SCO...
SCO...
SCO...
SCO...
SCO... ...om Chemical Accidents, 1989,
SCO... ...hort-Term Toxicity Tests for Non-Genotoxic Effects, 1990, 353 pp
SCOPE 42: Biogeochemistry of Major World Rivers, 1991, 356 pp
SCOPE 43: Stable Isotopes: Natural and Anthropogenic Sulphur in the Environment, 1991, 472 pp
SCOPE 44: Introduction of Genetically Modified Organisms into the Environment, 1990, 224 pp
SCOPE 45: Ecosystem Experiments, 1991, 296 pp
SCOPE 46: Methods for Assessing Exposure of Human and Non-Human Biota, 1991, 448 pp
SCOPE 47: Long-term Ecological Research. An International Perspective, 1991, 312 pp
SCOPE 48: Sulphur Cycling on the Continents: Wetlands, Terrestrial Ecosystems and Associated Water Bodies, 1992, 345 pp
SCOPE 49: Methods to Assess Adverse Effects of Pesticides on Non-Target Organisms, 1992, 264 pp
SCOPE 50: Radioecology After Chernobyl, 1993, 367 pp

Funds to meet SCOPE expenses are provided by contributions from SCOPE Committees, an annual subvention from ICSU (and through ICSU, from UNESCO), an annual subvention from the French Ministère de l'Environnement, contracts with UN Bodies, particularly UNEP, and grants from Foundations and industrial enterprises.

SCOPE 50

Radioecology after Chernobyl

SCOPE 50

Radioecology after Chernobyl

Biogeochemical Pathways of Artificial Radionuclides

Edited by

SIR FREDERICK WARNER

SCOPE Unit, Department of Chemistry and Biological Chemistry, University of Essex, Colchester, UK

and

ROY M. HARRISON

Institute of Public and Environmental Health, School of Biological Sciences, University of Birmingham, Edgbaston, Birmingham, UK

Published on behalf of the Scientific Committee on Problems of the Environment (SCOPE) of the International Council of Scientific Unions (ICSU)

by

JOHN WILEY & SONS

Chichester · New York · Brisbane · Toronto · Singapore

Published by John Wiley & Sons Ltd.
Baffins Lane,
Chichester,
West Sussex PO19 1UD, England

Other Wiley Editorial Offices

John Wiley & Sons, Inc., 605 Third Avenue,
New York, NY 10158-0012, USA

Jacaranda Wiley Ltd, G.P.O. Box 859, Brisbane,
Queensland 4001, Australia

John Wiley & Sons (Canada) Ltd, 22 Worcester Road,
Rexdale, Ontario M9W 1L1, Canada

John Wiley & Sons (SEA) Pte Ltd, 37 Jalan Pemimpin #05-04,
Block B, Union Industrial Building, Singapore 2057

Library of Congress Cataloging-in-Publication Data

Radioecology after Chernobyl : biogeochemical pathways of artificial radionuclides / edited by Sir Frederick Warner and Roy M. Harrison.
p. cm.—(SCOPE ; 50)
Includes bibliographical references (p.) and index.
ISBN 0 471 93168 3
1. Radioactive pollution—Environmental aspects. 2. Chernobyl Nuclear Accident, Chernobyl', Ukraine, 1986—Environmental aspects. 3. Radioecology. I. Warner, Frederick, Sir, 1910– . II. Harrison, Roy M., 1948– . III. International Council of Scientific Unions. Scientific Committee on Problems of the Environment. IV. Series: SCOPE report ; 50.
TD196.R3R3464 1993
574.5′222—dc20 92–26475
CIP

British Library Cataloguing in Publication Data

A catalogue record for this book is available
from the British Library

ISBN 0 471 93168 3

Typeset in 10/12pt Times from author's disks by Text Processing Department,
John Wiley & Sons Ltd, Chichester
Printed and bound in Great Britain by Biddles Ltd, Guildford, Surrey

International Council of Scientific Unions (ICSU)
Scientific Committee on Problems of the Environment (SCOPE)

SCOPE is one of a number of committees established by the non-governmental group of scientific organizations, the International Council of Scientific Unions (ICSU). The membership of ICSU includes representatives from 75 National Academies of Science, 20 International Unions and 29 other bodies called Associates. To cover multidisciplinary activities which include the interests of several unions, ICSU has established 13 Scientific Committees, of which SCOPE is one. Currently representatives of 35 member countries and 22 Unions, Scientific Committees and Associates participate in the work of SCOPE, which directs particular attention to the needs of developing countries. SCOPE was established in 1969 in response to the environmental concerns emerging at the time: ICSU recognized that many of these concerns required scientific inputs spanning several disciplines and ICSU Unions. SCOPE's first task was to prepare a report on Global Environmental Monitoring (SCOPE 1, 1971) for the UN Stockholm Conference on the Human Environment.

The mandate of SCOPE is to assemble, review, and assess the information available on man-made environmental changes and the effects of these changes on man; to assess and evaluate the methodologies of measurement of environmental parameters; to provide an intelligence service on current research; and by the recruitment of the best available scientific information and constructive thinking to establish itself as a corpus of informed advice for the benefit of centres of fundamental research and of organizations and agencies operationally engaged in studies of the environment.

SCOPE is governed by a General Assembly, which meets every three years. Between such meetings its activities are directed by the Executive Committee.

R. E. Munn
Editor-in-Chief
SCOPE Publications

Executive Director: V. Plocq-Fichelet
Secretariat: 51 boulevard de Montmorency
75016 Paris, France

Contents

Foreword

The explosion of Reactor No. 4 at Chernobyl on 26 April 1986 profoundly affected the world outside. News of the accident was not immediately given but reports of radioactive fallout came quickly from neighbouring countries, and action was taken to prevent contaminated food from being consumed. Some of the emergency measures had serious social and economic consequences which could have been mitigated with better information. This study has brought together the results of an international study on the ways in which radioactive materials released by human activities spread in the environment, but stops short of considering health effects. These are the concern of bodies such as the United Nations Commission on the Effects of Atomic Radiation (UNSCEAR) and the International Commission on Radiological Protection (ICRP), which evaluate the contribution received by humans.

This report is addressed to a wide community of scientists and others who study the environment. The Chernobyl accident had a positive aspect in providing experimental evidence which would not have been available otherwise. The direct scientific interest is important but the project has also been more widely followed. The universal interest in the environment needs to be fed with information if limited resources are to be used economically or difficult choices made. One of these is between burning fossil fuel with the effect on climate or using nuclear power with the possibility of an accidental release of radioactivity. There is also the risk of release from a nuclear weapon. The knowledge presented here should help in assessing what can be done to reduce effects on the environment or humanity.

The study of pathways included is based on radionuclides from the bomb tests, accidents and releases from re-processing. Contributions are made from other sources, such as fossil fuels, minerals, building materials and soils as well. These are pervasive and have not provided the identification available from individual locations.

The RADPATH study began in Moscow during a workshop in March 1988 to bring up-to-date the forecasts of environmental consequences of nuclear war (ENUWAR) made in SCOPE 28, a two-volume report from the Scientific Committee on Problems of the Environment of ICSU (International Council of Scientific Unions). A special session considered the improved models of movement in the atmosphere and their validation by measurements made on Chernobyl fallout. The opportunity was taken to visit the stricken reactor, the abandoned town of Pripyat and the surrounding country and marshes. At the SCOPE General Assembly in Budapest during June 1988, a proposal to begin this study was agreed with a request for the ENUWAR Unit at Essex University to carry it out.

A planning meeting later in 1988 at the Royal Society in London appointed a Steering Committee of experts who have organized a series of workshops and preparation of this report. It also identified other sources by which radionuclides are dispersed, such as nuclear weapons tests, accidents as at Windscale, Kyshtym and Three Mile Island and other escapes from the nuclear fuel cycle.

Three principal workshops were held in the UK—University of Essex, 1988 and 1991, and University of Lancaster, 1990. The latter afforded the opportunity to visit the nuclear fuel reprocessing plant at Sellafield. Other meetings with RADPATH participation were held in the USSR (Suzdal 1989, Gomel 1990, Zeleny Mys 1990 and Pushchino 1991), in Luxembourg 1990 through the International Union of Radioecologists and with BIOMOVS, Sweden 1990.

These meetings provided information from scientists with access to official sources of information as well as from academic and research institutions, but the non-governmental aspect of SCOPE work has been maintained. The contributions made by the International Union of Radioecologists and the Vernadsky Institute of Geochemistry have been considerable. The individuals who have given their time are identified, with their affiliations, in this report. There are problems in achieving consistency in the units used, especially curies and becquerels, because of old habits. The international collaboration made it impossible to impose uniformity.

Over the three years of RADPATH, the work of the Essex University Unit including the publication of Newsletters has been supported financially by grants from organizations which have also assisted the attendance of scientists at workshops and provided the facilities. They include the Royal Society of London, Commission of the European Communities, Fellowship of Engineering, the Leverhulme Trust, Wolfson Foundation and the Sigmund Sternberg Charitable Foundation. They have allowed Dr. Appleby to make a full-time contribution to all the RADPATH activities which has been invaluable.

Sir Frederick Warner, University of Essex, UK. *Chairman*
R. M. Harrison, University of Birmingham, UK. *Director*
L. Khitrov, Vernadsky Institute, Russia.
R. Kirchmann, International Union of Radioecologists, Belgium.
C. Lambert, Centre des Faibles Radioactivités, CNRS-CEA, France.
U. C. Mishra, Bhabha Atomic Research Centre, India.
T. Ohkita, Nagoya National Hospital, Japan.
A. G. Ryaboshapko, Scientific Committee on Problems of the Environment.
C. S. Shapiro, San Francisco State University and Lawrence Livermore National Laboratory, USA.

List of Contributors, Reviewers and Workshop Participants*

A. Aarkrog
Department of Radioprotection, Risø National Laboratory, Postbus 49, DK 4000 Roskilde, Denmark

J. P. Adloff
Laboratoire de Chimie Nucléaire, Centre des Recherches Nucléaires, BP 20, F-67037 Strasbourg Cedex, France

R. Allott
AEA Technology SRD, Wigshaw Lane, Culcheth, Cheshire WA3 4NE, UK

K. Andersson
Risø National Laboratory, DK-4000 Roskilde, Denmark

L. J. Appleby
SCOPE Unit, Department of Chemistry and Biological Chemistry, University of Essex, Wivenhoe Park, Colchester CO4 3SQ, UK

H. ApSimon
Air Pollution Group, Mechanical Engineering Department, Imperial College, Exhibition Road, London SW7 2BX, UK

S. I. Auerbach
Environmental Sciences Division, Building 1505, MS 6036, Oak Ridge National Laboratory, PO Box 2008, Oak Ridge, Tennessee 37831, USA

H. L. Beck
Environmental Measurements Laboratory, 376 Hudson Street, New York, NY 10014, USA

J. N. B. Bell
Centre for Environmental Technology, Imperial College, London SW7 2AZ, UK

* This list includes participants at workshops convened in the UK and those involved in the preparation and review of various sections of the book. Responsibility for the content of the report, nevertheless, rests with the principal authors. In view of the open nature of the study, some of those who participated or attended only some parts of a workshop may have been unintentionally omitted.

N. A. Beresford
Institute of Terrestrial Ecology, Merlewood Research Station, Windermere Road, Grange-over-Sands, Cumbria LA11 6JU, UK

E. J. Bradley
Environmental Investigation Group, NRPB Chilton, Didcot, Oxon OX11 0RQ, UK

S. B. Bradley
Environmental Protection Group, Building B433, British Nuclear Fuels plc, Sellafield, Seascale, Cumbria CA20 1PG, UK

D. P. Calmet
Laboratoire 501, Métrologie de l'Environment, Bois des Rames, 91400 Orsay, France

R. Carpenter
School of Oceanography, WB-10, University of Washington, Seattle, Washington 98195, USA

A. Chamberlain
1 Fifth Road, Newbury, Berks RG14 6DN, UK

S. Charmasson
CEA, Isremer, BP 330, 83507 La-Seyne-sur-Mer, France

A. Cigna
ENEA, I-13040 Saluggia VC, Italy

L. E. Coles
Pharmacy Department, Rookwood Hospital, Fairwater Road, Cardiff CF5 2YN UK

P. Coughtrey
WS Atkins Science & Technology, Woodcote Grove, Ashley Road, Epsom, Surrey KT18 5BW, UK

G. Desmet
Radiation Protection Programme, Commission of the European Communities, Rue de la Loi 200, B-1049 Brussels, Belgium

L. Devell
Studsvik Nuclear, Studsvik AB, S-61182 Nyköping, Sweden

M. Dickerson
Atmospheric and Geophysical Sciences Division, Lawrence Livermore National Laboratory, PO Box 808, L-262, Livermore CA 94550, USA

H. J. Dunster
19 Diamond Court, 153 Banbury Road, Oxford OX2 7AA, UK

J. B. Ellis
Middlesex University, Enfield, Middlesex, UK

S. Fowler
IAEA International Laboratory of Marine Radioactivity, 19 avenue des Castellans, MC 98000, Principality of Monaco

M. Frissel
RIVM, (National Institute of Public Health and Environmental Hygiene), PO Box 1, 3720 BA Bilthoven, The Netherlands

J. A. Garland
AEA, Harwell Laboratory, Oxfordshire, OX11 0RA, UK

G. Graziani
Joint Research Centre, Radiochemistry Division, 21020 Ispra (Varese), Italy

N. V. Grebenshikova
former USSR

P. Gudiksen
Atmospheric and Geophysical Sciences Division, Lawrence Livermore National Laboratory, PO Box 808, L-262, Livermore, CA 94550, USA

T. E. Hakonson
Los Alamos National Laboratory, Los Alamos, New Mexico 87545, USA

J. Hamilton-Taylor
Environmental Science Division, Institute of Environmental and Biological Sciences, Lancaster University, Lancaster LA1 4YQ, UK

W. Hanson
Hanson's Environmental Services, 1902 Yew St Rd, Bellingham, Washington 98226, USA

R. M. Harrison
Institute of Public and Environmental Health, School of Biological Sciences, University of Birmingham, Edgbaston, Birmingham B15 2TT, UK

C. N. Hewitt
Environmental Science Division, Institute of Environmental and Biological Sciences, Lancaster University, Lancaster LA1 4YQ, UK

F. O. Hoffman
Environmental Sciences Division, Oak Ridge National Laboratory, PO Box 2008, Oak Ridge, Tennessee 37831, USA

A. D. Horrill
Institute of Terrestrial Ecology, Merlewood Research Station,
Windermere Road, Grange-over-Sands, Cumbria LA11 6JU, UK

B. Howard
Institute of Terrestrial Ecology, Merlewood Research Station,
Windermere Road, Grange-over-Sands, Cumbria LA11 6JU, UK

J. Howorth
Environmental and Medical Sciences Division, Building 551, Harwell
Laboratory, AEA, Didcot, Oxon OX11 0RA, UK

M. Ivanovich
Isotope Geoscience Station, AEA Environment & Energy, Harwell
Laboratory, AEA, Didcot, Oxon OX11 0RA, UK

D. Jackson
British Nuclear Fuels plc, Sellafield, Seascale, Cumbria CA20 1PG, UK

S. R. Jones
Head, Environmental Protection, Building B434, British Nuclear Fuels plc,
Sellafield, Seascale, Cumbria CA20 1PG, UK

M. Kelly
Environmental Science Division, Institute of Environmental and Biological
Sciences, Lancaster University, Lancaster LA1 4YQ, UK

P. Kershaw
MAFF, Fisheries Laboratory, Lowestoft, Suffolk NR33 0HT, UK

L. Khitrov
Vernadsky Institute of Geochemistry and Analytical Chemistry,
Academy of Sciences, 19 Kosygin Street, Moscow 117975, Russia

R. Kirchmann
International Union of Radioecologists, 5 rue Cardinal Cardijn,
B-4680 Oupeye, Belgium

A. Konoplev
Institute of Experimental Meteorology, Typhoon Science and Production
Organization, Obninsk, Russia

T. Kosako
Research Centre for Nuclear Science and Technology, University of Tokyo,
Yayoi 2-11-16, Bunkyo-Ku, Tokyo 113, Japan

G. Kuznetsov
Vernadsky Institute of Geochemistry and Analytical Chemistry,
Academy of Sciences, 19 Kosygin Street, Moscow 117975, Russia

C. Lambert
AIMS Coordination Unit, Centre des Faibles Radioactivés, Laboratoire Mixte CNRS–CEA, Domaine du CNRS, Avenue de la Terrasse, F-91198 gif sur Yvette Cedex, France

L. J. Lane
Southwest Rangeland Watershed Research Centre, USDA-Agricultural Research Service, Tucson, Arizona 85719, USA

N. Lukyanov
RSMAS Program on Global Change, University of Miami, Rosensteil School of Marine and Atmospheric Science, 4600 Rickenbacker Causeway, Miami, FL 33149-1098, USA

S. P. Luttrell
Environmental Sciences Department, Pacific Northwest Laboratory, PO Box 999, Richland WA 99352, USA

A. B. Mackenzie
NEL/SURRC, Scottish Universities Research and Reactor Centre, Birnehill, E. Kilbride, Glasgow G75 0QU, UK

S. F. Malcolm
MAFF, Fisheries Laboratory, Lowestoft, Suffolk NR33 0HT, UK

T. Maruyama
Division of Physics, National Institute of Radiological Sciences, 9-1, Anagawa-4-chome, Chibashi 260, Japan

W. A. McKay
AEA, B 364, Harwell Laboratory, Didcot, Oxfordshire OX11 0RA, UK

Z. A. Medvedev
4, Osborn Gardens, Mill Hill, London NW7 1DY, UK

U. C. Mishra
Head of Training Division, Bhabha Atomic Research Centre, Trombay, Bombay 400 085, India

P. I. Mitchell
Department of Experimental Physics, Science Building, University College Dublin, Belfield, Donnybrook, Dublin 4, Ireland

C. Myttenaere
4 Place Croix du Sud, Labor. Physiologie Vegetale, 1348 Louvain La Neuve, Belgium

J. Naidu
535A Safety and Environmental Protection Division,
Brookhaven National Laboratory, Associated Universities Inc., Upton,
NY 11973, USA

K. Nicholson
AEA, Harwell Laboratory, Didcot, Oxfordshire OX11 0RA, UK

H. Nies
Bundesamt für Seeschiffahrt und Hydrographie, Laboratorium Sulldorf,
Wustland 2, D-2000 Hamburg 55, Germany

T. Ohkita
Nagoya National Hospital, Hirabari 4, Tenpaku-ku, Nagoya 468, Japan

B. Patel (deceased)
Health Physics Division, Bhabha Atomic Research Centre, Trombay,
Bombay, India

G. G. Polikarpov
Radiation and Chemical Biology Department, Institute of Biology of
Southern Seas, Academy of Sciences Ukraine, 2 Prospekt Nakhimova,
Sevastopol 335000, Ukraine

B. S. Prister
Ukrainian Branch of the All-Union Institute of Agricultural Radioecology,
7 Mashinostroiletnaya St, Chabany, Kiev, Ukraine

I. Rjabov
33 Leninsky Prospekt, Moscow V-71, 11701, Russia

W. Robison
Environmental Sciences Division, Lawrence Livermore National Laboratory,
PO Box 808, Livermore, CA 94550, USA

J. Roed
Environmental Science and Technology Department, Risø National Laboratory,
PO Box 49, DK-4000 Roskilde, Denmark

C. Ronneau
University of Louvain, Louvain-la-Neuve, Belgium

O. Rumianstev
Vernadsky Institute of Geochemistry and Analytical Chemistry, USSR
Academy of Sciences, 19 Kosygin Street, Moscow 117975, Russia

A. G. Ryaboshapko
Institute of Applied Geophysics, Glebovskaja Street, 20b Moscow, Russia

F. J. Sandalls
AEA Technology, Harwell, UK

P. Santschi
Professor of Oceanography, Department of Marine Sciences,
Texas A&M University of Galveston, Galveston TX 77553-1675, USA

W. R. Schell
University of Pittsburgh, Department of Environmental and Occupational Health, Graduate School of Public Health, Pittsburgh, Pennsylvania 15261, USA

C. S. Shapiro
Atmospheric and Geophysical Sciences Division, Lawrence Livermore National Laboratory, PO Box 808, L-262, Livermore CA 94550, USA

G. Shaw
Imperial College at Silwood Park, Department of Biology, Ascot, Berks. SL5 7PY, UK

J. Sinnaeve
Commission of the European Communities, DG XII-D-3, rue de la Loi 200, 1049 Brussels, Belgium

R. D. Small
Director for Thermal Sciences, Pacific-Sierra Research Corporation,
12340 Santa Monica Boulevard, Los Angeles, California 90025, USA

F. B. Smith
Meteorological Office, London Road, Bracknell, Berkshire RG12 2SZ, UK

J. S. Smitz
University of Liège, Geophysical Fluid Dynamics, Environmental Modelling Group, 4000 Sart Tilman, B5 Liège, Belgium

E. Steinnes
Department of Chemistry, University of Trondheim AVH, N-7055 Dragvoll, Norway

D. J. Swift
MAFF, Fisheries Laboratory, Lowestoft, Suffolk NR33 0HT, UK

U. Tveten
Institute for Energy Technology, PO Box 40, N-2007 Kjeller, Norway

T. Urabe
Computation Center, Nagoya University, Furo-Cho, Chikusa-ku, 464 Nagoya, Japan

J. van den Hoek
Laboratorium voor Dierenfysiologie, Haarweg 10, 6709 Wageningen, The Netherlands

P. Van Voris
Battelle Pacific Northwest Laboratory, PO Box 999, Richland, Washington 99352, USA

E. Voice
25 Miller Place, Thurso, Caithness KW14 7AL, UK

R. von Gunten
Laboratorium für Radiochemie, Universität Bern, Freiestrasse 3, CH-3000, Bern 9, Switzerland

A. Walmsley
Routine Releases Group, NRPB, Chilton, Didcot, Oxon OX11 0RQ, UK

Sir Frederick Warner
SCOPE Unit, Department of Chemistry and Biological Chemistry, University of Essex, Wivenhoe Park, Colchester CO4 3SQ, UK

B. T. Wilkins
Environmental Investigations Group, NRPB, Chilton, Didcot, Oxon OX11 ORQ, UK

D. S. Woodhead
MAFF, Fisheries Laboratory, Lowestoft, Suffolk NR33 0HT, UK

N. Zezina
Institute of Cell Biology and Genetic Engineering, Zabolotnogo Street 148, Kiev, GSP-22, Ukraine

Executive Summary

The movement of radionuclides in the environment has been studied for many years, with the principal objective of tracing the routes by which they accumulate in the food chain and become available for human consumption. The explosions in Reactor No. 4 of the nuclear power station of Chernobyl in the Ukraine provided a point source for distribution of radionuclides and a unique opportunity to trace the mechanisms by which they are distributed. The quantity and composition of the emissions are capable only of being estimated from environmental measurements, not by direct calculations from the composition of the reactor core and its mode of disintegration. Material deposited around Chernobyl and in neighbouring countries was traced through direct observation of radioactivity and radiochemical analyses through all stages from atmospheric dispersal to deposition and subsequent transfer. A major significance of Chernobyl was the relatively short-lived 'pulse' of emission which it caused. This has proved immensely valuable to the research community as a form of time marker from which the rates of subsequent processes may be estimated.

The basic properties of isotopes of interest and their sources are described in the opening chapter; the coverage is limited mainly to those whose release is due to human activities. A series of case-studies details the incidents which have been important for adding to knowledge of radionuclide behaviour within atmospheric, terrestrial, aquatic and urban compartments of the environment and the pathways through which transfer occurs. Major chapters then give detailed reviews of current knowledge of environmental pathways of radioactivity and recommendations for future research. This is followed by assessments of dosimetry and environmental effects.

Radionuclides which occur naturally in the environment have a pattern of distribution which is in dynamic equilibrium, with disturbances due to cultivation, earth movement and weathering. This equilibrium has had superimposed upon it the isotopes, some with comparatively short half-lives, which have arisen from mining, processing and accidents. The local disturbances allow the tracing of pathways through which the radionuclides are dispersed. Those which are most abundant and/or display the greatest activity provide most of the information for this study. These are principally $^{134,137}Cs$, ^{90}Sr, $^{103,106}Ru$, $^{141,144}Ce$, $^{129,131}I$ and $^{238,239,240}Pu$. Isotopes of other elements such as ^{110m}Ag, $^{127,129,132}Te$, ^{95}Zr, ^{95}Nb, $^{138,140}La$, ^{99}Tc and ^{241}Am are included but not treated in great detail.

1 SOURCES OF ENVIRONMENTAL RADIOACTIVITY AND ISOTOPES OF INTEREST

The artificial creation of radionuclides may result from physical processes involving nuclear fission, nuclear fusion and neutron activation. The most important source of artificially created radionuclides is neutron-induced nuclear fission. The chemical and physical forms of the active species determine deposition, migration and uptake of radioactivity by living organisms.

A variety of systems and processes may introduce radioactivity into the environment. Human activities involving nuclear weapons and the nuclear fuel cycle (including mining, milling, fuel enrichment, fabrication, reactors, spent fuel stores, reprocessing facilities and waste storage) are important, leading to significant creation and release of radioactivity. Human technology also releases pre-existing natural radionuclides which would otherwise remain trapped in the Earth's crust. The burning of fossil fuels (oil and coal) dominates direct atmospheric release of pre-existing natural radioactivity.

Since radionuclides are in some respects easier to study than stable nuclides there is an application in their use as environmental tracers. The most important tracers which are of artificial origin are the fission and activation products from nuclear tests. In addition, radioisotopes produced as a result of the bombardment of the Earth's atmosphere by cosmic rays provide useful tracers for studies of stratospheric–tropospheric exchange processes. Radon and thoron, and their associated decay products, are a second group of natural isotopes which may be utilized, for example, in studies of the atmospheric circulation processes of the lower atmosphere.

The physical and chemical form of radionuclides may vary depending on the release and transport condition in addition to the elements' properties. A general distinction can be made between gases, aerosols and particulate material. Particles with high activity concentration, known as 'hot particles', may result from atmospheric nuclear weapons tests or nuclear reactor accidents. This activity is diluted as material is transferred to soil and water directly or via vegetation and movement through other biota.

2 SOME OCCURRENCES OF SIGNIFICANT RADIOACTIVITY RELEASES TO THE ENVIRONMENT

The presence of radioactivity in the environment may be due to a variety of sources. One such source arises from routine releases which are made, for example, as a result of reprocessing activities undertaken at Sellafield, UK, and Cap de la Hague, France, or due to plutonium production operations at Hanford, USA. Besides planned discharges arising from operations associated with the nuclear fuel cycle, radionuclides may enter the environment as a consequence of accidental releases. Some important non-routine releases which have occurred include that due to an explosion in a high-level waste storage tank at Kyshtym, former USSR (1957), and those involving nuclear reactors at Windscale, UK (1957); Three Mile Island, USA (1979) and Chernobyl, USSR (1986). Additional releases have occurred upon re-entry of satellites powered by nuclear sources, such as SNAP 9-A (1964) and Cosmos 954 (1978). Whilst the Chernobyl accident is the most serious to have occurred in the history of nuclear reactor operation, other smaller releases are also of importance for the elucidation of the potential pathways of radionuclides in the environment.

3 RADIOACTIVITY IN THE ATMOSPHERIC ENVIRONMENT

The fate of radionuclides in the atmosphere is determined by various physical processes. The degree of dispersion and point of deposition depends on factors such as release parameters, the radionuclide form and meteorological conditions. Deposition of radioactive particles and gases to the ground may occur via dry deposition which includes direct gravitational settling, as well as turbulent transfer, or due to intervention of some form of precipitation (wet deposition). The involvement of precipitation processes may act as a concentration mechanism leading to areas of enhanced deposition. Entrainment of previously deposited material in the atmosphere (resuspension), due to the action of wind, mechanical agitation or falling hydrometeors, is another mechanism resulting in the presence of radionuclides with the potential for causing spread of contamination to adjacent clean areas and constituting an inhalation hazard. In some coastal areas a proportion of radionuclides discharged into the sea is blown back onto land as seaspray.

Measurements of the concentration of radionuclides in the air and on the ground are complemented by modelling studies, which require appropriate radiological and meteorological input data. A variety of air sampling methods may be used for measuring air concentrations of radionuclides, and a network of air monitoring stations is maintained in many nations. Many sites throughout the world have also been established to monitor the deposition of radionuclides on the Earth's surface. A major problem in most monitoring programmes lies in obtaining samples which are representative of the average deposition over a large area, as this is often highly variable on a small spatial scale. Techniques employed to acquire meteorological data as input to models depend on the release size. Data may be collected via on-site instrumentation or from a standard meteorological observing station, whilst at sea observations may be made either on board ships or gas and oil platforms, or using drifting buoys. Measurement of upper-air conditions are made using balloon-borne radiosondes. Aircraft and weather radar are other sources of valuable information regarding meteorological conditions and, in the case of the latter, to models tracking pollution.

A wide range of atmospheric dispersion and deposition models is available. Detailed modelling studies are an aid in understanding the role and interaction of specific microphysical and meteorological factors, whilst more general models are directly applicable to estimating the environmental consequences of radionuclides released to the atmosphere, which are of particular importance in guiding monitoring teams following accidental releases. The specific purpose for which a model is utilized governs the complexity of the model selected. It is, nevertheless, important that model simplifications, and limitations induced by the indeterminate nature of the atmosphere, should be recognized. Unique opportunities for model evaluation have been provided by the accidents which occurred at Windscale, Three Mile Island, and Chernobyl, where simulations of atmospheric transport have been compared with radiological observations.

4 RADIOACTIVITY IN TERRESTRIAL ECOSYSTEMS

The type of land and vegetation, and deposition mechanism, determine radionuclide behaviour in the terrestrial environment. The amount of a nuclide intercepted by surface vegetation depends on the surface area of leaves, roughness and botanical composition. Some small transfer of activity also occurs by translocation from the roots. Movement from roots and direct deposits on soil depend on vegetation type, soil geology and water movement. Grazing animals and cultivation cause redistribution in the topmost layer. Soluble materials may be removed from the soil by leaching due to the passage of water. Transfer back to vegetation, and thence to animals, occurs via root systems. A further transfer cycle is associated with mechanisms relating to metabolism, removal in food products, excretion and resuspension due to wind. The chemical similarity of potassium and caesium, and of strontium and calcium, has been a focus of studies concerning physiology of radionuclide uptake. As a result of bomb tests in Nevada and the South Pacific there is much information available for arid soils and movements in tropical islands. However this has little relevance with regard to Chernobyl fallout where effects in Arctic tundra ecosystems, particularly uptake by lichens, and in wetlands are of more importance. Knowledge is particularly needed on transfer of radionuclides in ecosystems such as forests and tropical agriculture.

5 RADIOACTIVITY IN THE AQUATIC ENVIRONMENT

The pathways of radionuclides in the aquatic environment (which is composed of both aqueous and solid phases and includes rivers, lakes, estuaries, shelf seas, deep oceans, ice sheets, glaciers, groundwater and ground-ice) are of importance since a major fraction of the Earth's surface is occupied by this compartment. The transfer of radionuclides present within the aquatic environment is affected by both non-living and living components. Radionuclide behaviour in the aquatic environment is determined by transport of the solution and solid phases, the chemical interactions between phases and their biological cycling. Of importance in determining radionuclide behaviour and its solid/solution distribution are the chemical characteristics of the water. An important contrast in the behaviour of radionuclides is observed between aerobic and anaerobic waters, and major differences in ionic composition which exist between saline water and freshwater (and in the latter case between alkaline and acid waters) are of significance in determining radionuclide speciation and behaviour. Furthermore, the solid/solution distribution of radionuclides in the aquatic environment is found to vary with grain size and composition of sediment grains, and bedrock grain surfaces. The nature of the grain surface, particularly the extent to which it is modified by inorganic (oxyhydroxide), or organic deposits, may also be of importance. Moreover, the potential mobility of radionuclide particulate fractions is determined by grain size. Transport of radionuclides within the aquatic environment not only leads to their redistribution, but results in dilution, fractionation and mixing. Accumulation of the particulate phase by sedimentation also affects the radionuclide distribution, with the radionuclide inventory in sediment deposits found to vary widely depending on radionuclide sources and deposit characteristics.

6 RADIOACTIVITY IN THE URBAN ENVIRONMENT

Since almost three-quarters of the population of the current developed world is located in urban areas, the study of radionuclides in towns and cities is of importance. The distribution pattern of fallout depends on the weather conditions (i.e. wet or dry) and variation in the degree of interception of fallout is found depending on the nature of the surface and the physico-chemical form of radionuclides. Large variability in surface contamination within cities may occur due to wash-off from impervious urban surfaces. The nature and rate of removal processes determines the retention behaviour of pollutants. In the outdoor environment wash-off probably governs weathering, whereas within houses cleaning constitutes the major removal process. Predicted dose reductions achievable in the urban environment through the application of various decontamination techniques reveal that digging gardens and defoliating trees should be of highest priority, based upon consideration of cost and effectiveness. Artificial radionuclides have been successfully employed in the study of individual urban environment pathways. Further applications lie in tracer studies of wet deposition, resuspension and wash-off, with applications to pollutant retention behaviour of greatest importance.

7 DOSIMETRY AND THE ASSESSMENT OF ENVIRONMENTAL EFFECTS OF RADIATION

A consequence of the presence of radionuclides in the environment is the potential for increased radiation exposure of living organisms. The impact of radionuclide releases on organisms may be assessed by consideration of the likelihood and extent of radiation exposure (i.e. radiation dosimetry) and potential consequential effects (i.e. the radiobiological response). With regard to characteristics influencing the maintenance of the population, for the majority of species (other than humans and rare/endangered wild species or valuable domestic animals) it is the population level rather than the individual response to radiation exposure which is considered to be of greatest importance. Nevertheless effects due to increased radiation exposure at, or above, the population level cannot appear without the occurrence of a clear response in individuals which comprise the population. Studies have revealed that no significant effects would be apparent in wild populations of terrestrial animals provided that the dose rate to the most exposed individuals did not exceed 0.04 mGy h^{-1}, or in terrestrial plants and aquatic plants/animals provided that the dose rates to the most exposed individuals did not exceed 0.4 mGy h^{-1}. With very few exceptions controlled waste disposal operations have not, and would not, cause these values to be exceeded. An exception is the deep ocean disposal of packaged low-level radioactive waste, where high dose rates could potentially be delivered to benthic organisms at the disposal site since the ultimate human exposure, which comprises the present basis for control, occurs far from the dumpsite and more than 100 years into the future. In the case of a major nuclear accident such as Chernobyl, although responses have been detected in pine trees, and to a lesser extent in animals, any serious risk to long-term survival of local populations is debatable. However, since nuclear accidents cause global concern, there is need of careful periodic assessment of safety systems and their upgrading to minimize probability of accident occurence.

8 FUTURE TASKS AND RESEARCH NEEDS

In summary, the findings contained in this volume constitute an authoritative body of basic scientific information utilizing practical experiences relating to the pathways of radioactivity, which may assist environmental managers and policy makers in achieving a swift and effective response to the assessment of any future (planned or accidental) releases. Wider applications of relevance lie, for example, in advancing knowledge generally of environmental processes and biogeochemical cycling of chemical elements or their compounds. The Chernobyl accident has provided a great impetus to work on this subject by highlighting the need for rapid, reliable predictive capabilities in the event of a nuclear accident. Such predictions of radionuclide behaviour will only be reliable if based upon a solid and comprehensive foundation of research. The tracking of contamination arising from Chernobyl has clearly illustrated areas where present knowledge is partially or even wholly inadequate.

Much fundamental and applied research remains to be done. In the atmospheric science field, much is still to be learnt about the physico-chemical nature of radionuclides originating from within a hot emission source, and its evolution during atmospheric transport of the various gaseous and particulate forms of airborne radionuclides. The Chernobyl accident demonstrated that large (> 10 μm) particles could be transported over long distances (> 100 km) within the atmosphere and these processes need to be better understood if the development of future accident scenarios is to be reliably predicted. Chernobyl also highlighted the importance of wet processes and orographic effects in deposition of radioactivity. Research is needed into the interaction of radioactive aerosol particles with cloud droplets and into wet deposition generally in convective and frontal systems. The high spatially variable nature of such deposition indicates a need for better understanding of radionuclide atmospheric dispersion in complex terrain and frontal systems, and of 'occult' deposition in cloud and mist. Present understanding of both terrestrial and marine resuspension processes is rudimentary and improved quantitative knowledge is sorely needed.

In the terrestrial environment, knowledge of radionuclide physico-chemical speciation is presently poor, particularly in relation to bioavailability to plants and animals. Chernobyl highlighted inadequacies in the knowledge of radionuclide behaviour in organic soils and further research is needed in this area, particularly in relation to long-term processes. Additionally, since much of the deposition in the near-field is in the form of particles, a better understanding is required of their behaviour, both in terms of physical migration of the particles themselves and leaching of the various radionuclides associated. The role of micro-organisms in the cycling of radionuclides and of agricultural practices upon radionuclide migration are inadequately understood. Radionuclide behaviour in forests and other natural and semi-natural ecosystems is poorly quantified as is the role of secondary sources of radionuclide contamination such as resuspension. Information is generally relatively abundant in relation to major fission products, but is lacking in respect of some 'exotic' species such as ^{110m}Ag. Numerical models of food-chain transfer are steadily improving, but require further validation, especially in regard of transfer parameters. Models for young animals require development and testing.

Research recommendations relating to the aquatic environment highlight the need for studies of freshwater pathways, especially within lakes, due to the sensitivity of the systems to pollution. Collaborative international efforts to study 'hot' zones require improved co-ordination and intensification. International radioecological studies of polar seas are also recommended. As with the terrestrial environment, the role of micro-organisms in biogeochemical cycling is not well understood. The role of the surface microlayer in aquatic systems in influencing radionuclide behaviour is comprehended inadequately and requires further study; improved techniques of sampling the microlayer will be necessary. In freshwater and estuarine systems the importance of interactions of radionuclides with colloids is well documented. Colloidal processes in marine systems however require further study, particularly in relation to size association and speciation of radionuclides.

In the urban environment, the major concern is to understand processes leading to contamination and radionuclide retention and to optimize decontamination. Use of artificial radionuclides will continue to be of value in the study of deposition, retention and clean-up processes, leading to the development of improved models.

A consequence of the rapid development of digital computers is that detailed and complex numerical models of multi-phase environmental transport, dispersion, deposition and food chain transfer can be constructed. These will of necessity become ever more sophisticated and embrace an ever wider range of biogeochemical pathways. One crucial need is for the results of field and laboratory research investigations of individual processes to be translated into compact but effective parameterizations for incorporation within the numerical models.

1 Sources

Co-ordinator: **C. S. Shapiro**

Contributors: **L. J. Appleby, L. Devell, U. C. Mishra, and E. H. Voice**

1.1 INTRODUCTION

Calculations of the transport of radionuclides through the atmosphere usually start with a source term. The source term provides the initial conditions from which the atmospheric pathway models begin. These initial conditions include a description of the isotopic composition and quantity of the released radionuclides, their physical and chemical forms and the spatial and temporal distribution of the release. In this chapter we examine the basic physical processes that create these nuclides, a description of the physical and chemical properties of significant radionuclides, and the systems and processes that result in their release into the environment.

1.2 BASIC PROCESSES: FISSION, FUSION AND NEUTRON ACTIVATION

The principal physical processes for artificially creating radionuclides are nuclear fission, nuclear fusion, and neutron activation. Radionuclides also include isotopes produced by natural processes. These include residual nuclides created before the Earth was formed, their daughter products and naturally produced radionuclides. We focus here on artificially created nuclides which dominate accidental releases. Later we shall examine natural radioactivity.

1.2.1 NUCLEAR FISSION

Neutron-induced nuclear fission is by far the most important source of artificially created radionuclides. This process produces the energy released in contemporary nuclear reactors. Fission also dominates the release of radionuclides in nuclear weapon explosions. The nuclides produced in fission and their decay products constitute a spectrum of hundreds of species with a range of half-lives that extends from fractions of seconds to billions of years. A primary fission product is usually the starting point of a sequence of decays, each daughter nuclide tending to have a half-life longer than its parent, until a stable nuclide is reached. Only some of

Radioecology after Chernobyl
Edited by F.E. Warner and R.M. Harrison
 Published by John Wiley & Sons Ltd

these fission nuclides are potentially significant to biological systems, because of their physical and chemical properties, and the quantities produced (fission chain yield). The actual distribution of fission yields depends on the fissionable nuclide (e.g. ^{235}U, ^{238}U, ^{239}Pu) and the energy of the neutron initiating the fission. A representative fission yield distribution for the isotope ^{235}U for a thermal spectrum of neutrons (thermal reactors) as well as for 14 MeV neutrons (from nuclear fusion of deuterium and tritium) is presented in Figure 1.1. The mass number distribution of the fission products is asymmetric, with the high yield isotopes having mass numbers that range from approximately 85 to 105, as well as from 130 to 150. The fission yields at the peaks are typically around 6 or 7 per cent, which includes the yields for the important isotopes strontium-89 and 90, iodine-131, and caesium-137. The total fission yield for all the isotopes in Figure 1.1 adds up to 200 per cent, reflecting the fact that two fragments are produced in each fission event. The relatively high fission yield of ^{137}Cs and ^{90}Sr, their half-lives of about 30 years, and the fact that caesium is chemically similar to potassium, and strontium is chemically similar to calcium, all combine to select these two radionuclides as potentially very significant to humans. Similarly, iodine-131 with its large yield, half-life of 8 days, and its affinity to the thyroid gland, is also very significant to humans.

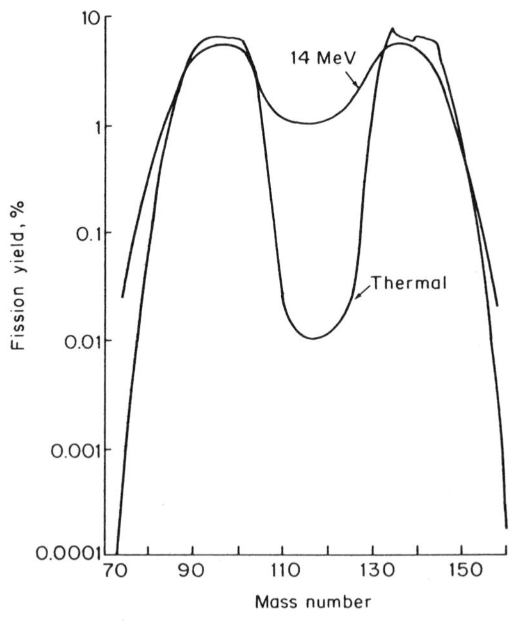

Figure 1.1 Percentage fission yield from thermal and 14-MeV neutron-induced fission of uranium-235 (Lamarsh, 1966; reproduced with permission of Addison-Wesley Publishing Company).

1.2.2 NUCLEAR FUSION

A second source of radionuclides is nuclear fusion. Here two nuclei of small atomic mass are combined (fuse) in an exothermic nuclear reaction. One such fusion reaction, important in the hydrogen bomb, is when deuterium and tritium combine to yield an alpha particle and a neutron.

$$^{2}_{1}H + ^{3}_{1}H \rightarrow ^{4}_{2}He + ^{1}_{0}n + 17.6\ MeV$$

In this reaction, 80 per cent of the 17.6 MeV of energy that is released is in the form of kinetic energy of the neutron (14.1 MeV). The initial reactant nuclei must collide at high energy (temperature) in order to overcome the coulomb repulsive force. Hence, to initiate a thermonuclear explosion, contemporary fusion weapons have a fission atomic bomb trigger. However in the laboratory, this reaction is much used for the production of fast neutrons.

1.2.3 NEUTRON ACTIVATION

A third source of man-made radionuclides is produced by the neutron activation process. This process can be important in both nuclear reactors and nuclear weapons. One reaction is:

$$^{14}_{7}N + ^{1}_{0}n \rightarrow ^{14}_{6}C + ^{1}_{1}H$$

This occurs when neutrons escaping from a weapon detonation interact with atmospheric nitrogen. The product nucleus, carbon-14, is also produced naturally in the atmosphere by the same reaction, from neutrons emanating from cosmic ray reactions. Many radionuclides produced in nuclear reactors come from the neutron activation processes. For example:

$$^{238}_{92}U + ^{1}_{0}n \rightarrow ^{239}_{92}U \rightarrow ^{239}_{93}Np + ^{0}_{-1}\beta$$

After a second beta decay, the $^{239}_{93}Np$ becomes $^{239}_{94}Pu$. This is the principal reaction process for producing plutonium-239. Further neutron absorptions produce

$$^{240}_{94}Pu,\ ^{241}_{94}Pu,\ ^{242}_{94}Pu$$

Another important neutron activation product is caesium-134. This radionuclide is not a direct fission product, but is produced by neutron activation of the stable fission product caesium-133. Hence, caesium-134 is produced copiously in reactors, but hardly at all in a weapon explosion. Measuring the ratio of caesium-134 to caesium-137 in a sampling of radioactivity is an important test to determine if the source of a radioactivity release is a nuclear reactor accident or a nuclear weapon explosion. Other important isotopes created by neutron activation in fission and fusion reactor structures are ^{55}Fe and ^{60}Co.

1.2.4 TIME DEPENDENCE OF RADIOACTIVE DECAY

Calculations of radionuclide transport that include quantitative estimates of activity concentrations and dose commitments must include radioactive decay with time. Figure 1.2 compares the overall time decay of the activity generated by the detonation of a 1 Mt fission weapon, with that present in a power reactor, a spent fuel storage pond, and a fuel reprocessing plant with a 10 year and 30 day high level waste storage facility. The relative contribution of each source depends on the time scale being considered. One can consider a hypothetical situation where the radio-

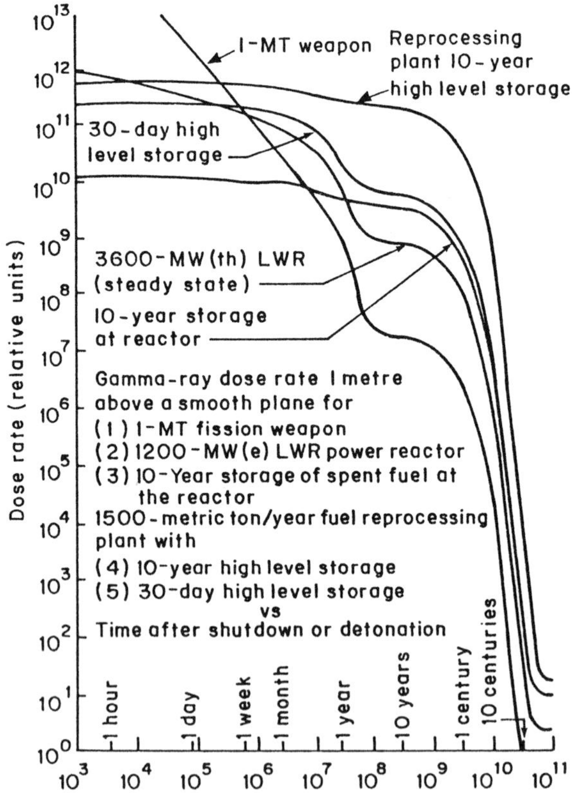

Figure 1.2 Gamma-ray dose rate area integral versus time after shutdown or detonation (Chester and Chester, 1976). The dose rate area integral is a measure of the radioactive inventory. (From Chester and Chester 1976; copyright 1976 by the American Nuclear Society, La Grange Park, Illinois).

activity from each source is spread over a square 1 km wide and examine the gamma dose rate in the centre at a height of 1 m. In the first few days, the higher activity of the nuclear weapon debris would dominate over the gamma radiation of the reactor. Likewise, gamma radiation levels from a light water reactor (LWR) are greater than that of 10 years worth of stored spent fuel for about one year after reactor shutdown. Subsequently, the spent fuel would be relatively more radioactive. Similarly, the gamma radiation from 10 years of spent fuel is greater than the radioactivity of a 1 Mt fission weapon after about two months because of the greater abundance of long-lived gamma emitters in the spent fuel.

1.3 ISOTOPES OF INTEREST

1.3.1 INTRODUCTION

The artificial radionuclides synthesized by humans range over all groups of the periodic table, and those released into the environment show a wide spectrum of behaviour. The RADPATH study focuses upon the more abundant and toxic isotopes.

Table 1.1 Radioactive nuclides, decay modes and half-lives

^{3}H	β	12.26 y	^{132m}I	β+IT	83 min
^{7}Be	EC	53.29 d	^{133}I	β	20.8 h
^{14}C	β	5370 y	^{133}Xe	β	5.25 d
^{24}Na	β	15 h	^{134}Cs	β	2.065 y
^{40}K	β	1.25×10^{9} y	^{136}Cs	β	13.1 d
^{60}Co	β	5.37 y	^{137}Cs	β	30.17 y
^{85}Kr	β	10.72 y	^{137m}Ba	IT	2.5 min
^{87}Rb	β	4.9×10^{10} y	^{140}Ba	β	12.76 y
^{89}Sr	β	50.52 d	^{140}La	β	40.3 h
^{90}Sr	β	29 y	^{141}Ce	β	32.5 d
^{91}Y	β	58.8 d	^{144}Ce	β	284.4 d
^{95}Zr	β	64 d	^{147m}Pm	β	2.62 y
^{95}Nb	β	53 d	^{210}Po	α	138.4 d
^{99}Tc	β	2.13×10^{5} y	^{210}Pb	β	19.4 y
^{103}Ru	β	39.24 d	^{222}Rn	α	3.82 d
^{106}Ru	β	372 d	^{231}Pa	α	3.27×10^{4} y
^{110m}Ag	β+IT	249.8 d	^{232}Th	α	1.41×10^{10} y
^{115m}Cd	β	44.6 d	^{235}U	α	7.04×10^{8} y
^{125m}Te	IT	58 d	^{237}Np	α	2.14×10^{6} y
^{125}I	EC	59.9 d	^{238}U	α	4.46×10^{9} y
^{127}Sb	β	3.84 d	^{239}Pu	α	2.41×10^{4} y
^{127}Sb	β	2.1 h	^{240}Pu	α	6537 y
^{129}I	β	1.6×10^{7} y	^{241}Pu	β	14.4 y
^{131}I	β	8.04 d	^{241}Am	α	432.2 y
^{131m}Xe	IT	11.92 d	^{242m}Am	α+IT	141 y
^{132}Te	β	78.2 h	^{244}Pu	α	8.1×10^{7} y

Principal nuclides include isotopes of the elements caesium, strontium, ruthenium, cerium, iodine and plutonium. Isotopes of other elements such as silver, tellurium, zirconium, niobium, lanthanum, technetium and americium, are included but not treated in great detail. Table 1.1 lists the radioactive nuclides with their half-lives in minutes, hours, days and years, and their principal modes of decay. Almost all simultaneously emit gamma or X-rays of many energies; those in the list which do not are ^{3}H, ^{14}C, ^{87}Rb and ^{90}Sr (all pure beta emitters). Most are fission products, some result from neutron activation, the first three are cosmogenic as well as human in origin, and six are primordial (see Sub-section 1.4.5).

The deposition, migration and uptake of radioactivity by living organisms is largely determined by the chemistry and physical form of the active species. The significant radionuclides are now considered individually.

1.3.2 CAESIUM

^{137}Cs ($t_{1/2}$ = 30 years) is a high-yield fission product, and stored highly-active waste and low level discharges generally contain large proportions of this isotope. Neutron activation of the fission product ^{133}Cs generates another active caesium isotope, ^{134}Cs ($t_{1/2}$ = 2 years), but since its production is proportional to the square of the neutron flux in the reactor, it is generally in a lesser proportion. Other sources of radiocaesium are nuclear weapons tests, and accidents such as those at Windscale and Chernobyl. The caesium released to the atmosphere at Chernobyl was associated with particles of a mean geometrical equivalent diameter of 0.5–1.0 μm. In rainwater it was mostly in colloidal form, with only a minor fraction as ionic.

The addition of ^{134}Cs or ^{137}Cs to soils as soluble compounds results in about 85 per cent rapidly becoming bound with a small fraction (about 15 per cent) remaining available for plant uptake. The availability of caesium in soils is related to the mineral composition. The subsequent movement and distribution is greatly influenced by the form of the overlying vegetation. It is believed that more than 95 per cent of ^{137}Cs remains in soluble ionic form upon the addition to seawater as a soluble compound. In freshwater, however, a large proportion of the total is expected to bind to suspended and deposited sediments.

Caesium is an alkali metal of Group I with properties similar to those of potassium and rubidium. As with all alkali metals, most of its salts are highly soluble in water. Despite the chemical similarities between caesium and potassium, they are not metabolized identically, and although potassium is often used to model caesium environmental transport, there can be more similarities between calcium and caesium.

1.3.3 STRONTIUM

Strontium is an alkaline metal of Group II, and its metabolic pathways tend to follow those of the other Group II elements calcium and radium. Radiostrontium is of environmental importance because of its tendency to concentrate throughout the volume of mineral bone in animals. The isotopes ^{89}Sr and ^{90}Sr are primary fission products with half-lives of 50.52 days and 28.8 years. Radiostrontium is generated in nuclear weapons explosions and is present in routine releases from nuclear facilities.

In both marine and freshwater environments strontium behaves primarily as a soluble element. In aquatic ecosystems ^{90}Sr is of greater importance than ^{89}Sr. On adding ^{90}Sr to seawater as a soluble compound, less than 1 per cent associates with suspended particulates or colloids. In freshwater, although the association with suspended particulates may be greater, it is unlikely to exceed 10 per cent of the total.

1.3.4 RUTHENIUM

Ruthenium, a transition metal, can exist in four valence states and forms numerous complexes. As a fission product it usually remains in fuel as a refractory, although at Chernobyl rapid oxidation formed the volatile oxides RuO_3 and RuO_4, which then condensed to aerosols or 'hot particles' of nearly pure ruthenium. The median diameter for particles bearing ruthenium in atmospheric aerosols was just less than 1 μm (ApSimon and Mahadeva, 1989). The primary ruthenium fission products are ^{103}Ru ($t_{1/2} = 39.3$ days) and ^{106}Ru ($t_{1/2} = 368$ days), discharging to the environment from fuel reprocessing and weapons detonations.

In soils, ^{106}Ru distribution and retention is largely dependent on the form (cationic, nitrosyl–ruthenium complexes, anionic) in which it is applied. Following aerial contamination of plants by ruthenium, 10 per cent retention may be expected. Reprocessing operations give rise to the production of nitrosyl–nitrato and nitrosyl–nitrito ruthenium complexes leading to nitrosyl-chloride in seawater. In discharges, particulate ^{106}Ru is bound to ferric hydroxides. Ionic ruthenium is in very low proportion in water, but up to 25 per cent may be present in colloids and as much as 60 per cent in particulates. A large proportion of ruthenium uptake in aquatic plants and animals is attributed to surface adsorption.

1.3.5 CERIUM

Cerium is the second of the lanthanide group of metals, rare in nature, but with two radioactive isotopes that are important fission products. The principal sources of ^{144}Ce ($t_{1/2} = 284$ days) and ^{141}Ce ($t_{1/2} = 32.5$ days) is fuel reprocessing. Other sources are from nuclear weapons tests and accidental releases, such as occurred at Chernobyl. At Chernobyl emitted fuel fragments contained cerium in concentrations close to the composition in the core. Fuel fragments carrying ^{141}Ce were not found to be sensitive to precipitation.

More than 95 per cent of ^{144}Ce added to soil is rapidly bound to exchange sites on clay minerals and to soluble, insoluble and colloidal organic complexes. It is likely that a very small amount of ^{144}Ce will be absorbed following deposition to the above-ground parts of plants. The addition of ^{141}Ce or ^{144}Ce to seawater leads rapidly to 70 per cent becoming associated with suspended sediments, over 25 per cent with colloidal material, and less than 5 per cent remaining in soluble ionic forms. There is marked association of cerium radioisotopes with particulate or sedimented materials in natural conditions. In living organisms cerium exhibits similar uptake and transfer behaviour as other lanthanides and actinides, and has a limited similarity with caesium and strontium.

1.3.6 IODINE

Iodine, a halogen of Group VII, is rare but widespread in Nature. Radioactive isotopes of iodine, ^{129}I and ^{131}I, are produced by nuclear weapons and reactor operation. Their half-lives are 1.57×10^7 years and 8.04 days. Around 6 per cent and 2 per cent respectively of the activity in boiling water reactors (BWR) and pressurized water reactors (PWR) liquid effluents can be due to ^{131}I. Several chemical and physical forms of iodine were released in the Chernobyl accident (particulate iodine, elemental iodine and organic iodides), the major fraction being present in gaseous form as elemental iodine and volatile organic iodides.

In most soils it may be assumed that 10 per cent of an addition of ^{131}I in soluble form remains available for uptake. Most elemental and organic iodine deposited on plants is directly absorbed by the foliage, whereas only about 10 per cent of particulates or inorganic compounds is absorbed by plants from solution. Iodide appears to be absorbed from water more effectively than iodate.

In the atmospheric dispersion of iodine, appreciably higher dry deposition of elemental, rather than particulate, iodine usually occurs, with dry deposition of the dominant organic iodide (methyl iodide) some two orders of magnitude lower than that of elemental iodine. In comparison with organic iodide, rapid washout of particulate and elemental iodine occurs. Iodine concentrates in the thyroid, absorption of the soluble form occurring through the skin, lung and alimentary tract.

1.3.7 PLUTONIUM

Neutron activation of uranium fuel in reactors leads to the production of the plutonium isotopes ^{238}Pu ($t_{1/2} = 87.7$ years), ^{239}Pu ($t_{1/2} = 2.4 \times 10^4$ years), ^{240}Pu ($t_{1/2} = 6.6 \times 10^3$ years) and isotopes of heavier elements with smaller abundances. The volatility of plutonium is very low and no significant releases occur during reactor operation, but environmental releases may occur from severe reactor accidents, as at Chernobyl. Further sources are controlled releases from research facilities, production facilities, experimental reactor operations, waste disposal and nuclear weapons testing.

In soils, the chemical and physical behaviour of plutonium is not well known. It seems that physical mechanisms dominate vertical transport as the chemical availability is rather low. In most soil types, soluble plutonium is absorbed rapidly.

Plutonium concentrations in plants depend on the species, the vegetation type, age and status, soil pH, cation exchange capacity, mineral and organic composition and physico-chemical form, and levels change with the duration of contamination. Redox, pH and organic ligands determine the solubility in water. Plutonium has a complex chemistry and most chemical forms are relatively insoluble, tending to adsorb onto sediments. At least four oxidation states Pu(III), Pu(IV), Pu(V) and Pu(VI) exist in aqueous environments, with the ions in a hydrated form. The solubility of plutonium in natural waters is low and mainly as Pu(V) or Pu(VI). The absorbed forms are usually in the Pu(III) and Pu(IV) state. Ionic, particulate, colloidal and pseudo-colloidal forms of plutonium are present in the water column. Over 90 per cent of plutonium is efficiently scavenged and removed from the water column within coastal environments in the short term where plenty of sediment is available. It is less efficiently scavenged and removed from the water column in mid-oceanic waters.

1.3.8 SILVER

Silver is a transition metal, rare but widespread in nature. In nuclear reactors it is both a minor fission product and an activation product. The radioactive nuclides of interest are ^{110}Ag ($t_{1/2}$ = 25 seconds), ^{110m}Ag ($t_{1/2}$ = 250 days) and ^{108}Ag ($t_{1/2}$ = 2.4 minutes). It is assumed that when silver enters soil a large proportion is rapidly adsorbed on the mineral fraction or complexed with insoluble organic compounds. Less than 2 per cent of the total in soil can be expected to be in soil solution.

1.3.9 TELLURIUM

Tellurium is an amphoteric element of Group VI, chemically related to sulphur and selenium. It is a minor fission product with five significant active isotopes and daughters: ^{126m}Te ($t_{1/2}$ = 109 days) decays to ^{127}Te ($t_{1/2}$ = 9.3 hours) plus ^{127}I (stable), ^{129m}Te ($t_{1/2}$ = 33.6 days) decays to ^{129}Te ($t_{1/2}$ = 73 minutes) plus ^{129}I ($t_{1/2}$ = 1.6×10^7 years), ^{132}Te ($t_{1/2}$ = 78.2 hours) decays to ^{132}I ($t_{1/2}$ = 2.3 hours). ^{132}Te has been measured in marine environments after nuclear detonations, mostly in particulate (including colloidal) form. Tellurium in the environment arises from accidents such as the Chernobyl release.

1.3.10 ZIRCONIUM

Zirconium is a refractory transition metal and is a major fission product. Nuclides of interest are ^{93}Ar ($t_{1/2}$ = 1.5×10^6 years) and ^{95}Zr ($t_{1/2}$ = 65 days). ^{95}Zr is widespread after weapons explosions and is a major gamma source in the mixed fission products from fuel reprocessing. Zirconium release to the environment also occurred during the Chernobyl accident from the extensive spreading of fuel fragments carrying such non-volatile elements. Zirconium is found in soil systems after nuclear detonations and near reprocessing plants. In most soils 95 per cent of the ^{95}Zr entering the system is rapidly adsorbed or precipitated. In water, over 95 per cent is expected to be bound to colloids, organic acids and suspended particles, leaving less than 5 per cent in solution.

1.3.11 NIOBIUM

A daughter product of ^{95}Zr is ^{95}Nb ($t_{1/2}$ = 35 days) which is released from reprocessing plants and nuclear weapons tests. An indication of ^{95}Nb chemistry in soil is provided by zirconium, although uptake of niobium to terrestrial plants occurs more readily than with zirconium. The behaviour of ^{95}Nb in water resembles that of ^{95}Zr but with a higher proportion bound by particulates, colloids and soluble organic complexes. The accumulation factor for sediment is higher for niobium than for zirconium.

1.3.12 LANTHANUM

Lanthanum, a transition metal element, is a fission product contributing to the long-lived radiation around Chernobyl. The principal active isotopes are ^{138}La ($t_{1/2}$ = 1.1×10^{11} years) and ^{140}La ($t_{1/2}$ = 40.3 hours). Lanthanum has a very low degree of availability in soils.

1.3.13 TECHNETIUM

Technetium, a transition metal element, no longer exists naturally on Earth as there is no stable isotope. It is a primary fission product, with the principal isotopes ^{99}Tc ($t_{1/2} = 2.1 \times 10^5$ years) and ^{97}Tc ($t_{1/2} = 2.6 \times 10^6$ years). Releases to the environment have occurred from weapons explosions and fuel reprocessing plants. The element is most stable as the very soluble pertechnetate anion. The form most likely to occur in fallout entering surface soils is Tc(VII). Soil type affects the extent to which technetium is fixed and the nuclide is very mobile in plants, transfer from roots to shoots and into developing fruit and seed tissues occurring readily. When entering waters as the pertechnetate ion, it can be assumed, in the absence of published data, that Tc remains mostly in a soluble form (Coughtrey *et al.*, 1983). It is assumed that on release of ^{99}Tc to the aquatic environment it remains available to all organisms.

1.3.14 AMERICIUM

Americium is an actinide element of which the most important isotope, ^{241}Am ($t_{1/2}$ = 432.2 years) forms from the decay of ^{241}Pu. It has been detected around waste disposal sites, operating power plants, and as a result of world-wide fallout. The isotope ^{242}Am ($t_{1/2}$ = 16 hours) also occurs in weapons fallout. A further product of weapons tests is ^{243}Am ($t_{1/2} = 7.4 \times 10^3$ years) but this isotope has not been detected in the environment. Oxides and hydroxides of Am(III) are relatively insoluble. The most likely chemical species of americium in soils is Am(III). The concentration factor of americium in vegetation depends on both plant and soil factors. Americium is generally assumed to exist in waters as the Am(III) oxidation state, and in freshwater and marine environments the soluble fraction of americium has been found not to be in the form of cations, but as negatively charged complexes. Americium hydroxide, resulting from the rapid hydrolysis of americium in solution, is insoluble in both freshwater and marine environments. These insoluble products of hydrolysis quickly form particulates or adsorb onto suspended particulates. Soluble complexes with organic ligands may maintain americium in solution. The association of americium with particulate material and sediments, controls its behaviour and distribution in the aquatic environment. In sediments, americium concentrations are generally highest where particle sizes are smallest. Americium deposits primarily on endosteal surfaces, and additionally on periosteal bone surfaces.

1.4 PROCESSES RELEASING RADIOACTIVITY INTO THE ENVIRONMENT

Radioactivity is released into the environment from a variety of systems and processes. We examine here the more important systems that lead to the significant creation and release of radioactivity. Man-made sources include nuclear weapons and the nuclear fuel cycle. The fuel cycle includes mining, milling, fuel enrichment, fabrication, reactors, spent fuel stores, reprocessing facilities and waste storage. Natural radioactivity includes primordial natural series decay products and cosmogenically generated isotopes.

1.4.1 NUCLEAR WEAPONS: FISSION, FUSION, ACTIVATION AND FUEL DISPERSAL

Modern nuclear weapons produce radioactivity from the fission, fusion and neutron activation process. The type and composition of a nuclear device markedly affects the kinds of radioactivity produced, while the location and size of detonation determine the quantity of radioactivity released to the biosphere. In a modern fission–fusion–fission weapon, the fission trigger may be fuelled either with highly enriched uranium or weapons grade plutonium. The fission trigger initiates the second stage fusion reactions. Fast neutrons from fusion then initiate the third stage fission. The third stage fission is normally fuelled with depleted uranium (^{238}U). Most of the fission yield usually emanates from this uranium. Table 1.2 contains approximations of the source term for these three processes. Most of the radio-

Table 1.2 Sources of radioactivity in nuclear explosions

Source	Isotope of interest	Strength of source
1. Fission fragments	Mass numbers 70–160	2.9×10^{23} fragments per kiloton of fission. Activity at 1 hour $= 1.5 \times 10^{10}$ Bq[a]
2. Fusion reaction products	Principally tritium ($^3_1$H) from reaction $^2_1H + ^2_1H \rightarrow ^3_1H + ^1_1H$	Depends on weapon design. Magnitude 10^4 Ci/kiloton[b] 10^{23} to 10^{24} atoms $^3_1$H/kiloton. Fusion energy release can vary from 0% to 99% of total
3. Neutron activation	Depends on explosion location and bomb case material. A principal isotope is carbon-14 from atmospheric nitrogen $^{14}_7N + ^1_0n \rightarrow ^{14}_6C + ^1_1H$	2×10^{23} neutrons liberated in fission and fusion per kiloton
4. Fissile materials	Plutonium-239	Depends on weapon design. Comes from non-fissioned weapons material plus neutron activation of uranium

a) Distribution and evolution of radioelements after nuclear explosion, Dupuis, M.J. (1970) *Bulletin of Information, Science and Technology*, **149**, 41–52.
b) Ericksson, E., *Tellus*, **17**, 118–130 (1965).

activity of interest comes from fission. An estimate of the activity of a specific isotope produced from fission energy release per kiloton of TNT explosive yield can be obtained as follows

$$A = \lambda N = \frac{0.693}{t_{1/2}}(1.45 \times 10^{23} Y) \cong 1 \times 10^{23} \frac{Y}{t_{1/2}} \text{ Becquerels.} \quad (1.1)$$

Here A is the activity of the isotope, λ is the decay constant, N is the number of atoms of the isotope produced in the explosion, $t_{1/2}$ = is the half-life in seconds, and Y is the fission chain yield (the fraction of the particular isotope out of all the isotopes and their daughter products produced in fission). For example, ^{137}Cs has a half-life of 9.5×10^8 seconds and a chain yield of 0.062 (6.2 per cent), resulting in an activity A of 6.7×10^{12} Bq/kT. The corresponding results for ^{131}I and ^{90}Sr are $A(^{131}I) = 4.4 \times 10^{15}$ Bq/kT and $A(^{90}Sr) = 6.3 \times 10^{12}$ Bq/kT. While the yields used in the above examples apply to thermal fission of ^{235}U, they can also be used as an approximation for the activity yields from fast neutron-induced fission of ^{238}U in nuclear weapons, and also for the yields from nuclear weapons fuelled with ^{239}Pu.

From these, it is easy to estimate the isotopic activity releases from Hiroshima (~15 kT), and from all the atmospheric weapons tests conducted between 1945 and 1963 (approximately 450 MT energy yield, of which approximately half was from fission, and half from fusion). Table 1.3 provides a comparison of the release of selected isotopes from weapon detonations and the Chernobyl and Windscale accidents.

The neutrons produced in fission and fusion can induce nuclear reactions that produce radioactive isotopes. One example of this neutron activation process, discussed earlier, is the reaction with atmospheric nitrogen producing carbon-14. Another is the production of iron-55 resulting from neutron activation with the bomb assembly materials ($\cong 10^{13}$ Bq per kiloton—Rotblat, 1981). For ground or near ground detonations, the escaping neutrons can induce activity in the ground, water or structures in the vicinity of the explosion. Some of the more important isotopes produced in the earth and in buildings are aluminium-28, manganese-56, and sodium-24. For explosions in salt water environments, sodium-24 and chlorine-38 can be important.

During a nuclear explosion, a considerable amount of un-fissioned uranium and plutonium is dispersed, constituting an additional injection of radioactivity into the atmosphere. In addition, neutron activation of uranium-238 can produce a considerable amount of additional plutonium. Rotblat (1981) suggests that a 10 Mt bomb with 5 Mt from fission can produce 250 kg of plutonium through activation. This suggestion, which should be considered speculative, is considerably more than the plutonium used in the trigger.

In almost all cases, the hazard of fallout of fission products from weapon explosions dominates over the neutron induced activity. The dispersed nuclear fuel normally poses a much smaller hazard, except in the case of a very low yield explosion, or in the case of an accidental non-nuclear explosion of a nuclear weapon.

Table 1.3 A comparison of radioactive releases from nuclear detonations and nuclear reactor accidents

	Radioactivity released (PBq)[a]			
Nuclide	Hiroshima	Weapon tests	Chernobyl	Windscale
^{137}Cs	0.1	1500	89	0.044
^{134}Cs[b]	—	—	48	0.0011
^{90}Sr	0.085	1300	7.4	0.00022
^{133}Xe	140	210 0000	4400	14
^{131}I	52	780 000	1300	0.59

[a] Decay corrected to 3 days after shutdown or detonation.
[b] ^{134}Cs is produced in reactors by neutron activation.
Gudiksen *et al.*, 1989; reproduced by permission of the Health Physics Society.

1.4.2 MINING, MILLING, ENRICHMENT AND FUEL FABRICATION

The nuclear fuel cycle includes mining, milling, enrichment processes, reactors, spent fuel storage, fuel reprocessing and radioactive waste storage.

In uranium mining, the principal effluent is radon-222. Release rates are about 1 GBq per ton of ore with a content of 1 per cent uranium oxide (U_3O_8). World production of mined uranium ore is about 40 kt per year. For the fuel needs of current design thermal reactors, this is equivalent to a release of about 20 TBq per gigawatt-year of reactor operation.

The extraction of the uranium from ore during milling typically leaves tailings which are a further source of atmospheric releases. Airborne emissions from a mill processing 2 kt of ore per day is roughly 1 GBq for ^{238}U, ^{230}Th, ^{226}Ra and ^{210}Pb. For ^{222}Rn, it is larger by a factor of a few thousand.

Fuel fabrication involves further processing of the uranium ore from the mill. These processes include purifying, conversion to uranium hexafluoride (UF_6), isotope enrichment of ^{235}U, conversion to metal fuel, and fabrication of fuel elements. Solid wastes produced in these fuel fabrication processes are trivial when compared to the mining and milling stages. Emissions of radionuclides in fuel fabrication are also relatively small.

1.4.3 REACTOR OPERATION

Nuclear reactors are copious producers of radioisotopes and the principal source of radioactivity in the fuel cycle. Each fission event produces two radioactive fission fragments, each of which undergoes radioactive decays, eventually producing a stable isotope. Hundreds of radioactive isotopes are produced. In addition, the intense neutron fluxes within the reactor create radioactive transmutations through neutron absorptions producing additional radioisotopes (neutron activation).

The actual inventory within a reactor core at any time depends on the type of reactor and its operating history. The primary factor is the power level as a function of time. The principal variable that determines the inventory of the longer-lived isotopes is the integrated power level (or energy output, sometimes called the burnup). The abundance of the shorter half-life isotopes approaches an equilibrium level proportional to the reactor power level. This can be understood qualitatively by a simple model of a single isotope that is being produced at a rate P, and simultaneously decaying with a decay constant λ and a half-life $t_{1/2}$

$$\frac{\mathrm{d}N}{\mathrm{d}t} = P - \lambda N \tag{1.2}$$

where N is the volume density of the isotope, and P is the rate of production per unit volume. The solution, assuming $N = 0$ at time $t = 0$, is

$$N = \frac{P}{\lambda}(1 - e^{-\lambda t}) \qquad (1.3)$$

We see that an equilibrium level (P/λ) is approached within a few half-lives (87.5 per cent of equilibrium level in 3 half-lives). If the reactor stops ($P = 0$), then a simple exponential decay ($e^{-\lambda t}$) ensues. Hence, in a typical reactor situation where the fuel elements are removed after about 3 years, we find the shorter-lived isotopes (half-lives of less than 1 or 2 years) approaching or reaching their equilbrium levels, and the longer-lived isotopes (half-lives > 5 years) building up their abundances in rough proportion to the time-integrated power level.

Table 1.4 contains an inventory of selected radioactive isotopes at 1 year and 5 years after startup, for a typical reactor operating at a power level of 3000 MW thermal (about 1000 MW electrical). Table 1.4 also contains the inventory for the same reactor after 1 year of continuous operation, for two different times after shutdown. One can see that power reactors normally contain many billions of GBq of activity. We have not listed in the table other isotopes that could be important for contamination problems in the very short term, or in the very long term, but have selected those deemed potentially important for human exposure.

In the process of normal operation of nuclear reactors some gaseous, liquid and solid wastes of low and intermediate activity are produced by fission and activation generated nuclides causing contamination of various materials, or due to releases from fuel or cladding surfaces. The type and amount of waste is determined by factors such as the reactor type, its design features, operating conditions and the integrity of the fuel.

Considering the generation of airborne radioactive wastes, three main particulate aerosol sources may be identified: emission of activated corrosion and fission products; involatile elements from radioactive decay of gases; and adsorption of fission-produced volatile radionuclides on suspended material present. Gaseous radioactive waste arises from the generation of volatile radionuclides, the most important of which are the halogens, noble gases, ^{3}H and ^{14}C. Filters are employed to remove particulate activity from contaminated gases and building ventilation air prior to discharge to the atmosphere via exhaust stacks. A combination of charcoal and particulate filters is being used to remove radioactive iodine from gas effluents. Decay-storage methods are used to reduce the quantities of noble radioactive gases discharged to the environment since these are mostly short-lived.

Gas-cooled reactors generate less liquid waste than reactors which are water-cooled and moderated, with significantly lower volumes of such waste. Virtually no liquid concentrates are generated by heavy water reactors (HWR), which operate mainly with once-through ion exchange techniques (Efremenkov, 1989). Active liquid wastes produced at nuclear power plants (PWR and BWR type) generally contain soluble and insoluble radioactive components (fission and corrosion products). Maintenance activities on plant piping and equipment generate liquid wastes from decontamination operations. In order that the bulk volume of aqueous waste

Table 1.4 Production and decay of important radionuclides produced in a reactor

Fission product		A (production)		B (decay)		
Isotope	Half-life (months)	1 year	5 years	100 days	1 year	5 years
^{89}Sr	1.7	4200	4300	1100	36	—
^{90}Sr	350	160	740	160	150	110
^{91}Y	1.9	5400	5500	1600	64	—
^{95}Zr	2.2	5500	5600	1900	110	—
$^{95}Nb^b$	1.2	5400	5600	3200	240	—
^{103}Ru	1.3	3400	3400	660	8.2	—
^{106}Ru	12	240	470	200	120	7.80
^{115}Cd	1.4	0.66	0.066	0.13	—	—
$^{125}Te^b$	1.9	3.8	15	4.3	4.0	1.40
^{127}Sb	0.13	87	87	—	—	—
^{131}I	0.27	2800	2800	0.44	—	—
$^{131}Xe^b$	0.40	28	28	0.16	—	—
^{132}Te	0.11	4100	4100	—	—	—
$^{132}I^b$	0.003	4100	4100	—	—	—
$^{133}I^b$	0.029	720	720	—	—	—
^{133}Xe	0.18	6100	6100	—	—	—
^{136}Cs	0.44	5.8	5.8	0.033	—	—
^{137}Cs	360	120	570	120	120	110
^{140}Ba	0.43	5700	5700	26	—	—
$^{140}La^b$	0.056	5700	5700	29	—	—
^{141}Ce	1.1	5300	5300	530	1.10	—
^{144}Ce	9.4	3000	4900	2300	1200	30
$^{147}Pm^b$	31	540	1800	530	440	150

Columns marked A represent the activity PBq (10^{15}dps)[a] after selected periods of continuous operation of a reactor at a power level of 3,000 megawatts thermal. Columns marked B represent the activity at specific times after shutdown or removal from a reactor that had been operating for 1 year

[a] Calculated using fission product yields.

[b] Daughter product.

IAEA, 1974; reproduced by permission of IAEA, Vienna.

may be released to the environment or recycled, waste treatment processes are employed. These treatments are based on four main techniques; evaporation, chemical precipitation/flocculation, solid-phase separation and ion exchange.

1.4.4 SPENT FUEL STORES, REPROCESSING, AND WASTE DISPOSAL

Most of the world inventory of the longer-lived radionuclides exists in spent fuel storage ponds, reprocessing plants and high-level waste storage facilities. Typically, commercial power reactor fuel elements are removed from the reactor after 3 years, and are kept in storage in 'swimming pools', usually on-site, allowing for shorter-lived activity to decay. In principle, after cooling sufficiently, the fuel elements are then shipped to reprocessing plants for the extraction of the uranium and plutonium. After extraction, the remainder is called high-level waste, and would normally be sent to a suitable waste storage facility. In practice, because of a limited reprocessing capacity, only about 5 per cent of spent fuel from commercial power reactors is reprocessed, in the UK at Sellafield, and in France at Cap de la Hague and at Marcoule. The remaining 95 per cent is kept in storage, usually near the reactor site. Japan has just begun operation of a reprocessing facility. The percentage for reprocessing from military production reactors is much higher. This limited reprocessing capability has the effect of confining these isotopes within the spent fuel, hence limiting them as a potential environmental source term. The noble gases (Kr and Xe) and iodine gas are however released under normal reprocessing conditions.

The current world total of nuclear generated electric power capacity is about 300 GW, and the annual production of nuclear electric energy is about 200 GW-year (the difference resulting from reduced operating levels and downtime). Hence, based on the inventory in a 1 GW reactor (Table 1.4), it is clear that the quantities being stored, and the rate of buildup worldwide, are enormous.

A survey of radioactive effluents from post-reactor sources has been evaluated by UNSCEAR (1988a). There, detailed inventories are kept of effluents from reactors, ponds, reprocessing plants and waste storage facilities. The isotopes of principle concern from these sources are: ^{3}H, ^{14}C, ^{60}Co, ^{85}Kr, ^{90}Sr, ^{95}Zr, ^{99}Tc, ^{106}Ru, ^{129}I, ^{131}I, ^{134}Cs, ^{137}Cs, ^{144}Ce, ^{147}Pm, ^{237}Np and the plutonium isotopes.

Reprocessing of spent fuel permits its separation into three basic components; uranium, plutonium and waste materials (fission products, actinides, cladding materials). Waste products account for only a small percentage of the total weight, the remainder being unused uranium and plutonium. Hence reprocessing enables resource conservation and ensures security of supply.

Since residual uranium, fission products and actinide elements are contained in the fuel upon discharge from the reactor, a period of on-site storage (in reactor ponds) is necessary to allow the decay of short-lived radioactivity prior to transport for reprocessing. The fuel is transported to reprocessing plants in specially designed containers (or 'flasks') which, in accordance with regulations based on International Atomic Energy Agency (IAEA) guidelines, are capable of withstanding the severest of accidents.

All industrial-scale fuel reprocessing plants have employed the Purex separation process, a solvent extraction method based on tri-*n*-butyl phosphate (TBP), or variants of it, following its development and use in the USA in 1954. After cooling, the first stage of reprocessing involves decanning, i.e. the removal of the fuels' cladding. More refractory fuel, ceramic UO_2 canned in stainless steel or zirconium alloy, is used in reactors with high power densities, e.g. PWR, BWR and AGR. This canning material cannot be easily cut or stripped from the fuel; hence chop–leach systems are used, whereby a shear is used to chop whole fuel assemblies. The chopped sections of fuel and cladding fall into a basket which is suspended in a dissolver containing nitric acid, effecting the removal of the fuel from the pins, leaving the cladding (or 'hulls'). The hulls are then withdrawn and may be treated prior to storage and disposal as intermediate-level waste. A series of TBP extraction and stripping cycles may then be used for the separation of the fuel into its constituent parts prior to eventual uranium and plutonium product finishing.

Uranium and plutonium constitute at least 96 per cent and up to 1 per cent respectively of the spent fuel, and may be recovered by reprocessing via extraction into the solvent. Almost all fission product activity remains in the aqueous phase, forming highly active waste which is stored following removal of any entrained solvent.

With regard to sources of low-level gaseous radioactivity associated with reprocessing activities, these arise principally from fuel-element shearing and dissolution processes. Considering radioiodines, most ^{131}I will have decayed as a result of storage procedures. In view of the amount which remains, and due to the presence of ^{129}I in reprocessing wastes, special chemical treatment is still required to remove these radionuclides. There has been no significant exposure due to stack releases of ^{85}Kr, moreover ^{3}H releases are small relative to amounts in liquid form. After treatment to minimize radionuclide concentrations, low-level liquid wastes, such as water from cooling ponds, may be discharged to the environment.

Leakage from spent fuel storage sites has been relatively low; there are extensive data on effluents from the major fuel reprocessing plants, with Sellafield and Cap de la Hague being major contributors. Information available regarding military facilities is more limited, but recent disclosures indicate major releases have occurred in the past from this source. Similar comments can be made about the high-level waste storage facilities. New information is now being released about a major accidental release in 1957 from a high-level waste storage tank at a military site in Kyshtym in the USSR. In size of activity this accidental release is second only to Chernobyl. New information from Hanford, Washington, in the USA, describes sizeable releases that occurred in the 1940s and 1950s from plutonium production reactor operations, fuel reprocessing and waste storage facilities.

Depending on the nature of the waste, various combinations of techniques may be used to achieve reductions in the volume and in the mobility of the radioactive content. Cutting, shredding and crushing treatments may be used to achieve volume reductions or as pre-treatments before compaction or incineration of solid low-level waste. Compaction may be used to reduce the volume of the waste, although this method does not improve waste properties with regard to longer-term management. Incineration, of suitable waste, enables very high volume and mass reduction, final packaging in containers for storage and disposal requiring no further conditioning treatments.

Intermediate-level wastes from reprocessing activities (such as cladding material, contaminated equipment, spent treatment materials) may be stored pending encapsulation and ultimate disposal. Encapsulation strategies have been developed for intermediate-level wastes, the nature of the waste determining the precise method of encapsulation used: for example, in-drum grouting may be used to encapsulate wastes such as hulls or swarf, whereas the addition of dry cement powder to materials like waste slurries may be employed.

The most significant fraction of wet-solid waste generated from the operation of nuclear power plants arises from spent ion exchange resin, with additional contributions to this category of waste from filter media and sludges (from pre-coated filters used to process liquid waste and decontamination corrosion products removed from liquid waste). Prior to final disposal of wet solids arising from liquid waste treatment, these must be converted to solid products using methods such as cementation, bituminization or incorporation into polymers.

Highly radioactive waste, in the form of a liquor, arises from the primary separation stage of fuel reprocessing. This high-level waste may be concentrated, by evaporation, for storage in cooled tanks prior to eventual disposal. It is desirable that such liquid wastes should ultimately be converted into solid form for storage and disposal. Hence, vitrification processes, based essentially on either metallic or ceramic melter techniques, are being employed for the conversion of high-level waste into borosilicate glass (Baehr, 1989).

^{85}Kr is a fission product of half-life 10.4 years and its main source is fuel reprocessing. It is produced in nuclear reactors to the extent of 4.38×10^{10} Bq per MWd(e) of fuel irradiated, but is mostly contained within the fuel rods and is released to the atmosphere during reprocessing of the fuel (for extraction of ^{239}Pu and other transuranics). Its total inventory in the global atmosphere is indicative of the amount of fuel of given irradiation reprocessed. ^{85}Kr is a rare gas and hence is not scavenged like aerosols. Unlike ^{14}C as CO_2, its solubility in water is low. Hence it is a very useful tracer for studying air mass exchange. Since the only sink for this tracer is radioactive decay, it distributes itself rather uniformly within the hemisphere. It is

Table 1.5 Normalized exposures of members of the public from solid waste disposal and globally radionuclides in effluents from the nuclear fuel cycle

Source	Normalized collective effective dose equivalent commitment (man Sv (GW a)$^{-1}$)
Mine and mill tailings and fuel fabrication (releases over 10^4 years)	150
Reactor operation	
Low-level waste	0.00005
Intermediate-level waste	0.5
Reprocessing solid waste disposal	0.05
Globally dispersed radionuclides	63
Total (rounded)	200

UNSCEAR, 1988a; reproduced with permission.

therefore used for studying air mass exchanges across the subtropical fronts and across the equator.

Table 1.5 puts some perspective on the size of these sources and the collective dose to the public. Mine and mill tailings and fuel fabrication dominate. The dose commitment in the table is integrated over 10 000 years; the dose in one human generation is quite small.

1.4.5 NATURAL RADIOACTIVITY

1.4.5.1 Primordial and natural series: the human release of natural radionuclides

While we focus in this book on the environmental dispersion of man's synthetic radionuclides, human technology also releases pre-existing natural radionuclides which otherwise would remain trapped in the Earth's crust. Most TENR ('technologically enhanced natural radiation' Gesell and Prichard, 1975) such as the mining and use of phosphatic fertilizers, monazite processing, uranium mining, and slag and pumice building construction, results in little emission of activity to the global atmosphere other than by incidental radon release (Eisenbud, 1987). For the direct atmospheric release of radioactivity, the burning of fossil fuels—coal and oil—dominates; all other anthropogenic emissions of natural activity to atmosphere such as ore smelting, deep-well hydrothermal schemes and mineral waters are negligible in comparison.

Natural radionuclides are present in the Earth's crust either because they are primordial, with half-lives comparable with the age of the Earth, or because they are continually generated by the decay of long-lived precursors.

Primordial radionuclides can be divided into non-series and series. Of the twenty or so natural non-series nuclides (radionuclides without progenitor on Earth and whose daughter is a stable isotope), only ^{40}K ($t_{1/2} = 1.25 \times 10^9$ years) and ^{87}Rb ($t_{1/2} = 5.0 \times 10^{10}$ years) are appreciable sources of activity, with average concentrations of 17 pCi and 2 pCi per gram of crustal rock (630 Bq and 74 Bq per kilogram).

The three significant series radionuclides, ^{238}U ($t_{1/22} = 4.5 \times 10^9$ years), ^{232}Th ($t_{1/2} = 1.5 \times 10^{10}$ years) and ^{235}U ($t_{1/2} = 7.1 \times 10^8$ years), each decays down a chain of short-lived radioactive daughters, the uranium series, the thorium series and the actinium series. The latter series is less important because ^{235}U now forms only 0.72 per cent of natural uranium.

The overall activity of a series radionuclide in a given rock depends not only on the total mass present but on how near the series is to radioactive equilibrium; that is, on how many of the short-lived highly active daughters are present. With the thorium series (parent ^{232}Th) equilibrium is established in less than a century. The actinium series (parent ^{235}U) requires half a million years, and the uranium series (parent ^{238}U) over a million years. Thus on geological timescales thorium is always at or near equilibrium and hence has a constant activity per unit weight of thorium, whereas uranium is often considerably out of equilibrium because of the long period needed to recover after changes in the local chemistry or the physical environment.

Two factors in particular perturb the equilibrium. Elements above radium in the decay chain (i.e. precursors of radium) are soluble in local sulphate-rich groundwaters and insoluble in chloride-rich, while radium and its daughter elements have the reverse character. In addition, the uranium series has ^{222}Rn as a member. This chemically unreactive gas has a half-life of 3.8 days, giving time to diffuse away from the parent or to transport in solution before decaying down the successive 11 members of the chain. The inert-gas members of the thorium series (^{220}Rn) and the actinium series (^{219}Rn), on the other hand, both have half-lives of seconds, not long enough to allow significant movement.

Hence a uraniferous deposit—and especially a porous sedimentary deposit—with a given uranium content can vary greatly in total activity per unit mass, and much of the activity can migrate to large distances from the parent isotope. In contrast, the activity of the thorium daughters mainly remains associated with the parent element, and is calculably constant with mass. The activity of the non-series ^{40}K is of course always closely proportional to the total potassium present.

Radioactivity associated with oil and gas fields

Most of the world's oilfields have formed over organic shales, whose uraniferous phosphate content can raise their uranium concentrations to 1000 ppm or more. The underlying porous sandstones contain brine, which continuously dissolves the ^{226}Ra ($t_{1/2}$ = 1620 years) and its daughters being generated by the uranium in the shale, carrying them to the oil and gas reservoirs above. As the uranium series is the major source of activity, individual geochemical circumstances make it impossible to predict the activity of the gas or oil even when the uranium content of the underlying beds is known.

The daughters of ^{222}Rn have half-lives of only seconds or minutes until ^{210}Pb ($t_{1/2}$ = 22.3 years) is reached, succeeded by ^{210}Po ($t_{1/2}$ = 138 days). With radium, these become the major contaminants of petroleum, and they also deposit in plant and pipelines from the radon swept out from a natural gas reservoir.

Initially the oil or gas from a new well is dry but soon water mixes with the outflow, forming an emulsion or an aerosol. This formation water contains many ions in solution, notably Group IIA cations Ca, Sr and Ba, with sulphate and carbonate anions. Temperature and pressure changes can then deposit a dense adherent scale in which radium (also Group IIA) and its daughters will co-precipitate. Most of the world's mature fields have operational problems from radioactive scale, with activities ranging from less than 1 Bq to over 15 000 Bq per gram, requiring health physics supervision, approved plant recovery procedures and active waste disposal. The main disposal route is to sea, and very few statistics of scale tonnage and activities are available. The most complete are from the UK sector of the North Sea, a field of exceptionally low activity. There, at least 10 per cent of the installations form scale and sludge above 1000 Bq g^{-1} in amounts ranging from 1 tonne to over

100 tonnes per year. Extrapolating globally, with some thousands of installations, 24 times the production (1988), and in general higher activities, a total ^{226}Ra scale discharge of 3.6 to 7.2 TBq (100–200 Ci) per year can be estimated.

Nuclides not depositing from the brine as scale are discharged in solution to seas or rivers. Activities in US production waters range from 3700 to 41 000 Bq m^{-3} ^{224}Ra, and up to 64 000 Bq m^{-3} ^{226}Ra, and up to 14 800 Bq m^{-3} ^{228}Ra. The mean of 172 German brines was 8800 Bq m^{-3} with a maximum of 286 000 Bq m^{-3} while Russian brines have been measured at 250 000 Bq m^{-3}. Worldwide, again one can only guess. In the North Sea 0.85 tonnes of formation water is discharged per tonne of crude oil produced (1989). Assuming this to apply globally, the 1988 world oil output of 3.0×10^9 tonnes was accompanied by 2.1×10^9 tonnes of water containing 2.1–21 TBq (57–570 Ci) of ^{226}Ra plus similar amounts of ^{224}Ra and ^{228}Ra.

Table 1.6 exemplifies the wide range of radon activities passing into natural gas distribution systems (the average figures are from UNSCEAR, 1977a). From the field outputs the apparent annual discharges to the atmosphere are deduced. The last column assumes all wells within the country to have equal production rates at the quoted average radon concentration, which is unlikely. No activities are quoted for the world's largest producer, the former USSR (770×10^9 m^3 in 1988), nor for the other 61 countries whose natural gas production statistics are known. World total output in 1988 was over 1900×10^9 m^3, which if averaging 1000 Bq m^{-3}, as seems possible globally, implies an annual ^{222}Rn release of 1900 TBq (50 000 Ci).

Accompanying the radon will be small amounts of pre-formed ^{210}Pb and ^{210}Po, present as aerosols. For North Sea gas, with radon averaging 35 Bq m^{-3}, ^{210}Pb and ^{210}Po each amount to 0.01–0.05 Bq m^{-3} at the land terminal, but fields with higher radon concentrations or less-effective filtering may emit more.

Table 1.6 Some natural radon gas activities and outputs

Source	^{222}Rn Range (Bq m^{-3})	^{222}Rn Average (Bq m^{-3})	Gas output (109 m^3 y^{-1})	Activity released (Ci y^{-1})	Activity released (TBq y^{-1})
North Sea (UK)	30–40	35	45	43	1.6
North Sea (ND)	40–1700	74	88	241	9
West Germany	100–4000	200	54	270	10
United States	185–54 000	1300	473	16 600	600
Canada	150–20 000	8700	98	23 000	900
Nigeria	40–60	67	3	5	0.2
Indonesia	55–92				

Radioactive emissions from coal burning

As the world's primary fossil fuel, 4.6×10^9 tonnes of coal are burnt each year (1987) with about 70 per cent used in large-scale power generation. All coals contain radionuclides of the uranium and thorium series, with the uranium as well as the thorium more nearly in secular equilibrium than in oil and gas fields. The equilibria established in the coal are destroyed in combustion as the chemical elements distribute according to their volatility over the range of ash particle sizes.

Radionuclides which have been measured in coal are ^{238}U, ^{234}U, ^{234}Th, ^{232}Th, ^{230}Th, ^{228}Ra, ^{228}Th, ^{228}Ac, ^{226}Ra, ^{214}Pb, ^{214}Bi, ^{212}Pb, ^{212}Bi, ^{210}Pb, ^{210}Po, ^{208}Tl and ^{40}K. Table 1.7 summarizes the range of activities for a wide variety of coals (Corbett, 1980; UNSCEAR, 1977a). The figures are roughly consistent with secular equilibrium. On burning, the non-volatiles remain in the ash, which for UK coals averaging 16 per cent implies a concentration by five or six times. Within the ash there is a partitioning between bottom and fly-ash with Th, U, and Ra fairly evenly distributed but ^{210}Pb and ^{210}Po enhanced in the finer fly-ash particles by as much as a further factor of six.

Table 1.7 Radionuclide concentration ranges in coal and ash (Bq kg^{-1})

	Coal	Bottom ash	Fly-ash
^{238}U	9–31	56–185	70–370
^{234}U	19	92	160
^{232}Th	9–19	59	81–174
^{228}Ra	6–20	18–78	63–130
^{228}Th	1–20	56–81	15–130
^{226}Ra	7–25	20–166	85–281
^{210}Pb	10–26	21–185	52–1813
^{210}Po	41	13–185	196–466
^{40}K	26–130	230–962	233–740

New coal-fired power stations in the United Kingdom now nominally retain 99.3 per cent of the fly-ash and at this rate the radioactive emission from a 1000 MW plant using coal of 16 per cent ash content would be:

^{222}Rn	7.5×10^{10} Bq y^{-1}
^{210}Pb and ^{210}Po	5.3×10^{9} Bq y^{-1}
^{40}K	3.7×10^{9} Bq y^{-1}
Th, U and Ra	5.3×10^{8} Bq y^{-1}

If all the world's coal-fired electricity generation were from such stations, the total annual emission would be over 1000 times the above, i.e. a total of 85 TBq (2300 Ci), with 5 TBq (140 Ci) of ^{210}Pb and ^{210}Po and 0.5 TBq (14.3 Ci) as Th, U and Ra. But few other countries achieve so low a fly-ash emission; parts of North America rarely exceed 97.5 per cent retention and most of the world falls far short of modern European and US standards. Indeed, it is probable that the majority of 'Third World' power stations use conventional cyclone separators only, releasing as much as 10 per cent of their fly-ash to the atmosphere. Worse still, in brick and pottery manufacturing aerosol retention world-wide is usually zero, and it has been suggested that the total release of ^{210}Pb and ^{210}Po from these industries could exceed that from power generation (Peute *et al.*, 1988). The global radon emmission is of course in direct proportion to coal tonnage consumed, i.e. 75 TBq (2030 Ci), since 100 per cent of this volatile gas is released whatever the fly-ash retention. But the total of thorium, uranium, radium, lead and polonium is likely to be at least ten times the above figures, i.e. 50 TBq (1400 Ci) of ^{210}Pb and ^{210}Po and 5 TBq (140 Ci) Th, U and Ra, totalling with the radon 130 TBq (3570 Ci) per year.

Environmental measurements around coal-fired power plants show levels of all the Table 1.7 nuclides to be up to 10 times the 'natural' concentrations, while long-term studies of glacial snow 150 km from an industrial centre found a 50-fold increase in ^{226}Ra concentration over 80 years (UNSCEAR, 1977a).

The global perspective

Large though these TENR emissions may seem, especially of the potentially hazardous ^{226}Ra and ^{210}Pb, they are by orders of magnitude smaller than global natural emissions (Robson, 1983), although on the local scale the man-made sources may dominate. One way of comparing impacts is to convert releases to the flow of ionization energy, which can differ by orders of magnitude for equal numbers of disintegrations of different nuclides. This takes into account the sum of energies of the parent and daughters and measures the true dose consequences in the environment (Jaworoski, 1982). Considering TENR emissions directly to atmosphere, radon from oil and gas wells greatly exceeds that from coal burning, but while the ^{226}Ra and ^{210}Pb from coal in Table 1.8 is that fraction going to atmosphere, almost all Ra and Pb from oil and natural gas is discharged to sea.

It is also of interest that one consequence of fossil-fuel burning, the Suess effect, *decreases* atmospheric specific radioactivity. The 6.7×10^9 tonnes of carbon released annually as carbon dioxide from coal and oil contains no ^{14}C, and its dilution into the atmospheric 6.6×10^{11} tonnes of carbon (containing 62 tonnes = 1.1×10^7 TBq or 285 MCi of ^{14}C at equilibrium) implies simplistically a reduction of atmospheric ^{14}C concentration by 1 per cent per year. In actuality, because of ocean and biosphere involvement, there was a decrease in atmospheric ^{14}C concentration by 3–4 per cent from this cause between 1850 and 1950 (Aitken, 1978). After that a rise of at least 50 per cent was seen from nuclear weapons testing, but since 1970 the fall probably continues.

Table 1.8 Relative emissions calculated as radiation energy flow, (J s^{-1})

	Nuclear power including mining, 1981	Coal burning in 1980	Natural gas and oil	Natural sources
^{222}Rn	53 000	3000	75 000	120 000 000
^{226}Ra	0.22	10	20	1100
^{210}Pb	8	26	18	19 000

1.4.5.2 Cosmogenically produced isotopes

Radioactivity is easily measured in very small quantities and hence serves as a very useful tracer for atmospheric circulation.

Radioisotopes used as atmospheric tracers are both of natural and man-made origin. Amongst natural sources are cosmogenic and terrestrial origin isotopes whereas man-made sources comprise fission and activation products from nuclear weapons tests and the nuclear fuel cycle including nuclear power stations and fuel reprocessing.

Cosmogenic radioisotopes are produced by cosmic ray bombardment of the Earth's atmosphere. The important cosmic ray produced isotopes are ^{3}H, ^{7}Be, ^{14}C, ^{22}Na, ^{32}P, ^{33}P and ^{33}S. These isotopes are produced by spallation reactions with the atmospheric nitrogen, oxygen and argon, by high-energy cosmic rays, mostly protons and neutrons. Spallation reactions are those in which nuclei are split into smaller nuclei due to the very high energy of the bombarding cosmic ray particles. Table 1.9 lists some of the cosmic ray produced isotopes, their production modes and range of concentrations observed in the air and rainwater. Production is maximum in the upper atmosphere at high latitudes where cosmic ray particles are maximum. Consequently, these isotopes are good tracers for studying stratospheric–tropospheric exchange processes. Detailed calculations are available on the expected concentrations of these isotopes at various latitudes and altitudes (Lal *et al.*, 1958).

The second type of natural isotopes which can be used as atmospheric tracers are radon and thoron and their decay products. The soil always contains varying amounts of uranium-238, thorium-232, uranium-235 and their daughter products. The radon (^{222}Rn), thoron (^{220}Th) and actinon (^{219}Rn) gases formed in the decay of uranium-238, thorium-232 and uranium-235 series, respectively, escape through the pores of the soil and diffuse into the atmosphere.

Actinon, although present in the atmosphere, is in such small quantities that it is not of much significance. Even for radon and thoron, their emanation over the oceans is negligible in comparison to that over the land due to the low radioactivity content of sea waters and lower emanation rate of radon and thoron as compared to continental areas. The daughter products of the gases, radon and thoron, are solids and condense on the fine aerosols always present in the atmosphere forming radioactive particulates. Since these radioactivities, both gaseous and particulate, orginate from the soil, their concentrations generally decrease with height and at high altitudes only the longer-lived isotopes are found in measurable concentrations. The behaviour of these radioactive materials in the atmosphere thus can be used for the study of the atmospheric circulation processes of the lower atmosphere.

Fission and activation products dispersed as global fallout from nuclear tests are the most important, and to a lesser extent are those from reactor accidents and other releases from peaceful uses of nuclear energy such as fuel reprocessing. Dispersal patterns of these isotopes depend on the radioactivity release site and height, magnitude or amount of release and meteorological and other factors of the release event. High-yield tests push the bomb cloud into the stratosphere. In the case of low-yield tests, the dispersal of the radioactivity is by tropspheric wind systems.

Table 1.9 Radionuclides produced by cosmic rays

Primary radio-nuclide	Atmospheric half-life	Production mode	Approximate concentration range	
			Air ($Bq/10^3\ m^3$)	Rain ($Bq/10^3 l$)
^{3}H	12.3y	$^{14}N(n, ^{12}C)^{3}H$	0.020–0.200	200–4000
^{7}Be	53 d	Spallation (N and O)	0.080–30	400–4000
^{10}Be	1.6×10^{6} y	Spallation (N and O)	—	$4 - 40 \times 10^{-5}$
^{11}C	20.3 min	Spallation (N and O)	—	—
^{14}C	5730 y	$^{14}N(n, p)^{14}C$	—	40
^{18}F	1.8 h	$^{18}O(p, n)^{18}F$		
		Spallation (Ar)	—	
^{22}Na	2.6 d	Spallation (Ar)	$(2 - 8) \times 10^{-4}$	0.004–0.04
^{24}Na	14.7 h	Spallation (Ar)	$(2 - 8) \times 10^{-3}$	4–40
^{28}Mg	20.9 h	Spallation (Ar)	—	0.2–1.20
^{31}Si	2.6 h	Spallation (Ar)	—	0.8–8
^{32}Si	104 y	Spallation (Ar)	$(4 - 20) \times 10^{-5}$	$(4 - 20) \times 10^{-3}$
^{32}P	14.3 d	Spallation (Ar)	0.008–1.0	20–200
^{33}P	25.3 d	Spallation (Ar)	0.008–1.0	20–200
^{34m}Cl	32.2 min	Spallation (Ar)	—	4.0–32
^{35}S	87.5 d	Spallation (Ar)	0.008–1.0	20–200
^{36}Cl	3×10^{5} y	$^{35}Cl(n, \gamma)^{36}Cl$	—	—
^{37}Ar	35 d	$^{37}Cl(n, p)^{37}Ar$	—	—
^{38}Cl	37.2 min	Spallation (Ar)	—	200–3000
^{38}S	2.8 h	Spallation (Ar)	—	20–200
^{39}Ar	269 y	$^{38}Ar(n, \gamma)^{39}Ar$	—	—
^{39}Cl	55.6 min	$^{40}Ar(n, np)^{39}Cl$	—	400–4000
		& spallation (Ar)		
^{41}Ar	1.8 h	$^{40}Ar(n, \gamma)^{41}Ar$	—	—
^{81}Kr	2×10^{5} y	$^{80}Kr(n, \gamma)^{81}Kr$	—	—

In both cases the latitude and season of testing affect the dispersal and distribution of the radioactivity. The radioactive releases from nuclear tests and reactor accidents act as 'point sources' specific of the location and time, unlike natural radioactivity which is a 'spread continuous source'. This difference is of significance in their use as a tracer. For most tracers the isotope is present on very small (micron size) particulates which follow the air mass motions.

1.5 PHYSICAL AND CHEMICAL FORMS, AND BIOSPHERIC TRANSFERS

Radionuclides appear in various physical and chemical forms depending on condition of release and transport and on the properties of the elements involved. A general distinction can be made between gases, aerosols and particulate material. Gases may be inert, like the noble gases, or reactive like elemental iodine, interacting with aerosols or other surfaces. Particulates carrying radionuclides may result from an initial release of particles, or from a condensation to form solid particles. Radionuclides may also condense on natural aerosols. Particles with a high specific activity, called hot particles, can result from nuclear weapon detonations and nuclear reactor accidents.

1.5.1 GENERAL ASPECTS

Once artificially created radionuclides are released, their subsequent evolution is determined by their physical and chemical properties through biogeochemical processes. It is convenient to categorize these nuclides into chemical groups that share similar chemical volatility, transport properties, and aerosol characteristics such as particle size and settling velocities. For nuclear explosions, these differing group characteristics produce the phenomenon known as fractionation, in which the mix of isotopes within fallout changes with distance and time as the fresh fission products move through the atmosphere. For reactor accidents the fraction of core inventory released depends on the accident scenario as well as the chemical properties. In Table 1.10 a listing of the chemical groups is presented along with a compar-

Table 1.10 Estimated fractional release of radionuclides available for long-range transport from Chernobyl (Gudiksen *et al.*, 1989) compared with the USAEC reactor safety study (WASH-1400, 1975) and the Windscale reactor accident (Clarke, 1974)

WASH–1400 (%) Chemical group	Chernobyl estimate	Range estimate	Best estimate	Windscale estimate
Noble gases				
Xe, Kr	100	50–100	90	100
Halogens				
I, Br	80	50–100	90	20
Alkali metals				
Cs, Rb	40	40–90	80	20
Tellerium group				
Te, Se, Sb	10	5–25	15	20
Alkaline earths				
Ba, Sr	1.2	2–20	10	0.2
Noble metals includes Ru, Rh, Pd, Mo, Tc	0.9	1–10	3	2.0
Rare earths and actinides includes Y, La, Ce, Pr, Nd, Pm, Sm, Eu, Np, Pu, Cm	0.2	0.01–1	0.3	0.2
Refractory oxides				
Zr, Nb	0.2	0.01	0.3	0.2

ison of the estimated fractional releases in the Chernobyl and Windscale accidents. Also included in the Table is an estimate of release fractions made in the US Atomic Energy Commission reactor safety study WASH-1400 (1975). These fractional release estimates illustrate how important the physical and chemical properties are in determining the source term for a reactor release.

1.5.1.1 Noble gases

Noble gases are emitted from many nuclear installations. Short-lived isotopes like ^{87}Kr, ^{88}Kr, ^{133}Xe and ^{135}Xe can be released from nuclear power plants when fuel pin cladding fails. Some plants have decay systems to limit such releases. Fuel reprocessing plants emit significant amounts of ^{85}Kr ($t_{1/2} \approx 10.4$ years). Releases also occur from nuclear weapon testing and in nuclear accidents.

The noble gases with the exception of Xe are chemically inert and show little adsorption on surfaces at environmental temperatures. They can be trapped in cooled activated carbon filters and later released by warming. Such techniques have been carried out for analytical purposes in the vicinity of nuclear power plants and after nuclear weapons testing.

1.5.1.2 Iodine

Halogens are volatile elements of Group VII. Bromine and iodine are important fission products and may be released from defective nuclear fuel elements. They are very chemically reactive, soluble in water and strongly adsorbed on filters and surfaces. The halogen of importance in this study is iodine, which has the isotopes: ^{129}I, $t_{1/2} = 1.7 \times 10^7$ years; ^{131}I, $t_{1/2} = 8$ days; ^{132}I, $t_{1/2} = 2.8$ hours; ^{133}I, $t_{1/2} = 20.8$ hours. The most important isotope is ^{131}I, which can appear in atmospheric releases as elemental iodine, or adsorbed on particles, or as caesium iodide and other organic forms. It also forms volatile organic compounds, such as methyl iodide.

A distinction can be made between normal operation and accidents. Iodine penetrates particulate filter during normal operation when gaseous elemental iodine is present. For this reason charcoal filters are added to particular filters in the off-gases of nuclear facilities. Even then some iodine can penetrate the filters when volatile organic compounds such as methyl iodide are present, needing special filters.

The substantial environmental impact of iodine released in a nuclear power plant accident has generated extensive research on the question of iodine forms. These very much depend on the accident conditions. A sudden overheating of the fuel to temperatures of 1800–2000 °C or more will release substantial amounts of iodine. Initially the iodine is released as atoms which react chemically with other elements to form, for example, caesium iodide. Water present in the reactor containment will trap caesium iodide, which is highly soluble, as are several other forms of iodine. Particulate iodine may settle on surfaces. Only minor amounts of organic iodides are expected to form in a LWR core melt-down accident.

Having been released from the reactor plant, the behaviour in the environment is very much dependent on the composition of iodine in the atmospheric plume. Elemental iodine and particulates will wash out quite efficiently in rainfall and also adhere to ground and vegetation surfaces during dry conditions. The organic iodides are more inert and their dry deposition on surfaces is two orders of magnitude lower than elemental iodine. During atmospheric transport the composition of iodine in a plume may change, depending on its interaction with aerosols and moisture. It is of interest that uptake of organic iodides in lungs after inhalation is as high as the uptake of elemental iodine, in contrast to the different behaviour when in contact with other surfaces.

1.5.1.3 Tritium

Tritium is formed in small amounts by spallation reactions of cosmic rays with oxygen and nitrogen in the stratosphere. Hydrogen bomb testing has added substantial amounts to the natural background level, as have releases from nuclear power stations and reprocessing plants. In the future, nuclear fusion plants will also contribute. Tritium can appear mainly as hydrogen gas or as water, or can be incorporated into organic materials. Hydrogen gas will be oxidized to water in the atmosphere with a half-life of some years. A fast oxidation of tritium to water can occur in the upper soil layer by bacterial activity. The radiotoxicity of tritium is small compared to many other radionuclides, as its beta energy is low, and it widely distributes in the environment by its chemical identity with hydrogen.

1.5.1.4 Carbon-14

Besides continual generation in the atmosphere by cosmic ray interaction, the beta emitter carbon-14, $t_{1/2} = 5600$ years, is emitted by nuclear power plants and fuel reprocessing facilities. Concern for long-term global collective doses has initiated special studies of carbon-14 releases. Measurements on BWRs and reprocessing plants show that their carbon-14 is emitted almost exclusively as carbon dioxide, whereas from PWRs there could be substantial fractions as carbon monoxide, methane or other hydrocarbons.

1.5.1.5 Semi-volatile elements

Typical radionuclides in this group are caesium and tellurium. Of special interest for radiation protection are caesium-137 (half-life 30 years) and caesium-134 (half-life 2 years). The long-lived caesium-137 is of major concern in ground contamination after a nuclear accident as well as in nuclear weapon testing. Nuclear power plants have air filters to prevent significant releases during normal operation.

Caesium is more volatile than strontium, another semi-refractory element, and its migration in nuclear fuel during operation is much higher. Release fractions of strontium are therefore generally lower than those for caesium.

During atmospheric transport, caesium, like most radionuclides, will appear as aerosols, or adsorbed on to aerosols. In contact with water, part of the caesium can be dissolved and appear in ionic form, available for uptake into living organisms.

1.5.1.6 Non-volatile elements

The nuclear fuel matrix containing uranium, some transuranium elements, and fission products such as cerium, zirconium and ruthenium, has a high melting point and is in practice non-volatile. The elements may be released to the environment only under special conditions, such as that of an explosion creating small particles which can be carried by the winds. Only minor amounts of such aerosols are released during normal operation. Activation products, such as ^{60}Co from construction materials in the reactor, may be more common in accident releases than fission product aerosols.

1.5.2 SIGNIFICANT ELEMENTS IN NUCLEAR ACCIDENTS

The few nuclear accidents which have resulted in significant atmospheric releases are individual in cause and consequence, and do not allow many general conclusions about the forms of radionuclides released. In the Windscale accident in 1957 the graphite fire released substantial amounts of iodine. The 1957 Kyshtym accident in the Urals, an explosion in a high-level waste storage tank of fission products from reprocessing, released about 2×10^{17} Bq of strontium and caused significant ground contamination during plume transport. Most of the caesium had previously been extracted from this waste. In the TMI-2 accident in 1979 most of the released fission product from the core was retained within the containment, and only minor amounts of noble gases and iodine were released to the environment. The Chernobyl accident was unique in that the high release fraction to the environment included a significant part of the fuel matrix and a release period of 10 days, which resulted in a wide dispersion of material of changing composition over a wide area.

Estimated release fractions of the Chernobyl core inventory to the environment appear in Table 1.10. The fractions were very high because of the severe core damage and lack of containment or mitigation processes. A surprising phenomenon was the tremendous release of fuel material of a range of particles sizes to the environment. These ‘hot particles’ were spread over wide areas not only during the initial phase but also during the second release period several days later. They had various compositions but fall into two main categories; (a) fuel fragments, including fission products depleted in the volatile and semi-volatile elements iodine and caesium; (b) particles containing predominantly ruthenium. The largest particles were deposited close to the release point but hot particles of the size of up to several microns were found even in Scandinavia. The volatile and semi-volatile nuclides ^{131}I, ^{103}Ru, ^{132}Te and ^{137}Cs were carried on particles with a median diameter of

0.3–0.9 μm (sampled in Germany and Scandinavia). There are indications that iodine was associated with particles in the lower part of this range. ^{90}Sr was apparently carried by somewhat larger particles.

Iodine appeared in both gaseous and particulate fractions and the composition was variable with time. Part of the gaseous iodine was organically bound as methyl iodide, which has a lower deposition velocity compared with elemental or particulate iodine. The unique characteristics of the plume affected the subsequent behaviour in the environment. For example, non-volatile elements carried by fuel fragments had a higher dry deposition velocity than volatiles such as caesium carried by other types of particles. On the other hand, washout of caesium was more efficient during rainfall. The properties of the fallout material have been studied with special emphasis on ionic and colloid fractions for caesium, as such fractionation results in part of the caesium fallout being less available for uptake in living organisms.

2 Case-studies of Significant Radioactive Releases

Co-ordinator: **L. J. Appleby**

Contributor: **S. P. Luttrell**

2.1 INTRODUCTION

This chapter examines specific data which are available as a consequence of various planned and accidental releases of radioactivity to the environment. The case-studies presented do not comprise an exhaustive treatment of all incidents involving radioactive discharges, but include examples considered to be of particular significance, especially with regard to elucidation of environmental pathways. Releases associated with waste management operations, due to plutonium production and fuel reprocessing activities, are examined in connection with planned discharges. A number of case-studies relating to non-routine releases, arising from nuclear warhead production (Kyshtym) and reactor operations (Windscale, Three Mile Island, Chernobyl) are then examined chronologically in detail, and information relating to the re-entry of nuclear powered satellites (SNAP 9A and Cosmos 954) is presented. Another significant potential source of radioactivity in the environment exists due to the practice of ocean disposal of packaged radioactive waste. Further discussion of this issue appears in Chapter 5.

2.2 SELLAFIELD

The Sellafield complex, located on United Kingdom's west coast in Cumbria (see Figure 2.1), is the largest of British Nuclear Fuel's (BNF) sites and has been involved with the reprocessing of spent fuel since 1952. The current reprocessing plant at Sellafield began operation in 1964, replacing the earlier plant, its main function being to reprocess fuel from Magnox reactors. A new fuel handling plant concerned with the receipt, storage and decanning of Magnox fuel was commissioned in 1985 (primarily to increase the storage capacity and throughput of spent fuel for reprocessing), the year in which the site ion exchange plant (SIXEP) began operation. BNF has also built a large vitrification plant (for the conversion of highly active waste into solid glass blocks) and store at Sellafield. The enhanced actinide removal plant (EARP) and thermal oxide reprocessing plant (THORP), a new facility to reprocess fuel from more modern nuclear power stations in Britain and over-

Radioecology after Chernobyl
Edited by F.E. Warner and R.M. Harrison
 Published by John Wiley & Sons Ltd

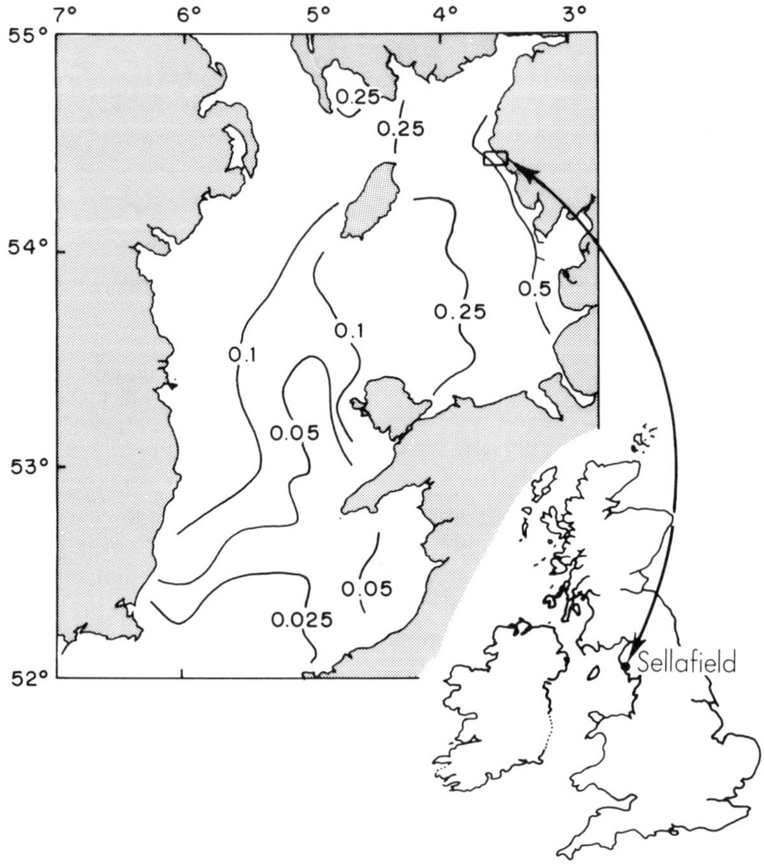

Figure 2.1 The concentration of ^{137}Cs (Bq kg^{-1}) in filtered water from the Irish Sea, April 1987 (from Hunt, 1988. Copyright © British Crown copyright, 1988.)

seas, are scheduled for commissioning in the early 1990s. Apart from reprocessing and waste management activities, Sellafield is also the location of the world's first industrial-size nuclear power station, Calder Hall.

Considering releases of radioactivity to the environment, discharges from Sellafield are subject to authorization (by appropriate UK regulatory bodies), inspection, monitoring and assessment. The authorization is kept under review, subject to advice from national and international organizations, with due account of any discharge reductions achieved. Six main categories of effluent from reprocessing operations at Sellafield may be identified; pond purge liquours (from pond storage of spent fuel) and arisings from dissolution and solvent extraction processes which include medium active salt-free concentrate, medium active salt concentrate, highly active effluents, low active effluents and spent solvent (Horsley and Howden, 1990).

Another source of environmental activity arises from the disposal of materials originating from within controlled areas and the immediate working environment of personnel on site (which have not been in contact with process materials), comprising solid low-level waste. These wastes are disposed in shallow trenches at Drigg, 6 km from Sellafield.

With regard to the aquatic environment, activity concentrations in the Drigg stream, into which the solid low-level waste disposal trenches are presently drained, have not, on average, exceeded 5 per cent of the authorized limits (1.5×10^5, 2.0×10^6 and 1.0×10^9 Bq m^{-3} respectively for alpha, beta and ^3H: Donn, 1990). Moreover, whilst most (over 99 per cent) of the radioactivity in waste generated from fuel reprocessing operations is, in fact, presently stored on site, there are routine discharges of low-activity liquid waste (containing a variety of radionuclides) into the north-east Irish Sea from Sellafield. This occurs via pipelines which terminate 2.1 km beyond the low water mark. The fuel element storage ponds and the reprocessing plant are responsible for the most significant liquid radioactive discharges. The peak beta release from Sellafield for any year was 9×10^{15} Bq in 1975, a peak in the alpha release occurring in 1973 with a discharge of 1.8×10^{14} Bq (mostly in the form of plutonium). The introduction of new plant (e.g. SIXEP and the salt evaporator) has caused discharges to the aquatic environment to decrease considerably in recent years. During 1988 discharges from the Sellafield pipelines involved the release of 8.1×10^{13} Bq total beta activity and 2.1×10^{12} Bq total alpha emitting radionuclides (British Nuclear Fuels, 1989a). Further reductions in discharges will be achieved through EARP, and the stringent effluent standards of THORP will lead to only a limited contribution to site liquid discharges of isotopes which are radiologically significant.

Radionuclides discharged to the aquatic environment may be accumulated by aquatic plants and animals, which may subsequently form human foodstuffs, terrestrial livestock feeds or crop fertilizers. Important pathways of exposure arise from fish and shell-fish consumption. Previously, harvesting of *Porphyra* used in the manufacture of laverbread, was significant; however in recent years this has not been harvested in the immediate vicinity of Sellafield. Additional pathways to the terrestrial environment (and humans) result from grazing farm animals in intertidal zones, from the inhalation of water vapour and gases which arise from the water or from external gamma-ray exposure over sediments. Radiocaesium and transuranic radionuclides are considered the more important with regard to contributors to exposure from discharges to the aquatic enivronment (Hunt, 1988). The concentration of ^{137}Cs in filtered water from the Irish Sea in April 1987 is shown in Figure 2.1. Monitoring programmes have revealed that the decline in discharges is associated with an approximately parallel decrease in radionuclide concentrations in foodstuffs.

Hence, it is apparent that radionuclide pathways may be varied and in considering dose to humans, as an indicator of the radiological impact of Sellafield releases, it is important to take into account the different contribution of radionuclide concentrations in various foodstuff tissues. The mechanisms by which doses are received by members of the public, as a result of discharges from the Sellafield plant, are continuously reviewed. Radiation exposure of the general public from Sellafield discharges as a consequence of fish consumption is primarily due to radiocaesium whereas a wide range of radionuclides contribute to radiation exposure of shell-fish consumers. Since the radiation dose to fish and shell-fish consumers varies according to the product of the mass of the foodstuff consumed and the concentration of radioactivity contained, a range of annual doses may be expected. Thus, a 'critical group' approach is applied, based upon the identification of groups of individuals exposed to the highest radiation doses. The committed effective dose equivalent to the local sea-food consuming critical group was just under 4×10^{-4} Sv in 1988, which is within the International Commission on Radiological Protection's (ICRP) recommended principal dose limit of 1×10^{-3} Sv y^{-1} for members of the public. In recent previous years exposures are considered to have remained within this principal limit, whilst over a period of a few years prior to this exposures exceeded 1×10^{-3} Sv y^{-1}, although remaining within the ICRP subsidiary dose limit of 5×10^{-3} Sv y^{-1}. Since dose-rates greater than 1×10^{-3} Sv y^{-1} did not occur over a sufficiently long period to exceed the 1×10^{-3} Sv y^{-1} average lifetime exposure, the dose limitation objectives of ICRP will be achieved. For typical fish-eating members of the public the average dose in 1988 was around 1×10^{-5} Sv.

As noted above, radiation exposure of the public may also arise as a result of the uptake of gamma-emitting radionuclides by intertidal sediments. At present these gamma dose rates are primarily attributable to caesium. It is found that radioactivity is most readily absorbed by fine-grained muds and silts (such as are often found in estuaries and harbours) than by coarser-grained sands (e.g. on open beaches). Monitoring of the period of time spent in intertidal areas at various locations has been undertaken to identify members of the public subject to the highest external exposures. This group comprises boat-dwellers located along the intertidal area of the coastline boardering the north-east Irish sea, who received lower radiation doses than the critical group for sea food in 1988 (British Nuclear Fuels, 1989a).

Considering non-aquatic pathways, the release of treated ventilation air from the process plants and the Calder Hall reactors is primarily responsible for atmospheric discharges, leading to low concentrations of radioactivity in the air and in foodstuffs such as milk and vegetables as a result of deposition onto soil or vegetation. Radiation doses to the critical group due to discharges via these atmospheric pathways were less than 2×10^{-4} Sv in 1988. The critical group which is most affected in the case of gaseous discharges has been indentified as children drinking milk from local farms (British Nuclear Fuels, 1989c). Further studies have revealed no excess airborne radioactivity (and accordingly a negligible dose impact) arising from the Drigg site. Direct radiation from the plants, rather than from discharged waste, is mainly due to gamma radiation. This direct radiation mostly affects a small group of people near Sellafield, who received an estimated dose of less than 2×10^{-4} Sv in 1988 (British Nuclear Fuels, 1989a).

In conclusion, routine discharges from the entire nuclear industry actually account for less than 0.1 per cent of the radiation received by the UK population: The UK collective dose commitment (truncated at 500 years) due to Sellafield discharges represented less than 50 per cent of the total collective commitment for the nuclear fuel cycle in 1986 (Berry and Coulston, 1989).

2.3 CAP DE LA HAGUE

Cogéma's Cap de la Hague centre, located on the Normandy coast of France (see Figure 2.2), commenced reprocessing spent fuel from commercial graphite/gas reactors in 1966. The complex was completed 10 years later by a high-activity oxide

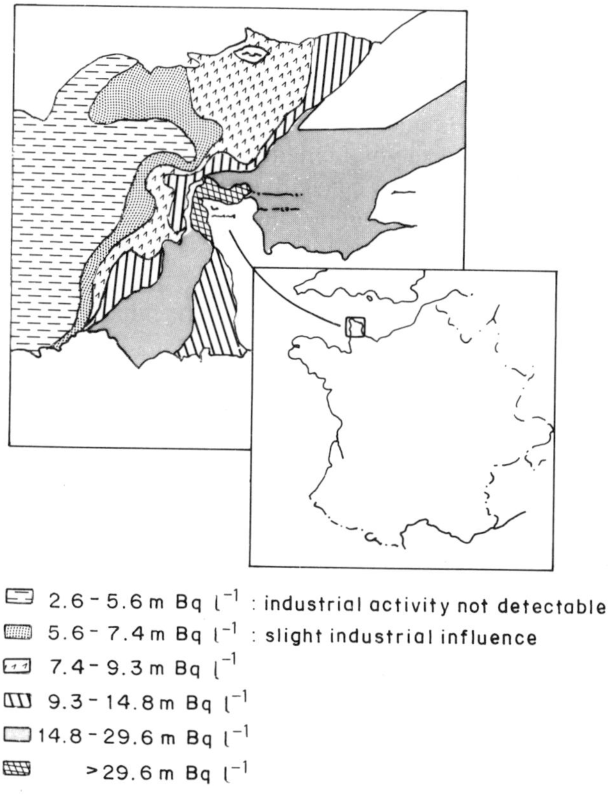

Figure 2.2 Location of Cap de La Hague and the distribution of ^{137}Cs in the locality in 1983 (from Gueguenlat *et al.*, 1988; reproduced by permission of Elsevier Science Publishers, Ltd).

head-end facility to provide for reprocessing light water reactor (LWR) fuel. Almost 3000 t of LWR fuel has been reprocessed at the plant's Usine Plutonium-2 (UP2)/400 facility to date (Rougeau and Lallement, 1990), with commissioning of the UP3 facility, which has a nominal capacity of 800 t $year^{-1}$, during 1990. Work is also being undertaken to increase the capacity of UP2 to 800 t $year^{-1}$. The capacity of Cap de la Hague is intended to reach 1600 t y^{-1} by 1992. With regard to the transportation of spent fuel, since 1975 over 5000 t of LWR spent fuel has been shipped to this complex, from western Europe and Japan, without incident (Aycoberry and Rougeau, 1988).

The amount of waste arising from reprocessing spent graphite/gas fuel is decreasing and being replaced by that due to LWR fuel reprocessing. Considering waste management techniques at Cap de la Hague, high-level radioactive wastes generated from reprocessing operations undergo a period of on-site storage, prior to incorporation in borosilicate glass (through vitrification techniques), further interim storage, with eventual deep geological disposal. Depending on the specific nature of low- and intermediate-level wastes various containment procedures may be employed utilizing, for example, bitumen, cement or polymer resins, or metal boxes or casks (Beniston *et al.*, 1987; Rougeau and Lallement, 1990). Shallow land (near-surface) disposal (e.g. at Centre de la Manche) is then used for containerized short-lived wastes, whereas long-lived wastes will be disposed in deep geological formations. Another technique employed, for low-activity liquid and gaseous effluents, is environmental dispersal.

Considering routine discharges of low-level radioactive wastes from Cap de la Hague, one of the principal releases concerns liquid effluents discharged to the aquatic environment. The release of liquid effluents is subject to very strict controls, with specification of maximum permissible activities. Various effluent treatment processes may be utilized (based upon e.g. techniques such as co-precipitation and ion exchange) before discharge to the sea occurs. Statutory analyses are undertaken prior to the release of liquid wastes to assess the specific activity of radionuclides and the total activity of the release.

The absolute amount of activity discharged annually from Cap de La Hague is, in fact, much lower than from Sellafield. It should also be noted that the composition of nuclides released from the two plants differs. At Cap de la Hague, besides tritium, the principal radionuclides released include; ruthenium, caesium, strontium and antimony. Routine discharges into coastal waters made by Cap de la Hague, up to 1985, involved total releases of approximately: 1.019×10^{16} Bq, ^{3}H; 4.905×10^{15} Bq, ^{106}Ru; 9.4×10^{14} Bq, ^{137}Cs; 7.55×10^{14} Bq, ^{90}Sr; 3×10^{12} Bq $^{238}Pu + ^{239/240}Pu$ and 1×10^{15} Bq, ^{125}Sb (Pentreath, 1988). The total discharge of caesium isotopes from Cap de la Hague is some 3×10^{13} Bq annually. The distribution of activity in the Channel due to ^{137}Cs is shown in Figure 2.2. Since the mid-1970s, however, the ^{90}Sr contribution to Cap de la Hague's efluents has exceeded that due to ^{137}Cs, and peaks in the discharges of ^{90}Sr are found to occur every four to five years. The activity ratio for $^{90}Sr/^{137}Cs$ has, in recent years, been greater than 1.0, and in 1983 the ratio was of the order of 12. A maximum in the release of ^{137}Cs from Cap de la Hague occurred in 1971 whereas a maximum release of ^{90}Sr occured in 1983.

Discharges from the reprocessing operations of the Cap de la Hague plant give rise to contamination effects in the North Sea, contributing to concentrations of plutonium isotopes, americium and curium. The contribution of caesium isotopes contained in Cap de la Hague's liquid effluents is, however, small. Examination of transuranic elements (e.g. plutonium and americium), in 1987, revealed very low activity concentrations in the southern part of the North Sea, which is attributed to the fact that the dominant fractions are buried in sediments in the discharge area of the Cap de la Hague plant (Nies, 1990). The ^{134}Cs activity is too small to be detected at a distance, although the ratio $^{134}Cs/^{137}Cs$ is significantly higher for Cap de la Hague than for Sellafield (Gueguenia *et al.*, 1988). Considering the ^{137}Cs distribution in the North Sea, values in the range 2.7–12 Bq m^{-2} were found in 1987 (Nies, 1990).

Regarding gaseous effluents, the dominant nuclide is ^{85}Kr, which gives rise to a dose of 3×10^{-4} Sv at the skin (for 800 t of fuel reprocessed) within the maximal fallout area of Cap de la Hague's stack (Lafaille, 1980). In the terrestrial environment, it is considered that releases of gaseous effluents may give rise to contamination at the site, as a result of the passage of radionuclides into the subsoil (Lafaille, 1980). The gaseous effluents released by the stack and the plant's dozen secondary outlets are, therefore, monitored continuously. No difficulties are experienced in the implementation of this standard monitoring procedure.

With regard to non-routine releases of radioactivity, on 6 January 1981 an incident involving the ignition of solid waste in an air-cooled, steel-lined, concrete radioactive debris storage silo occurred, leading to the escape of radioactive vapour from the silo. It was announced by Cogéma that the radioactivity did not exceed a fifth of the permissible dose, although the following day Cap de la Hague's Hygiene and Security Committee revealed that in parts of the site radiation reached maximum levels. In addition, a claim was made by the unions (which was undisputed by Cogéma) that at the plant's medical centre, situated some 200 m from the main road, levels reached a value 10 times higher than those permitted in the plant (Lloyd, 1981). It was, nevertheless, concluded that the incident did not constitute any serious risk to the employees or the local population.

A working group on Irradiated Fuel Management, which was chaired by Professor Castaing of the French Academy of Sciences, has undertaken a review of the exposure of Cap de la Hague's personnel to ionizing radiation, and the impact of radioactive waste discharged during normal operation, which does not reveal any serious reasons for concern (ATOM, 1983). Hence, by considering personnel exposure statistics, since no significant incidents have occurred, an indication of the safety performance of Cap de la Hague may be obtained. The maximum dose in 1975, of 5.09×10^{-3} Sv, decreased to 1.56×10^{-3} Sv in 1986, which may be compared with the maximum allowable dose of 5×10^{-2} Sv (Aycoberry and Rougeau, 1988).

2.4 KYSHTYM

The possibility of the occurrence of an accident involving the release of radioactive material in the eastern Urals region of the Soviet Union during the late 1950s was first widely publicized by Zhores Medvedev in 1976. It was suspected that a release had occurred in the Kasli area of Chelyabinsk province. Examination of Soviet radioecological literature revealed a series of studies in which the location of large experimental areas remained unidentified, but which involved the investigation of soil and biota typical of the Kasli area (Medvedev, 1979; Auerbach, 1990). In response to concerns expressed with regard to the occurrence of such an accident, Soviet specialists issued reports to the International Atomic Energy Agency (IAEA) in 1989. Subsequently excerpts from a previously classified 1974 Soviet report were published, from January 1990, in *Energiya*, and additional papers about the accident were presented in 1990 at the Commission of the European Communities (CEC)/International Union of Radioecologists (IUR) Luxembourg Seminar. However, Trabalka and Auerbach (1991) note that a self-consistent data set is not presently available to permit an authoritative analysis of the accident.

It has recently been confirmed that on 29 September 1957, at 16.20 local time, a chemical explosion took place at Kyshtym in a storage tank containing 250 m^3 of high-level radioactive waste, generated as a result of plutonium production operations. The explosion has been attributed to the ignition of acetate–nitrate concentrate following the failure of a crude cooling system. This is reported to have led to the release of 7.4×10^{17} Bq of activity to the atmosphere. Whilst most of this release was deposited on the ground close to the site of the explosion, a plume of finer particulates (containing about 10 per cent of the activity released) was carried to a height of 1 km. This radioactive cloud was transported to the north and north-east towards the towns of Kamenets-Ural'skij and Tyumen (see Figure 2.3). The largest contribution to the total activity of the release (accounting for over 60 per cent) was due to ^{144}Ce+^{144}Pr. Additional contributions were from; ^{95}Zr + ^{95}Nb (over 20 per cent), ^{90}Sr + ^{90}Y (over 5 per cent), ^{106}Ru + ^{106}Rh (over 3 per cent) and ^{137}Cs (over 0.03 per cent). Detection of ^{89}Sr, ^{147}Pm, ^{155}Eu and plutonium in the release has also been reported (Nikipelov, 1989). The principal source of radiation dose to biological surfaces over the first year after the accident was ^{144}Ce + ^{144}Pr, whilst ^{90}Sr was the principal contributor to long-term exposure.

The resulting terrestrial distribution of radioactivity may be considered to comprise two phases associated with initial radioactive fallout and wind migration (Ternovskij *et al.*, 1989). Deposition of virtually all radioactivity from the plume occurred within 11 hours along a $\geq$ 300 km path from Kyshtym (see Figure 2.3) leading to the contamination of an area which extended over some 2×10^4 km^2 (Trabalka and Auerbach, 1991). Effects of wind resuspension with regard to redistribution of contaminants were most apparent during the first few days after the accident, but did not lead to significant changes in the trace boundaries. By 1958 wind transfer caused the redistribution of less than 1 per cent of the original radioactive

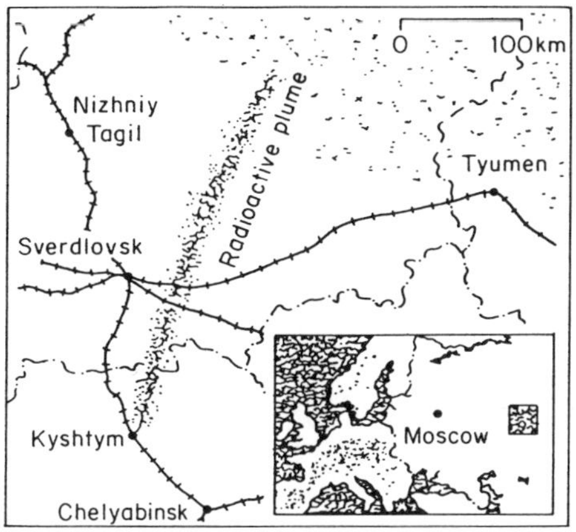

Figure 2.3 Location of Kyshtym area and radioactive plume (from Dickson, 1989).

fallout, the formation of the radioactive trace being essentially completed. The stability of the geographical distribution of contamination was demonstrated by further measurement and sampling programmes.

The distribution of radioactive fallout in the environment was subsequently affected by biogeochemical processes. Radionuclides were rapidly redistributed through the soil profile due to vertical migration and leaching processes. In the spring following the accident most activity (90–95 per cent) was concentrated in the turf, with 0.5–1.5 per cent in living plants and 5–10 per cent in the mineralized portion of the soil. Maximum radionuclide movement was observed in turfy podzol soils with minimum movement occurring in leached chernozem soils. A rise in the accumulation of ^{90}Sr in natural herbaceous plants 6–12 years after the contamination event was apparent, due to vertical migration from turf into the mineralized portion of the soil, with a subsequent decrease in accumulation. Root uptake became more important over this period, the contribution of ^{90}Sr from soil by this route increasing to 95 per cent in the last decade, from 10 to 20 per cent in the year after the accident. The ^{90}Sr contamination levels in the above-ground biomass of trees fell, over a period of 20–25 years, to 0.02 and 7 per cent respectively for pine and birch, compared with 80–90 per cent in 1957, as a consequence of biogeochemical migration processes.

As a result of the release, over the two year period following the accident effects of radiation damage were apparent to varying degrees, depending on the radiation dose, in both trees and herbaceous vegetation (Trabalka and Auerbach, 1991). All pine trees, within some 20 km^2 in which the dose to needles exceeded 30–40 Gy, perished by autumn 1959, whilst doses in excess of around 200 Gy to herbaceous vegetation and meristem buds of birches (involving an area of some 5 km^2) killed these plants and trees. Lasting effects arising from radiation exposure of wildlife have not been recorded, due to the potential for migration of animals from less affected or uncontaminated areas to replace those lost due to acute exposure, whilst the survival and fitness of exposed populations is not expected to be significantly affected as a consequence of sub-lethal effects. Regarding farm animals, loss of animals exhibiting acute radiation sickness symptoms is reported to have begun within 9–12 days from those sites nearest the accident. However, at greater distances from the accident no mortality was apparent in farm amimals over a six-month period, whilst upon removal from such contaminated areas no diffences in comparison with control animals were found.

With regard to the aquatic environment, half periods for the elimination of radionuclides from lake water varied from 1 to 24 days for ^{144}Ce, and from 780 to 1100 days for ^{90}Sr. Contamination densities in the more highly contaminated rivers ranged from 4×10^3 to 28×10^3 times higher than pre-accident levels immediately after the accident. However, radioactive decay, absorption by bottom deposits and natural migration led to reductions by a factor of 150 in river contamination by 1958, causing a decrease in the contamination level in lakes by a factor of 20–30. Over a 25 year period the ^{90}Sr concentration in lake-water within the contaminated region decreased by a factor of 30. Similar reductions of 20–30 per cent were apparent in the upper layer of bottom deposits (0–0.05 m), with reduction by a factor of 35 in fishes (Ternovskij *et al.*, 1989).

During the autumn/winter seasons following the accident fish received doses of 40 Gy in the most heavily contaminated lakes. Nevertheless, although diminished reproduction was apparent in the vulnerable herbivorous fish species (e.g. carp and goldfish) a few years after doses to eggs exceeded lethal levels (in excess of 10 Gy), by 1960 no ecological effects were observable. No effects on plankton, invertebrates or aquatic plants have been detected.

Immediate actions to mitigate the consequences of the accident involved the evacuation of 1154 people from four villages, three of which were located in the most affected zone (≥ 37 T Bq km^2 ^{90}Sr), within 7–10 days. An additional 9580 inhabitants located in areas with progressively lower levels of activity were moved out over a period of 250–670 days. Thus, in total almost 11 000 people were evacuated from a 700 km^2 area in which contamination exceeded 74–148 GBq km^{-2} ^{90}Sr. Additional measures taken to mitigate the circumstances of the accident included: decontamination of portions of agricultural land, monitoring of agricultural produce (with the rejection of produce exceeding accepted norms), controlled agricultural production through specialized state farms with introduction of restrictions on the use of contaminated areas, reorganization of agriculture and forestry. Besides ploughing 200 km^2 of land at the head end of the radioactive cloud using standard procedures during 1958–1959, deep ploughing (to a depth exceeding 0.5 m) was carried out on 62 km^2 of land during 1960–1961 (Nikipelov, 1989; Medvedev, 1990). Whilst quantities of grain, potatoes, meat, milk, vegetables and other produce were rejected as unfit for consumption, effective control over all produce was not ensured due to administration problems and delays in the introduction of such measures.

The implementation of a hydrological diversion and isolation system for various purposes which may include remediation of contamination from the 1957 accident has been noted (Trabalka and Auerbach, 1991). An area of 170 km^2 which was most highly contaminated following the accident has, nevertheless, remained unsuitable for human habitation, agriculture or forestry and is designated for research purposes as a radioecological reserve.

Although average external exposure doses preceding evacuation are reported to have reached 0.17 Sv, with effective dose equivalents of 0.52 Sv (1.5 Sv to the gastro-intestinal tract) it should be noted that significant uncertainties are associated with these figures (Trabalka and Auerbach, 1991). It is suggested by Nikipelov (1989) and Buldakov *et al.* (1989) that these doses may be doubled to account for the non-uniformity of contamination density and exposure conditions. Inhabitants of the area exhibiting measurable contamination located outside the most contaminated zone were estimated to have received effective radiation dose equivalents of about 1–10 per cent in excess of that due to natural background radiation over the 30 year period following the accident. No significant radiation effects have been revealed by studies, which have been undertaken over the past 30 years, of evacuated and resident exposed populations.

2.5 WINDSCALE ACCIDENT

The first substantial publicized release of radioactive material from a nuclear reactor accident occurred in October 1957 at the UK Atomic Energy Authority's Windscale Works, Sellafield, Cumbria, the location of which is shown in Figure 2.4. The accident to the Windscale No. 1 reactor (an air-blast cooled, graphite moderated, reactor designed for the production of military plutonium) was caused by overheating during a procedure to release Wigner energy (*Nuclear Engineering*, 1957; Command 302, 1958). This led to can failure, resulting from an exceptional temperature rise, uranium oxidation and subsequently a graphite fire which affected some 150 fuel channels. In order to contain the fire the fuel channels adjacent to those which were red-hot were discharged, thereby creating a fire break. An attempt to extinguish the fire using CO_2 proved ineffective; thus on 11 October water was injected into the core, and was applied for 24 hours until the core was cold. Recent assessments have revealed that at the time of the accident the reactor was also being used as an irradiation facility to produce ^{210}Po from bismuth.

The Windscale accident resulted in the release of radioactivity, mainly isotopes of the noble gases and volatile elements, into the atmosphere from the 125 m reactor stack. The release took place over the period from approximately noon on Thursday 10 October, 1957, to noon on Friday 11 October (Command 302, 1958). Crick and Linsley (1982, 1983) estimated that the principal nuclides released during the accident were; ^{137}Cs (2.2×10^{13} Bq), ^{131}I (7.4×10^{14} Bq), ^{210}Po (8.8×10^{12}Bq), ^{239}Pu (1.6×10^{9} Bq), ^{106}Ru (3×10^{12} Bq), ^{90}Sr (7×10^{10} Bq), ^{132}Te (4.4×10^{14} Bq)

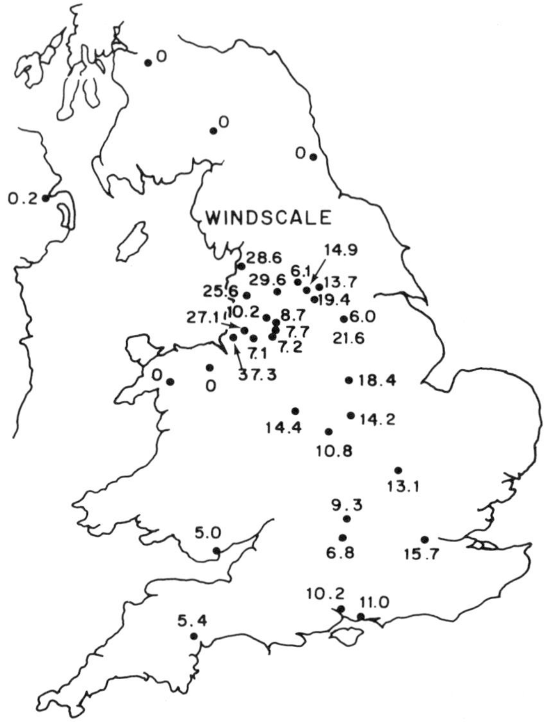

Figure 2.4 Time integral of air concentration of ^{131}I Bq days m^{-3} (from Stewart and Crooks, 1958; reproduced by permission of UKAEA. Copyright © 1958 Macmillan Magazines Ltd).

and ^{133}Xe (1.2×10^{16}Bq). Dunster *et al.* (1958) estimated that about 3×10^{12} Bq of ^{89}Sr was also released. Chamberlain and Dunster (1958) concluded, from radioautographic examination of filter paper and other fallout-exposed surfaces, that most of the radioiodine was in gaseous form or absorbed on very small particles. There was found to be little dissemination of particulate uranium oxide, the amount of uranium disseminated being considered negligible (Chamberlain, 1981; Chamberlain and Dunster, 1958).

The meteorological conditions over the period of the accident were examined by Crabtree (1959) and Stewart and Crooks (1958) who found that this information provides a qualitative explanation of the deposition and air concentration data. Analysis of airborne dust samples collected on routine industrial monitoring filters located in the UK and Europe enabled the passage of cloud, the direction of which was ultimately south-easterly, to be traced. The cloud was detected in Mol, Belgium, at 19.00 GMT on 11 October and at Frankfurt, Germany, and Sola, Norway, on 12 and 15 October respectively. The highest activity in the portion of the cloud which reached Belgium, Holland and Germany was found to be only

6 per cent of the London value (Crabtree, 1959). Results from long-range measurements of the time integral of ^{131}I air concentration are provided in Figure 2.4. This demonstrates that the radioactive plume avoided north Wales but diffused to cover a wide front in the southern regions of the UK.

Considering deposition, radioactive dust was deposited on grass mainly by turbulent diffusion and impaction because little rain fell while the cloud passed over England (Stewart and Crooks, 1958). The mean value of the deposition velocity for ^{131}I in northern and southern England were 0.30 cm s^{-1} and 0.11 cm sec^{-1} respectively. The deposition velocities of ^{137}Cs and ^{103}Ru were less than 15 per cent of that of ^{131}I. Marked deposition of ^{131}I on high ground was observed (Chamberlain and Dunster, 1958; Dunster *et al.*, 1958), the pattern of strontium on grass in the Windscale area being found to be virtually the same as that for ^{131}I on grass. The disappearance of ^{131}I and ^{137}Cs from grassland after the Windscale accident was examined by Booker (1958) revealing field loss half-lives, T_b, of 13 and 10 days respectively for the two nuclides. It was noted that the field loss of ^{137}Cs and other nuclides from grass after the accident appeared to slow down after mid-November 1957 (Booker, 1958).

Whilst the fallout after the accident of 1957 comprised mainly dry deposition of sub-micrometre particles, some larger particles were found near the reactor (Chamberlain and Dunster, 1958). Ground surveys of the deposition of particulate matter in the vicinity of the Windscale works revealed that deposition extended for some 4 km south-south-east from the site. Particles in the size range 20–500 μm were found with total beta activities for individual particles of 37–(4.81×10^3) Bq. Such particles were not considered to contribute appreciably to public health problems.

With regard to contamination of aquatic systems, analysis of drinking-water samples collected from reservoirs and streams revealed that concentrations of radioiodine and other radionuclides were not found to exceed the maximum permissible concentration permitted by International Commission on Radiological Protection (ICRP).

The determination of levels of strontium isotopes in samples of produce such as potatoes, cabbage, kale, turnips and lettuce taken from within and outside the restricted area around Windscale is reported by Dunster *et al.* (1958). Results from farms at 3.2 and 12.8 km south-east of the Windscale site revealed concentrations in turnips of 5.55 and 1.48 Bq ^{90}Sr g^{-1} Ca, and in potatoes of 4.81 and 0.93 Bq ^{90}Sr g^{-1} Ca, respectively. Lower concentrations were found in other samples such as cabbage and kale. None of the levels measured represented any danger.

Considering transfer of radionuclides to animals, the level of ^{131}I in the thyroids of unselected sheep were examined by Van Middlesworth (1958), who observed that 10 days after the Windscale accident the average level was 5.18×10^2–5.55×10^2 Bq g^{-1} for London sheep thyroids, compared with concentrations averaging 37 Bq g^{-1} during August/September 1957. Comparing results from sheep slaughtered in London with those of sheep slaughtered in Tennessee reveals that the former received less ^{131}I from Windscale than the latter received from distant nuclear weapons tests.

It was apparent, from radiation surveys and air sampling activities, that by the evening of Friday 11 October, no external radiation hazards or significant inhalation problems existed. However a possible marginal level of milk contamination was recognized on the basis of gamma radiation survey data. Measurements to analyse the passage of radioiodine and radiostrontium activity into milk from contaminated pasture land indicated that contamination was due essentially to ^{131}I, the release having involved a preferential radioiodine content rather than a normal distribution of mixed fission products. The highest measured ^{131}I level during the survey was 5.18×10^4 Bq l^{-1} in a sample collected on 13 October 1957 from a farm 16 km from Windscale, lying in the direct path of the plume. Further studies of milk were undertaken elsewhere, for example in Cardiganshire revealing a value on 13 October less than 1 per cent of the highest level found in the Windscale area. Consequently, restrictions on the consumption of milk in regions where the radioiodine activity exceeded levels of 3.7×10^3 Bq l^{-1} were introduced, involving over an area of about 500 km^2 by 14 October. Derestriction of all regions in this area had occurred by 23 November. A survey of ^{137}Cs in milk during late October 1957, when gamma spectrometry analysis was feasible due to the reduction in ^{131}I activity, revealed a similar pattern to that of ^{131}I. However, the concentration of ^{137}Cs in the area of maximum concentration corresponded to a level of 3.7×10^2 Bq l^{-1}.

A retrospective assesment of the wider implications of the Windscale accident was recently undertaken by the National Radiological Protection Board (Crick and Linsley, 1982, 1983). The collective thyroid dose and collective effective dose equivalent were calculated to be 2.6×10^4 and 2×10^3 person-Sv respectively in Europe.

2.6 THREE MILE ISLAND

The first nuclear power plant accident occurred in 1979 at the US Three Mile Island Unit 2 (TMI-2), a 900 MWe pressurized water reactor, located in Pennsylvania. The events leading to this loss-of-coolant type accident are attributed to the combined effects of equipment malfunctions and operator error. The sequence commenced when feedwater pumps supplying the steam generators tripped or shut down at 4.00 a.m. on 28 March 1979 (Kemeny, 1979) with consequent automatic shutdown of the steam turbine and generator, and reactor. Increases in the temperature and pressure of the reactor coolant resulted, due to the heat from decay fission products in the core, causing a pressure relief valve to open (as designed); however this subsequently failed to close automatically. Provision of cooling water via three emergency feedwater pumps did automatically occur; however two valves inadvertently remained closed, thereby preventing water from reaching the reactor, with loss-of-coolant occurring due to the pressure relief valve which had remained open. In order to cool the system high-pressure injection pumps then automatically began to operate but these were deactivated by operators, who were unaware of the continued water release occurring through the open relief valve. A temperature in excess of 2273 K was reached as water turned to steam, causing cladding failure and melting of some 50 per cent of the fuel, thereby releasing fission products to the reactor vessel and coolant system (Gerusky, 1988). More than two hours after the accident began it was realized that the pressure relief valve (through which coolant drained to a tank in the reactor building with subsequent pumping, upon overflow to a sump, to an auxiliary building) had remained open, and a manually operated backup valve was closed, thus preventing further coolant loss. Eventual cooling of the core was achieved subsequently, via the operation of the high-pressure injection pumps, bringing the accident under control.

Very little radioactivity was actually released to the environment, despite the transport of a large amount of radioactive material from the core into the containment, since reactor vessel and containment structure remained intact. The amount of radioactivity released into the environment was of the order of 10^{17} Bq, and consisted mainly of the noble gases ^{133}Xe, ^{133m}Xe and ^{135}Xe. The detection of a relatively small amount of radioiodine, well below levels apparent during preceding nuclear weapons fallout episodes, was also reported. It has been calculated that less than 1.11 TBq of ^{131}I was released in the course of the accident.

Environmental monitoring undertaken in the locality of TMI following the accident included analysis of samples of milk, air, water, produce, soil, vegetation, fish, and river silt and sediment. These studies revealed that little or no radioactivity was present in the environment apart from the noble gases, and it has been concluded that the accident did not contribute significantly to increased levels of radioactivity in local foodstuffs. Increased environmental radionuclide concentrations were observed only due to ^{131}I in cow's milk, goat's milk, on-site non-drinking water, on- and off-site air and ^{137}Cs in fish following the accident (Gerusky, 1988).

The highest doses received as a consequence of the accident were considered to be to people living within a 3.2 km radius of the plant. An estimated dose of some 0.5 Sv was received by one person located on the islands in the vicinity during the first few days of the accident, with 260 people, mostly located on the east bank of the local river, receiving 0.2–0.7 Sv. The estimated collective dose to the population within an 80 km radius of the TMI plant was 20 person-Sv (President's Commission on the Accident at Three Mile Island, 1979). The exposure of nearby residents due to subsequent controlled venting of ^{85}Kr present in the reactor building to the atmosphere was considered to be trivial.

2.7 CHERNOBYL

The most serious accident to have occurred in the history of nuclear reactor operation, was in April 1986 at the Unit 4 reactor of the Chernobyl nuclear power plant (NPP) located in the Ukraine, USSR (see Figure 2.5). This reactor was one of the 15 Soviet RBMK (Reactor Bolshoi Moschnosti Kipyashiy—Large-power boiling reactor) type, featuring a graphite-moderated light-water-cooled system in which vertical zirconium alloy pressure tubes contain the fuel pins. Cooling water flows, under high pressure, upwards through these tubes and boils, producing steam which is used to drive turbines.

The accident at Chernobyl occurred as a result of a power excursion during an experiment to test the ability of turbine generators to supply energy for a limited period, in the event of an external power failure, until standby diesel generators could be brought into operation. During procedures intended to reduce the reactor power to the required 700–1000 MW_{th} (thermal) level, there were strong violations of operating procedures (including the over-riding of all in-built safety systems) and the reactor was brought to low power. Operator mis-handling caused the reactor power to reach 30 MW_{th}; consequently the control rods were withdrawn from the reactor (beyond safe limits) to increase the power (to 200 MW_{th}). Inherent RBMK design weaknesses result, in such low-power régimes, in a positive void coefficient, i.e. increased neutron flux and power resulting from the conversion of water to steam voids in the coolant circuits, producing more heat and steam voids.

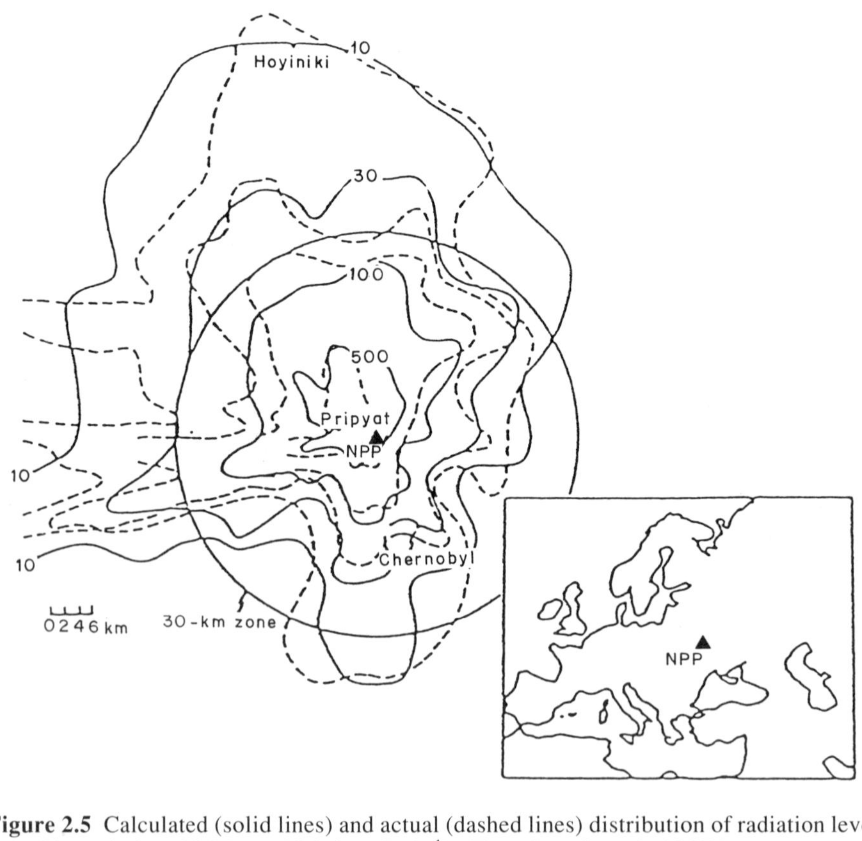

Figure 2.5 Calculated (solid lines) and actual (dashed lines) distribution of radiation levels near Chernobyl on 29 May, 1986 (in μ Sv h^{-1}). (From Israel *et al.*, 1987.)

The fact that additional circulation pumps were started, in order to permit the experiment to be repeated if required, caused flow limits to be exceeded, leading to flow cavitation and noticeable vibration. Hence an uncontrollable growth in power, resulting from rapid void formation in the coolant flow in a large part of the core combined with the positive void coefficient, began on 25 April 1986 at 21.23 GMT (26 April, 01.23 h local time), the reactor power reaching 100 times normal full power within four seconds. This power excursion led to fuel rupture, succeeding explosions causing the movement of the 1×10^3 t reactor cover plate, which cut all reactor cooling channels, and exposure of the core to the environment. Ejection of material containing spent fuel, enriched with noble gases and volatile nuclides of iodine, tellurium and caesium, into the atmosphere occurred during the initial explosions, with a continuing release due to the subsequent fire in the graphite moderator. Efforts to suppress radionuclide releases and to prevent further fission, involved covering the core with more than 5×10^3 t of boron, dolomite, sand, clay and lead. This caused the emissions to be decreased substantially for a few days; however the blanketing of the core led to increased temperatures and a further release peak from 1 to 5 May. Emissions were terminated on 6 May upon pumping a nitrogen coolant through tunnels constructed under the core.

At the end of the 10-day release period following the accident some 2×10^{18} Bq of activity, of condensible radioactive fission and transuranium activation products, was present in the environment, representing some 3–4 per cent of the core inventory. The composition of fission products in the release resembled that in the fuel, excepting the preferential release of the more volatile nuclides. Besides a total release of noble gases, some 20 per cent of iodine available in the core was released (6.7×10^{17} Bq), with releases of around 10 per cent of the inventory of caesium (1.9×10^{16} Bq ^{134}Cs, 3.7×10^{16} Bq ^{137}Cs) and about 3 per cent of rare earths and actinides.

Of the fraction of the core ejected to the atmosphere, 0.3–0.5 per cent was deposited on the site, 1.5–2 per cent was deposited within 20 km, the remainder being dispersed beyond 20 km. The deposition pattern close to the site was dominated by gravitational settling of coarser material released over the first five days. With regard to dispersal beyond the Chernobyl region, during the first day the radioactivity became segmented, with sections heading toward Scandinavia and central Europe, and easterly across Asia to Japan, the North Pacific and the North American continent. The plume reached Scandanavia on 27 April, where the highest levels of deposition outside the USSR were found. Radioactivity became incorporated in a developing frontal precipitation system the following day, causing localized deposition peaks (greater than 1×10^5 Bq m^{-2}) in parts of central Scandinavia. The release spread south-westward across Europe from Poland, and a section then turned northward to the United Kingdom, other material travelling eastward across the Soviet Union and southward to Turkey and Greece, much material eventually moving across the North Sea and Scandinavia thereby leaving Europe (ApSimon *et al.*, 1988, 1989). By mid-May levels of the order of 1×10^{-3} Bq m^{-3} were observed in Hong Kong and North America. Transport of the release to the southern hemisphere was not apparent. A component of Chernobyl fallout within the Soviet Union and various European countries (e.g. Sweden, Poland, Greece) comprised hot particles. Two basic categories of hot particles have been identified: those comprising mixed fission products (originating from the fuel) and those associated with a few elements only (such as Ru, Ce and Sr isotopes). An important feature of the observed contamination pattern was the patchy nature of the deposition of radioactive material, which was often concentrated in a few small areas within countries. This was found to be highly correlated with local variations in amounts of rainfall during the passage of the plume (ApSimon *et al.*, 1988).

As a consequence of the accident a 30 km radius zone around the power plant was evacuated during the first few days following the release. An area of more than 2×10^4 km^2 outside this zone has been designated as contaminated. Radiation levels in the Chernobyl region shortly after the accident are shown in Figure 2.5. Agricultural land comprised about a quarter of the contaminated territory, consequently restrictions have been placed on food production (a total ban on produce where ^{137}Cs contamination levels exceed 1.5×10^{12} Bq km^{-2}) and agricultural techniques have been altered on land contaminated to intermediate levels. Various measures to reduce contamination at the reactor site were also implemented, including activities such as collection and disposal of loose fuel fragments, removal of topsoil, application of special solutions to effect decontamination and the containment of the damaged Unit 4 within a concrete 'sarcophagus'. Extensive monitoring of radioactivity, especially within and near the 30 km zone, has been undertaken by Soviet scientists: caesium isotopes generally dominate persistent radiation levels, with contributions from Sr, Pu, Zr, Nb, Ru and La isotopes.

Considering ecological damage within the Soviet Union, it has been reported that deciduous trees were considerably less affected than pines (the most susceptible species to radiation damage and mutation) although, in the first year, over-sized leaves were observed in species such as oak. There is no evidence that the effects observed in *Pinus* or any other species involve the genome. With regard to small mammals and birds, the recovery of these has been recorded (Clough, 1990; Sokolov *et al.*, 1990).

Contamination of river water and the Kiev reservoir in the vicinity of the accident was due initially to jets of volatile and aerosol products, with subsequent possible contamination as a result of precipitation washing radionuclides from the surface. The total amount of radioactivity in the water was calculated to be 7.4×10^{15} Bq, with less than 0.05 per cent attributable to ^{90}Sr (Israel and Petrov, 1988). A system of protective and filtering dams was established on waterways feeding the Kiev reservoir during the winter of 1986/1987 in order to preclude the possibility of increased contamination from spring flood influxes. Contamination levels were not found to increase above those of the previous autumn, the level of ^{137}Cs falling by a factor of 20, to 1 Bq Bq l^{-1}, in the period July 1986–May 1987. It has been concluded that during the period following the accident contamination levels did not exceed prescribed limits acceptable for drinking.

Considering collective effective dose commitments, the collective whole-body dose to the 1.35×10^5 Soviet evacuees is estimated to be 1.6×10^4 person-Sv. In the European Soviet Union the collective effective dose equivalent is in the range 2×10^5–2×10^6 person-Sv, whilst in the European Community the value is 8×10^4 person-Sv (Clarke, 1989).

2.8 THE HANFORD SITE, WASHINGTON STATE, USA

The Hanford Site is located in a rural region of south-eastern Washington State in the USA. The Site, which is operated by the US Department of Energy (DOE), was acquired by the US government in 1943 for the construction and operation of facilities to produce plutonium for the atomic weapons programme. From 1945 through 1960, Hanford's primary mission was the production of plutonium for national defence and management of the wastes generated by associated chemical processing operations. The current mission no longer includes defence, but rather waste management, environmental restoration, advance reactor development, and research and development.

The Hanford Site is underlain by a thick unconfined aquifer contained in alluvial and glaciofluvial sediments and a series of confined aquifer systems isolated by a sequence of basalt flows. The following discussions pertain to the unconfined aquifer because essentially all contamination resulting from liquid waste disposal occurs in this aquifer. The depth to groundwater ranges from less than 1 m near the Columbia River to approximately 105 m in the centre of the site about 15 km from the river. The thickness of the aquifer ranges from less than 1 m where the basalt rises above the water table to more than 200 m.

Recharge to the unconfined aquifer is both natural and artificial. The vast majority of recharge occurs from operational liquid-waste disposal, which has created groundwater mounds near the 200 Area location (Graham *et al.*, 1981). These mounds have altered the aquifer's local groundwater flow pattern, shown on Figure 2.6, which is generally from the recharge areas in the west to the discharge area, primarily the Columbia River, which is hydraulically connected to the unconfined aquifer. A representation of the water table at the Hanford Site is constructed and reported on at least annually.

Operations on the Hanford Site have resulted in large volumes of waste water being discharged to the ground (and soil column) through cribs, ditches and ponds. These discharges have contaminated the groundwater and have greatly influenced groundwater flow and contaminant movement in the unconfined aquifer beneath the Site. Discharge of waste water to the ground at Hanford began in the mid-1940s and reached a peak in 1955. After 1955, discharge to the cribs declined because of improved treatment of waste streams and deactivation of various facilities.

Approximately 23.7 billion litres of liquid effluent were disposed to the ground in the 200 Area during 1988. This value is indicative of the magnitude of previous

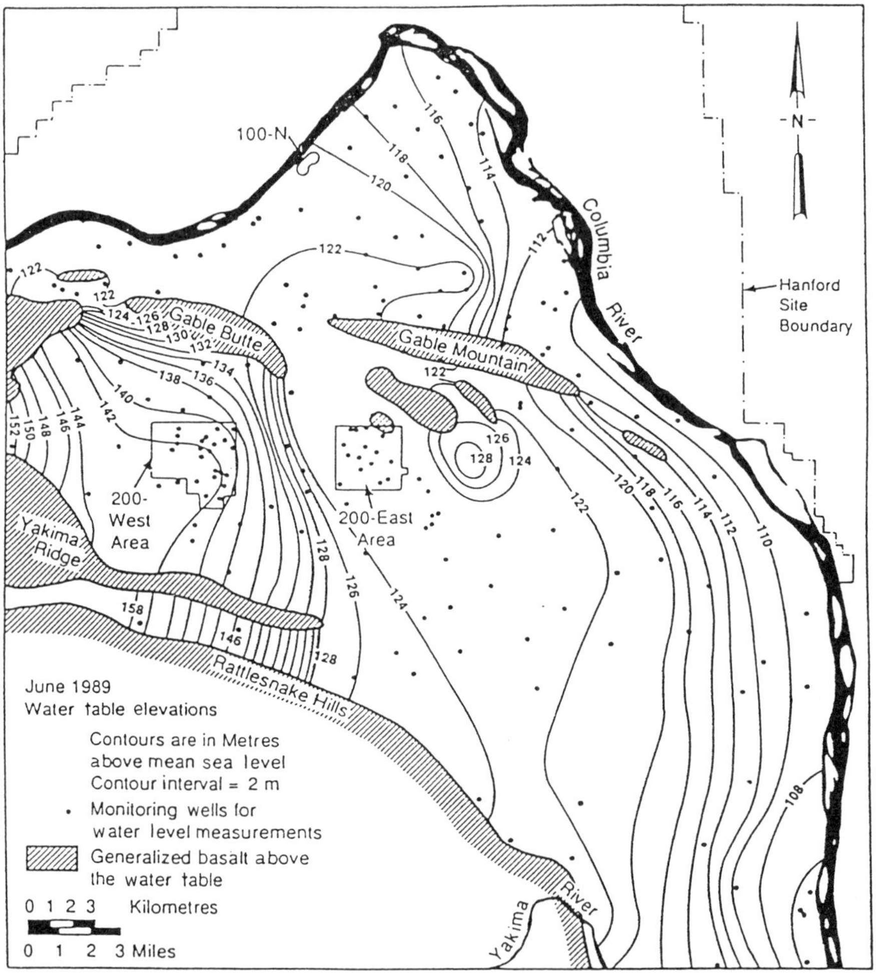

Figure 2.6 Water table map with location of wells. (From Jaquish & Bryce, 1990.)

years' discharges to ground. Approximately 0.8 billion litres of liquid effluent were disposed to ground in 1989. Radionuclides inventoried in the waste streams in recent years include, ^{3}H, ^{90}Sr, ^{103}Ru, ^{106}Ru, ^{113}Sn, ^{129}I, ^{137}Cs, ^{147}Pm, U (total), ^{238}Pu, $^{239/240}Pu$, ^{241}Am, gross alpha, and gross beta.

2.9 SOVIET SPACE REACTOR RE-ENTRY

The re-entry of the nuclear reactor powered Cosmos 954, a Soviet satellite, occurred on 24 January 1978 causing radioactive debris to be spread over a 1000 km path from Great Slave Lake to Barker Lake upon its disintegration over Canada (Tracy *et al.*, 1984). At this time the reactor core was estimated to contain some 3.11 TBq of ^{90}Sr, 181 TBq of ^{131}I and 3.18 TBq of ^{137}Cs. The retention of 75 per cent of the original material in the upper atmosphere has been reported. Samples of air, water and foodstuffs did not reveal any detectable contamination.

2.10 SNAP-9A RE-ENTRY

Satellite nuclear auxiliary power (or 'SNAP') units utilize the decay heat of radionuclides to supply electrical energy for space satellite equipment. The re-entry of a navigational satellite containing such an isotopic power unit (known as SNAP 9A) occurred on 21 April 1964, over the Indian Ocean, after failure to attain orbital velocity. The heat of re-entry caused volatilization of the SNAP 9A device which contained 629 TBq of ^{238}Pu. The deposition of around 95 per cent of the plutonium intially injected was estimated to have occurred by the end of 1970. The device's re-entry resulted in a 50 year dose commitment of approximately 0.36 Sv to the respiratory lymph nodes of the world's population due to exposure between 1961 and 1968 (Shleien *et al.*, 1970).

3 Atmospheric Pathways

Co-ordinator: **R. M. Harrison**

Contributors: **H. M. ApSimon, H. Beck, L. Devell, M. Dickerson, J. A. Garland, G. Graziani, P. Gudiksen, W. A. McKay, U. C. Mishra, K. W. Nicholson, F. B. Smith and C. S. Shapiro**

3.1 INTRODUCTION

The atmosphere can provide rapid dispersal and deposition of radionuclides out to very long distances from a release. Some of the processes involved in such transport and deposition are described in succeeding sections. Measurement and numerical modelling methods are outlined subsequently.

3.2 ATMOSPHERIC PROCESSES

3.2.1 ATMOSPHERIC TRANSPORT PROCESSES AND TRACERS

Radionuclides released into the atmosphere are subjected to a variety of physical processes that determine their eventual fate (Figure 3.1). Some of the most important processes, from a dose assessment point of view, are associated with atmospheric dispersion and subsequent removal of the radioactivity from the atmosphere. Dispersion of radionuclides injected into the planetary boundary layer (the lowest few kilometres of the atmosphere) occurs as a result of both the transport (advective) and mixing (diffusive) properties of the atmospheric circulations that occur on local to global scales. The radionuclides are initially dispersed by local scale circulations that occur over time scales of seconds to minutes and over horizontal dimensions of up to a few kilometres. Dilution and mixing is induced by turbulent eddies and wind shear. In the vertical these eddies are limited in size to the mixing layer, whereas in the horizontal they extend up to the synoptic scale of large weather systems.

The advective component (the mean transport of radionuclides) is determined from measurements of winds at the surface and aloft collected by various wind measurement systems operated as part of local, regional or global scale monitoring networks. Meteorological forecast models use such data to produce analysed windfields over a region and forecast future development. The winds observed at a particular location may be thought of as consisting of a mean wind component

Radioecology after Chernobyl
Edited by F.E. Warner and R.M. Harrison

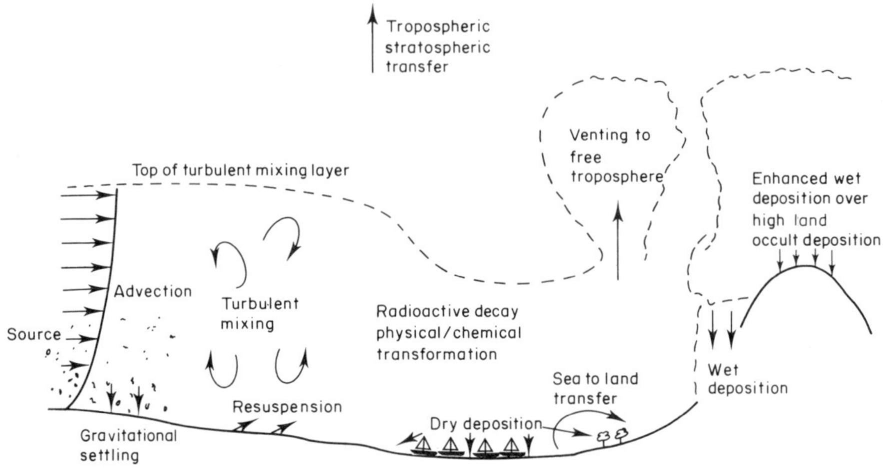

Figure 3.1 Schematic diagram of processes influencing radionuclides in the atmosphere.

and a superimposed fluctuating component (commonly referred to as eddies) that can be related to the diffusive (turbulent) properties of the flows. Wind fluctuations larger than the cloud tend to transport the entire cloud (advection) downwind rather than diffuse it, while eddies that are about the same size as the cloud will cause the cloud to grow rapidly and dilute the radioactivity. Thus, atmospheric turbulence consists of a full spectrum of eddies that range from distances of thousands of kilometres down to molecular dimensions. There is a continuous transfer of kinetic energy from the largest eddies to the smallest eddies where the energy is dissipated. The turbulent intensity of the atmosphere is primarily dependent upon the complexity of the underlying terrain, the wind shears as a function of height, and the vertical temperature distribution in the atmosphere. Thus, one expects greater turbulent intensities over complex terrain than over flat terrain, and during periods when strong solar heating causes vertical temperature instabilities.

Much research effort has been focused on relating the diffusion of atmospheric pollutants to measurements of various meteorological variables. In particular, boundary-layer turbulence is often related to vertical temperature gradients in the atmosphere, and to the variability of the horizontal wind directions and speeds. Since the theoretical foundations of these relationships are generally poorly understood, it is necessary to rely on empirical relationships based on data acquired from meteorological field experiments. An example of these relationships is the Pasquill–Gifford curves which describe the rate of diffusion of a pollutant released into the boundary layer. These curves may be considered to give a measure of the mechanical component of turbulence due to frictional drag near the surface and the thermal component due to the vertical heat fluxes.

The radioactivity is removed from the atmosphere by two principal mechanisms: precipitation scavenging and dry deposition. Precipitation scavenging is the removal of particulate matter and gases from the atmosphere by various types of precipitation. The process involves incorporation of the radioactivity into the rain water and the subsequent deposition of the material onto the surface of the Earth. The deposition rates, which determine the residence time of the material in the atmosphere, can markedly affect the downwind surface and airborne concentration patterns. The removal of radioactive particles and gases from the atmosphere by precipitation scavenging depends on complicated microphysical and microchemical processes that are functions of the conditions both within and outside natural cloud bearing layers. Some of these are nucleation scavenging, gas and particle diffusion to cloud and rain droplets, aerodynamic and electrostatic capture, and thermophoresis and diffusiophoresis. Removal from the atmosphere by dry deposition is important within the surface layer where the airborne radioactivity may come in contact with the surface by a variety of mechanisms that include diffusion, gravitational settling, impaction, interception, electrostatic effects, diffusiophoresis and thermophoresis. Since these processes are extremely complicated and poorly understood, it is usual to simulate them by means of a deposition velocity, which is defined as the deposition flux divided by the air concentration (see Section 3.2.3).

Material deposited on the surface may undergo resuspension due to wind stresses and by mechanical disturbances as a result of human activities. This resuspension process may continue to occur over a contaminated area for long periods of time. The physical processes involved in the resuspension process are also poorly understood and difficult to quantify. Resuspension studies are often based on the concept of a resuspension factor, which is the ratio of the airborne concentration to the surface concentration. These factors, however, may range over many orders of magnitude, and therefore are of limited use for predicting the inhalation exposure due to a resuspended contaminant. For numerical modelling purposes, the resuspension rate is often preferred. It is a function of various physical processes such as particles moving in saltation and surface creep. Since the rates are believed to depend on the soil type, particle size, surface wetness, surface winds, and atmospheric stability, they also vary over several orders of magnitude. Thus, our ability to predict resuspension rates is limited due to our lack of understanding of the physical processes involved and also the great variability of observed rates. The concepts above are expanded in the succeeding sections (see Sections 3.2.5 and 3.2.6).

Radioisotopes may be used as tracers of atmospheric processes and much has been learned from observations of their behaviour. The movement of radioisotopic tracers is controlled by the meteorological processes in the atmosphere; a distinction can be made between stratospheric and tropospheric processes. Radioactive tracers confined within the troposphere (e.g. activity of low-yield nuclear tests, reactor accidents, etc.) are subjected to tropospheric wind systems only. Tracers reaching the stratosphere are initially controlled by the stratospheric processes and later, after transport to the lower part of the atmosphere, by tropospheric meteorology. Thus, these tracers move with the wind systems of both the regions.

Wind systems, both in the stratosphere and troposphere, are predominantly zonal, i.e. west to east or east to west. The tropospheric winds are predominantly west to east at higher latitudes with the wind speeds increasing with height up to the level of the jet stream. Moreover, intensity of the jet stream varies with location and height. At lower latitudes the winds are often easterly. However, the zonal winds are modified by the presence of cyclones and anticyclones resulting in the wind system 'meandering' and having north–south components. In addition, at lower levels, regional wind systems are observed at certain locations, e.g. the monsoons of the Indian Ocean region, were due to the uneven heating of land and sea; we observe wind systems changing directions with the seasons. Meridional transport can also take place by circulations of the Hadley cell type in the troposphere at low latitudes. In this circulation pattern, winds are raised near the equator, move towards the poles and descend to the Earth's surface in the sub-tropical regions. Movement towards the equator compensates for the poleward shift at higher altitudes. Radioactive debris from nuclear tests of France and China have been used to trace the Hadley cell circulations.

Radioactive tracers are subject to the regional wind systems when they reach the lower levels of the atmosphere. They can be studied to give information on the meteorological phenomena described above. Some of the applications in which radioactive tracers have been employed are:

1. estimation of stratospheric residence time and diffusion within the stratosphere,
2. stratospheric–tropospheric exchange rates and their seasonal and geographic dependence,
3. planetary scale zonal air mass movements within the troposphere and tropospheric residence time estimation,
4. movement across the equator and regional scale movements,
5. rain removal of atmospheric aerosols and scavenging within and below clouds,
6. dry deposition velocities on various surfaces (surrogate and natural) and their dependence on particle sizes,
7. growth of particles from formation to deposition.

Residence times of stratospheric particulates have been estimated using measurements of radioactive tracers carried high into the atmosphere by megaton yield nuclear tests. The residence time is calculated assuming exponential removal of the material by meteorological processes. Since removal at lower levels of the atmosphere is much more rapid than in the stratosphere, the decrease in stratospheric levels is estimated by collecting samples of long-lived tracers ^{137}Cs and ^{90}Sr at ground level and following their year-to-year decrease in activity at similar periods of the annual cycle (to even out seasonal variations in stratospheric–tropospheric exchange). Confirmation is obtained by high-altitude sampling.

Radioisotope tracers from low-yield tests are excellent tracers for wind systems in the troposphere. A study of the detection dates of radioactive debris indicated that for the middle and high latitude tests, the movement has been predominantly by the westerly winds, possibly under the influence of jet streams. For example, Chinese test debris reached Japan in 1–3 days, the USA in 3–6 days and Central Europe in 7–8 days; the global circulation time was 10–14 days. The fast zonal movement due to the jet streams results in the activity distribution being a maximum in the latitude of testing and decreasing at higher and lower latitude; the geographical distance is not relevant.

Regional air mass circulations, like the monsoon wind systems of the Indian subcontinent, have been studied using isotopic tracers. The source of the Indian summer monsoon has been traced using debris from the low-yield French tests in Polynesia in the southern hemisphere. Many of these tests were in the summer period of the northern hemisphere. A study of the travel times of debris from the French test series of 1968 to various locations in the globe showed detection early on the west coast of India (in about 16–22 days). This is similar to the travel time of the Chinese test debris in spite of its being in the same hemisphere. A comparison of the levels of activity also showed that the west coast region of India receives more activity, up to an order of magnitude higher compared to other stations of the northern hemisphere. This study confirmed that a significant southern hemispheric component exists in the Arabian Sea branch of the monsoon. This type of study would have been very difficult by conventional meteorological means as few stations operate over the oceans of the equatorial regions to collect the required data continuously.

The levels of radon activity can also be used as an indicator of air mass mixing. This is based on the fact that air masses which travel a long distance over the oceans have low radon activity as the emanations over the oceans are much less as compared to that over the continents. Hence an abrupt increase in radon over oceanic areas indicates the presence of continental air.

The tropospheric aerosol residence time, i.e. the mean time spent by aerosols in the atmosphere before deposition on the surface, can also be measured using the ratios of radon daughters. This is based on the principle that in a parent–daughter series of radioisotopes, the growth of the daughter is affected by the non-radioactive removal rate constant. The required condition is that the radioactive half-life should be of comparable magnitude to or longer than the 'removal rate' for the relative concentrations of parent and daughter to be significantly affected. $^{210}Bi/^{210}Pb$ ratio studies can provide this information, there being no source of ^{210}Bi in the troposphere other than the parent ^{210}Pb. Residence times of about a week for the lower troposphere have been obtained, the value increasing slightly at higher levels.

The values of tropospheric residence times have also been estimated from the decrease of tropospheric fission product debris but this may be in error as it involves meridional diffusion and other uncertainties.

3.2.2 TRANSFORMATION PROCESSES AND AEROSOL BEHAVIOUR

The physical and chemical form of a radionuclide affects its behaviour in the atmosphere and may itself be subject to transformation during atmospheric transport. As indicated in Chapter 1, the radionuclides may be emitted in various forms, including:

1. an inert gas, lost from the atmosphere only by natural radioactive decay (but possibly producing radioactive daughter products with different properties);
2. a more reactive species such as elemental iodine-131 which may interact physically or chemically with other species, for example by adsorption on the surface of the ambient aerosol;
3. as small nuclei-mode particles (*ca.* 0.01 μm diameter) which interact with the ambient aerosol;
4. as in the case of the actinides and more refractory nuclides from Chernobyl, attached to coarse fuel particles which sediment out rapidly.

Thus radionuclides may enter the atmosphere in a wide range of physico-chemical forms and may be modified during transfer, particularly by interaction with the ambient aerosol. The range of travel, and depositional properties, are crucially dependent upon the physico-chemical form. These matters may be illuminated by considering the typical size distribution of the atmospheric aerosol—an idealized aerosol size distribution appears in Figure 3.2; it comprises three modes, as follows:

1. *Nuclei mode*: these very small particles are the product of vapour condensation and gas-to-particle conversion processes. They represent the primary particles formed by this route, and are only a transient species.

   Aerosol formed from elements of moderate volatility (e.g. Cs) in a high temperature accident (e.g. Chernobyl) will be released largely as vapour, with subsequent rapid condensation to form aerosol in the nuclei mode. Elements released in gaseous form which are subsequently converted to less volatile forms (e.g. iodine) also associate substantially with the nuclei mode due to the large surface area/volume ratio of very small particles.
2. *Accumulation mode*: since nuclei mode aerosol is inherently unstable with respect to growth mechanisms such as coagulation and vapour condensation, there is a tendency for growth, leading to formation of accumulation mode particles.

   Coagulation rates for nuclei mode particles are high, favouring formation of accumulation mode, by which time particle number densities have typically fallen to *ca* 10^4 cm^{-3} or less, and further coagulation is very slow. The other factor favouring accumulation mode aerosol is the inefficiency of sink processes in this size region. For smaller particles, both dry and wet deposition are favoured by the high diffusivity of the aerosol, whilst for larger particles inertial and gravitational forces are operative and lead to efficient dry removal. Thus the accumulation range is at a minimum in the efficiency curve both for dry and wet deposition, as neither scavenging within or below the cloud layer are effective (see Section 3.2.4).

3. *Giant, or mechanically generated mode*: the mass size distribution of aerosol in ambient air shows a minimum typically at 2–3 μm; particles above this size are termed 'giant', or 'mechanically generated'. The latter name reflects the fact that primary aerosol in this size is generated predominantly by mechanical attrition mechanisms, rather than from gas or vapour. Thus particles arising from mechanical break-up of the reactor core at Chernobyl were largely in this size range, as are particles of resuspended soil. Radioactivity associated with larger particles need not arise by this mechanism, as both coagulation and condensation processes can lead to attachment of radioactive substances to large airborne particles. Condensation of vapour can cause formation of particles with a stable interior and a radioactive external coating.

   Giant particles are subject to rather rapid gravitational settling, as well as being removed effectively by processes depending upon their inertia (see Section 3.2.3).

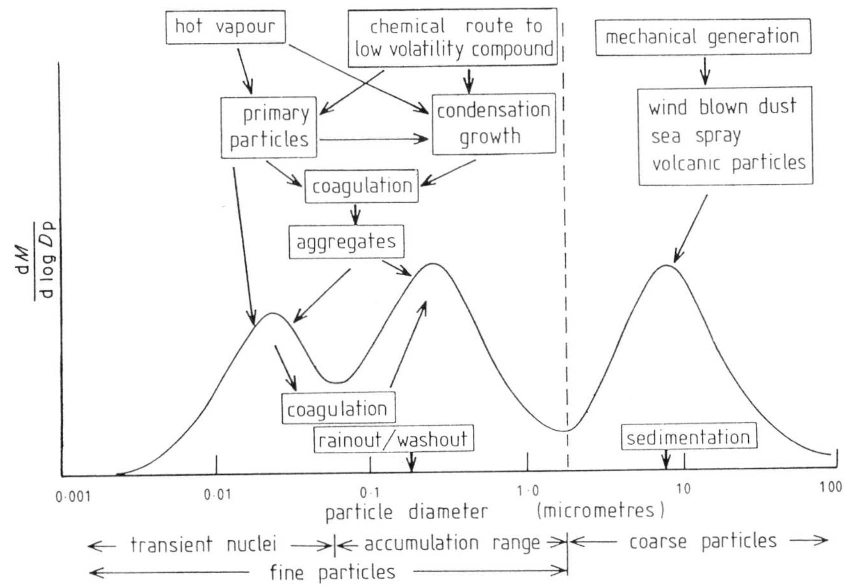

Figure 3.2 Schematic diagram of the size distribution (expressed as mass per increment in log particle diameter) and formation mechanisms for atmospheric aerosols (adapted from Harrison, 1990).

Aerosol coagulation processes affect particles of all sizes, although they are most rapid for small particles at high number densities. A plume of radioactive particles mixing with ambient air will lead to coagulation of radionuclides with the ambient aerosol. This will both alter the size distribution of the aerosol and affect its physico-chemical behaviour. For instance, when a hydrophobic nuclei mode radioactive particle coagulates with a larger hygroscopic particle in the ambient aerosol, its properties will be modified by incorporation into the accumulation mode, and by the presence of hygroscopic material which will influence its ability to act as a condensation nucleus.

Measurements of the size distribution of airborne radionuclides reflect the physical processes described above. Aerosol generated in nuclei mode is rapidly lost by deposition or transformation to larger sizes. It is thus normally observed only close to a source. Giant mode particles are also usually rather locally generated (including by resuspension), although atmospheric turbulence processes can keep them airborne for appreciable distances (Pye, 1987); tens, hundreds or exceptionally thousands of kilometres. Knowledge of the occurrence of very large particles (i.e. >8 μm diameter) is restricted due the difficulties of collecting them from the atmosphere with conventional air samplers. Their range of travel is likely to be very short, although there is evidence from Chernobyl of very large 'hot' particles depositing tens or hundreds of kilometres from source.

Thus in the Chernobyl emission, whilst approximately 27 per cent of the volatile elements Cs and I was deposited within 80 km, 60 per cent of the more refractory Zr and Ce, expected to be associated with large particles, was deposited within this range (USSR State Committee, 1986). The major part of the activity reaching Western Europe was associated with particles of <2 μm (Jost *et al.*, 1986; Winkelmann *et al.*, 1987); larger particles containing a higher proportion of refractory elements reached Scandinavia (Devell *et al.*, 1986; Persson *et al.*, 1987).

Accumulation mode aerosol is the vector of efficient long-range transport of aerosol. It has a very low deposition velocity (see Section 3.2.3), typically 0.01–0.1cm s^{-1}, leading to a lifetime with respect to dry deposition of 12–120 days in a boundary layer of 1 km depth. In practice, lifetimes of 7–30 days are typical.

Aerosol particles which experience high relative humidities are liable to hygroscopic growth, a process crucial to cloud formation and precipitation scavenging. The particles which grow most rapidly by absorption of water comprise soluble salts. Very small particles (< *ca.* 0.1 μm), even when soluble, do not take up water except at appreciable supersaturation owing to the high energies associated with their surface curvature (the Kelvin effect). Thus even water-soluble accumulation mode particles require a small supersaturation to grow into cloud droplet sizes (*ca.* 10 μm) at which their dissolved salt concentration is very low and hence the suppression of vapour pressure at their surface is negligible. Supersaturations of up to *ca* 3 per cent occur in the updraughts associated with convective cloud formation and under these conditions many water-soluble particles act as cloud condensation nuclei. Small water-insoluble particles will not condense water under these conditions, but the presence of a water-soluble component to the particle, perhaps as a result of coagulation with the background aerosol after emission, may lead to incorporation of both the hydrophilic and hydrophobic sections of the particle being incorporated in a cloud droplet. There has been some speculation that the presence of water vapour in the Chernobyl emissions deriving from the reactor coolant circuits may have contributed to an enlargement of particles in the plume upon cooling, with consequent tendency to more rapid deposition.

Aerosol particles can sometimes acquire electrical charges which profoundly influence their behaviour. Normally, atmospheric aerosol has little associated charge (it is in a Boltzmann charge equilibrium), but there has been speculation that aerosol emitted from Chernobyl may have been charged. The presence of charge can lead to considerably altered properties, for example either enhanced or decreased coagulation rates dependent upon whether charging is bipolar or unipolar. It has been suggested that the frequency of electrical storms may have been increased due to charged radioactive aerosols arising from Chernobyl emissions.

3.2.3 DRY DEPOSITION

It is convenient to distinguish 'wet' deposition mechanisms, which involve the intervention of rain or some other form of precipitation, from those that result from the direct interaction of airborne material with the surface. The latter collectively constitute 'dry' deposition. Thus, dry deposition is continuous, whereas wet deposition is intrinsically episodic. Deposition cleans the atmosphere but places radioactivity on the surface. Atmospheric radionuclides result in doses to humans and other organisms via external exposure and by inhalation; deposited radioactivity may cause external radiation exposure and may enter food chains over extended periods. Often, deposited radioactivity contributes a larger total dose than direct exposure from the air, and the rate of the deposition processes determines the distribution and intensity of the contamination of terrestrial and aquatic ecosystems and of the radiation exposures that result.

Studies of the transport from the atmosphere to the surface have been reviewed by Chamberlain (1991), Garland (1980), Garland *et al.* (1987) and Nicholson (1988b). Both gases and particles are transported to the immediate vicinity of the surface by turbulent mixing (see Section 3.2.1 above). Close to the surface is an air layer where turbulence is suppressed and transport is dominated by molecular diffusion (for gases and vapours) or a combination of Brownian diffusion, inertial effects and sedimentation for particles. At the surface specific mechanisms for removal depend on physical, chemical and physiological properties of the surface as well as the depositing material. Thus, deposition involves a sequence of steps, and the overall rate of deposition cannot exceed that of any step.

Deposition rates are usually described using the deposition velocity,

$$v_g(z_1) = \frac{F}{C(z_1)} \tag{3.1}$$

where F is the flux density of the depositing material to a unit area of land including any vegetation or other structure it supports, and $C(z_1)$ is the concentration at a height z_1 where C may be measured or specified.

It is convenient to call the reciprocal of v_g the resistance to deposition

$$r(z_1) = \frac{C(z_1)}{F} = r_a(z_1) + r_b + r_s \tag{3.2}$$

Here $r_a(z_1)$ is the resistance for transport by turbulence to the vicinity of the surface; it can be estimated using the experience of agricultural micrometeorologists in studying the exchange of water vapour, heat and momentum at extensive crop surfaces. An additional resistance, r_b, is found to be necessary in such studies to account for observed behaviour; it can be loosely identified with transport very close to the surface. Additional resistance due to the interaction at the surface itself is represented by r_s.

r_a varies with wind speed, surface roughness and the atmospheric stability (e.g. see Garland 1980). Typical values at moderate wind speeds are 30 s m^{-1} for forest and 70 s m^{-1} for grassland, but values are often several times higher in the low winds common at night. r_b is comparable in magnitude with r_a. r_b varies as (diffusion coefficient)$^{-2/3}$, and is roughly 100 times larger for typical small particles (0.3 μm diameter) than for typical gas. Although r_s is known to be negligible for particles below about 5 μm diameter, it is the controlling resistance for most gases, and generally field measurements are necessary to establish its value.

The noble gases, krypton, xenon and radon, are not very soluble and do not react with common surfaces. There may be some loss at water surfaces, but this is unlikely to be significant, and for practical purposes r_s is infinite and the deposition velocity zero. Radioisotopes of these gases are thus lost predominantly as a result of radioactive decay.

Iodine has a number of isotopes of radiological significance of which the chief, ^{131}I, contributed an important fraction of the dose following the Windscale and Chernobyl accidents (see also Section 3.5.2 and 3.5.3). The dose due to this short-lived isotope arises through its deposition to vegetation, and its prompt and efficient transfer via cows to dairy products. Several airborne forms have been postulated including the element (I_2 or possibly I), methyl iodide (CH_3I), hypoiodous acid (HOI), $IONO_2$, I_2O_2 and iodine attached to particles (Jenkin *et al.*, 1985). Field measurements indicate a mixture with similar proportions of a readily absorbed gas (I_2, I or HI) a less absorbable gas (CH_3I or possible HOI) and particulate I on particles of 0.1 to 2 μm diameter. Studies of deposition show that I_2 is absorbed via stomata and on the leaf cuticle; in field conditions r_s is small and for grass $v_g \sim 1$ cm s^{-1}. Forest might have a v_g several times higher, but this would be of little practical importance. CH_3I has a deposition velocity about four orders of magnitude smaller, and particulate iodine about two orders smaller than I_2. Thus, the deposition rate is largely determined by the fraction of the reactive form. In practice, the net deposition velocity for the mixture appears to be about 0.3 cm s^{-1}.

The long-lived ^{14}C is released from nuclear fuel during reprocessing as $^{14}CO_2$, and has also been enhanced as a result of nuclear weapon tests. It is incorporated into vegetation during photosynthesis and permeates all living organisms and food chains. In the long term, all biotic carbon is formed with practically the same isotopic composition as the atmospheric CO_2. Following a short release, a deposition velocity might be useful in estimating contamination, but v_g would depend on the rate of photosynthesis, and might be of order 0.05 cm s^{-1} during daylight, but zero at night.

The deposition of airborne particles is strongly influenced by particle size (see Nicholson, 1988b, for a review of available information). Table 3.1 shows the Brownian diffusion coefficient and sedimentation velocity for particles spanning the range of interest. In normal wind speeds turbulent velocities of order 0.3 m s^{-1} are common in the boundary layer: particles bigger than 100 μm diameter clearly could not be supported by these turbulent motions, and would fall out more or less along trajectories determined by their sedimentation velocities and the mean wind. (In stronger winds, or in the organized motions in convective storms, a higher threshold would be appropriate). Particles of around 30 μm and smaller would be readily mixed throughout the boundary layer, and particles of intermediate size would be transported to varying degrees by the turbulent flow, while subject to rapid deposition by sedimentation and inertial impaction.

Particles smaller than about 30 μm would be conveyed to the vicinity of the surface by turbulent motions assisted by gravity according to particle size. Within the air layers close to surface elements where turbulence is damped, mechanisms which cause particles to deviate from the flow become important. Brownian motion becomes significant for particles of ~0.1 μm and smaller (Table 3.1), and leads to an increase of deposition velocity for smaller sizes (Figure 3.3). Particles of 2 or 3 μm and larger have stopping distances comparable in magnitude with the smaller surface elements in vegetation canopies (Table 3.1). In consequence, they have suf-

Table 3.1 Microphysical behaviour of small particles suspended in air[a]

Particle diameter (μm)	Brownian diffusion coefficient ($mm^2\ s^{-1}$)	Sedimentation velocity ($mm\ s^{-1}$)	Stopping distance[b] (mm)
0.01	0.052		
0.03	5.6×10^{-3}		
0.1	6.8×10^{-4}	8.7×10^{-4}	2.6×10^{-5}
0.3	1.3×10^{-4}	4.2×10^{-3}	1.3×10^{-4}
1	1.3×10^{-5}	0.035	1.1×10^{-3}
3	8.0×10^{-6}	0.28	8.4×10^{-3}
10		3.0	0.093
30		27	0.84
100		300	9.3

[a] For particles of density 1000 kg m^{-3}.
[b] The distance travelled by a particle projected into still air with an initial velocity of 0.3 m s^{-1}.

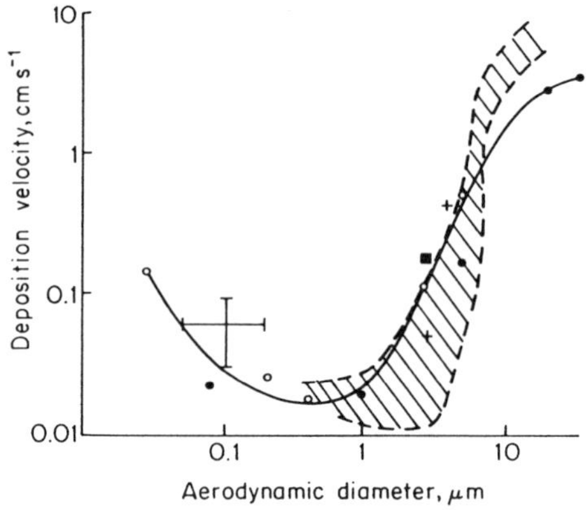

Figure 3.3 Experimental data from several sources, on deposition of small particles to grass. The symbols and hatched area represent the distribution of field and wind-tunnel data. (Copyright AEA.)

ficient inertia to carry them to grass blades and similar obstacles. Inertial impaction explains the rise in deposition velocity up to about 10 μm. The further rise for larger particles is chiefly due to the increase in sedimentation. Particles larger than ~5 μm in diameter may bounce at surfaces, if both particle and surface are dry. This may sometimes offset the contribution of impaction to deposition. However, typical deposition velocities of 10 μm particles to grass are above 2 cm s^{-1}, and deposition to forest may be several times higher.

None of these mechanisms is effective in the size range 0.1 to 2.0 μm. Here deposition is due to the interception of particles following the flow of air past surface elements. Particles which touch the surface adhere and are deposited. The microscopic roughnesses constituted by leaf hairs and other surface structures may increase the surface area of vegetation and may contribute to this effect, but the deposition rate in this region is expected to be of order 0.03 cm s^{-1}. Some field measurements indicate much larger deposition velocities, but there are a number of possible sources of difficulty in the interpretation of these data, and the general behaviour of submicron particles in the atmosphere is consistent with the value indicated (Nicholson, 1988b). Measured deposition velocities of several nuclides determined in Sweden after Chernobyl appear in Figure 3.4. The more volatile elements, Cs and I, are in the accumulation mode and show the lowest values.

Water covers large areas of the Earth's surface, and deposition to water is important as a cause of contamination of aquatic systems as well as decontamination of the atmosphere. Water surfaces are smooth aerodynamically, even when waves are generated by strong winds. Generally, deposition velocities for particles of 0.1 μm to 10 μm diameter would be predicted to be smaller than those for vegetated surfaces, and would not much exceed the sedimentation velocity through much of the range. However, most atmospheric particles are hygroscopic, and they are expected to grow by accretion of water vapour, in the shallow layer of air close to the sea

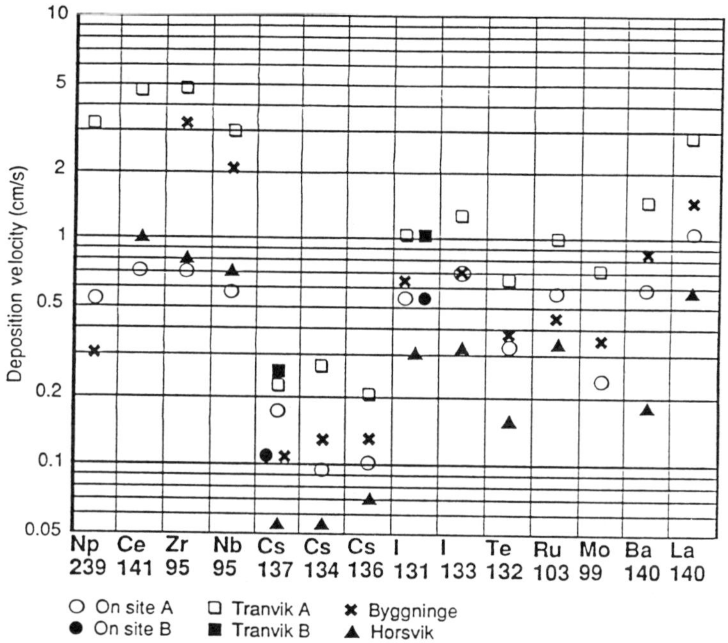

Figure 3.4 Dry deposition on various grass surfaces at Studsvik for 28–30 May. Iodine is corrected for the presence of 70 per cent gaseous iodine (from Devell 1991; reproduced by permission of the Swedish Radiation Protection Institute).

surface where the air is near saturated with vapour. If the particles grow quickly enough to reach their equilibrium size at saturation within this layer, the deposition velocity for most particles of practical interest should increase to ~1 cm s^{-1} (Slinn and Slinn, 1980).

3.2.4 WET DEPOSITION PROCESSES

It was recognized well before Chernobyl, that in the event of a major nuclear accident, precipitation processes could act as a concentration mechanism leading to areas of enhanced deposition (e.g. ApSimon and Simms, 1986). Not surprisingly therefore the major problems after Chernobyl coincided with areas of high rainfall, especially over hilly and mountainous regions of Europe. Many countries found a correlation between deposition, D_{dep}, and amounts of precipitation scavenging the radioactive cloud:

$$D_{dep} = \int W.XJ\,dt \qquad (3.3)$$

where W is the dimensionless 'wash-out ratio', X the concentration in air near ground level and J the precipitation rate. Observations of bomb fall-out and radioactive aerosols from Chernobyl indicate that typical values of W for small aerosols (~1 μm diameter) commonly lie in the range between 100 and 1000 Bq per litre per Bq m^{-3} mean concentration in the air. Alternatively a 'wash-out model' is envisaged, whereby the rain falls through the radioactive material from above, uniformly removing radioactive material from a vertical column of air with an efficiency dependent on the rainfall rate at the ground, and depositing it instantaneously at the surface. This leads to the definition of a wash-out coefficient expressed as an integral over the raindrop spectrum

$$\Lambda = \int N(a).E(a).\pi a^2.V(a).\mathrm{d}a \tag{3.4}$$

where $N(a)$ is the number density of raindrops of radius a, $E(a)$ is their collection efficiency for the pollutant, and $V(a)$ is their terminal velocity. The difference beween the two approaches is that the latter depends on the vertically integrated air concentration, whereas the former depends on what is likely to be measured, namely the time-integrated air concentrations near ground level.

These simple techniques provide a straightforward approach to mapping wet deposition. For example in the UK, Smith (Clark and Smith, 1988) used wash-out ratios and observed rainfall amounts to map deposition in considerable detail and pick out hot-spots of $^{137}Cs/^{134}Cs$ from Chernobyl. In an emergency situation such methods are dependent on detailed rainfall data being readily available. ApSimon *et al.* (1989) illustrated the potential of weather radar observations of rainfall in conjunction with a wash-out model to give a very rapid assessment within a matter of hours in any future emergency. The use of such remote sensing techniques is expanding steadily, especially in Europe.

However, not surprisingly, quite large variations are observed in the scavenging efficiency of precipitation. For example near Gavle in Sweden, which was one of the areas subject to the highest levels of deposition of $^{137}Cs/^{134}Cs$ from Chernobyl outside the USSR (~100 kBq m^{-2}), the hot-spot areas not only coincided with the highest rainfall, but the efficiency of removal in the precipitation was an order of magnitude higher than in some other less contaminated areas of the country (Persson *et al.*, 1987). Similarly in the UK, for example, the problem areas for sheep farming coincided with upland areas, again subject to much enhanced deposition.

3.2.4.1 Effect of physical and chemical form of radionuclides

To appreciate the efficiency of removal in precipitation systems we first have to consider the chemical and physical form of the radionuclides. In a major accident a whole range of radionuclides may be released; and their characteristics may depend on the exact nature of the accident—for example on core temperatures and volatility. Thus after Chernobyl there were two distinct components of the release; the first similar in nature to particles of fuel (with some depletion of the volatiles possibly); and the second due to volatilization and condensation of nuclides. The latter is a form of gas-to-particle conversion, and often results in rather small aerosols ($\ll 0.1$ μm), which quickly attach to ambient aerosols, particularly in the accumulation range (~0.1 up to 1 or 2 μm). A special case is ^{131}I which is so volatile that it is still partly in the gaseous state at ambient temperatures and pressures (for example about two-thirds of the total ^{131}I was observed as gaseous iodine, and one-third attached to ambient aerosols after Chernobyl). This also introduces complexities into the rain chemistry.

The character of the radioactive aerosols determines the probability that they will readily nucleate water droplets when cloud first forms. As air becomes supersaturated there is a competitive process between the atmospheric particles, with the largest soluble particles being the most successful at capturing the available moisture. The unnucleated dry aerosols remain in the interstitial air. This is very important because the probability of capture by a falling raindrop is generally very much higher for a nucleated cloud droplet than for a bare dry aerosol.

3.2.4.2 Orographic enhancement

A simple example of this is provided by the feeder–seeder mechanism over hills and upland areas. The situation occurs when there is some general rain from overlying 'seeder' cloud; air forced to rise over the hill cools and becomes saturated forming the 'feeder' cloud. Rain falling from the seeder cloud picks up additional moisture from the feeder cloud, leading to heavier rainfall over and downwind of the summit region. However the order of magnitude difference in capture efficiency of cloud droplets relative to bare aerosols can lead to far more dramatic enhancement, and hot spots of deposition over a hill-top for radionuclides attached to efficient cloud condensation nuclei. The exact location of hot spots is very complex, depending on meteorological and topographical factors. The feeder–seeder mechanism has been extensively studied by Choularton and co-workers (Hill *et al.*, 1986).

3.2.4.3 Occult deposition

Even in the absence of a seeder cloud and precipitation, hill-cloud can lead to very much enhanced deposition. The effective deposition velocities for cloud droplets can easily be in excess of 10 cm s^{-1}, especially over rougher surfaces such as forests, and with high wind speeds. This is more than two orders of magnitude higher than typical dry deposition velocities for bare sub-micron aerosols over, say, grassland. Studies of acid deposition have indicated that direct deposition from cloud can provide a significant proportion of the total annual sulphur deposition in sensitive areas, even at moderate altitudes of 500 m or less above sea level; also fog droplets can be very efficiently captured by obstacles such as bushes, hedges or isolated trees (ApSimon *et al.*, 1991b). Sea-fog or morning mist may also enhance deposition over low-lying ground.

3.2.4.4 Convective shower systems

Much of the wet deposition of $^{137}Cs/^{134}Cs$ across Europe from Chernobyl occurred in convective shower systems. For simplicity we shall consider a shallow short-lived shower. In such a shower material is continuously drawn into the updraught from the boundary layer of the atmosphere—see Figure 3.5. As the air rises it condenses, forming cloud droplets which coalesce, eventually forming rain droplets large enough to fall down despite the upward current of the updraughts. The whole system acts rather like a giant vacuum cleaner filtering a large volume of boundary layer air, and depositing a proportion of the moisture and pollutant material in a small area below the updraught with the precipitation. The first rain which falls has scavenged a full column of undepleted material in the updraught and is often substantially more contaminated than later precipitation, which has often reached a higher altitude and colder temperatures (with ice-phase processes) where a greater proportion of the vapour has condensed out. If incorporated in cloud droplets the

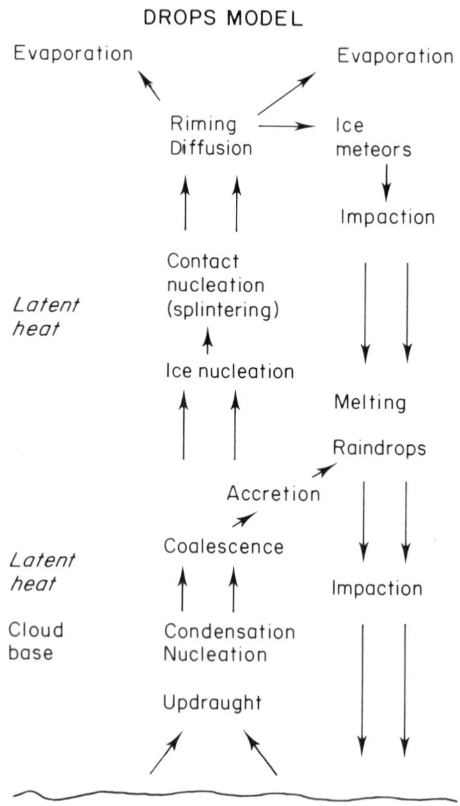

Figure 3.5 Schematic diagram of processes included in DROPS model

radioactivity will be more efficiently deposited than interstitial aerosol, when factors such as the aerosol size will be significant. As much as half the radioactive aerosols entering the storms will generally be vented from the top of the storms. This still reduces concentrations at ground level but is an important mechanism for transferring contamination to higher levels in the free troposphere.

In a shallow shower system with rain falling through the updraught, evaporation of the rain and hence cooling of the air eventually quenches the updraught, and precipitation generally ceases within about half an hour. More intense and persistent convective systems tend to be deeper; also wind-shear leads to displacement of the falling precipitation relative to the updraughts, so that evaporative cooling occurs in air outside the column (forming a downdraught of colder air) and does not quench the storm. The complex intertwined updraught and downdraughts in a thunder-cloud lead to large local wind changes and the characteristic anvil shape of the cloud. Ice particles are formed in different ways. As air falls further below freezing, say −10 °C, direct nucleation can take place on small ice nucleating aerosols; the latter are quite different from the cloud condensation nuclei, being insoluble and often minerals or clays. These small ice nuclei can then nucleate super-cooled droplets and raindrops forming larger ice nuclei, and resulting in conversion of water to ice phases. Further growth of ice particles can occur by such processes as diffusion of vapour and riming (collection and freezing of droplets, etc.). When diffusion occurs the vapour may be supplied by evaporation of water droplets which can lead to resuspension of the bare aerosols, reducing scavenging efficiency. Riming, on the other hand, transfers the radioactivity to the large ice meteors which are likely to precipitate leading to deposition.

The vacuum cleaner effect of convective storms drawing in large volumes of air and depositing radioactivity in precipitation over a relatively narrow area was evident in many locations across Europe after Chernobyl, but was particularly acute in the vicinity of Gomel some 140 km from the accident. In this region of Byelorussia hot spots occurred at levels in excess of 1500 kBq m^{-2} where precipitation had fallen. ApSimon *et al.* (1991a) show that the concentrated deposition from a simple summer storm, generating about 5 mm of precipitation (amounting to a deposition of 1.2 kBq m^{-2} for air contaminated with 1 Bq m^{-3} of ^{137}Cs aerosol), is sufficient to account for the observed levels, given the estimated release rates of Sedunov *et al.* (1990).

3.2.4.5 Frontal systems

The full complexity of warm and cold cloud processes is also encountered in frontal systems. These are very likely to be important in an accident situation, passing over northern mid-latitudes on average every two to three days. In such situations warm air from the south is drawn northwards, over-riding colder air, and giving rise to a conveyor belt of ascending air in the warm sector ahead of the cold front. Such systems are however much more complex to model. In particular it is extremely difficult to deduce the trajectories of material during passage close to the warm and cold frontal surfaces from analysed windfields produced by forecasting models. Such windfields are often only specified every three to six hours (or even 12-hourly), during which time the frontal surfaces can move 100 km or more, and the discontinuities in winds at the frontal surfaces are poorly resolved by the spatial resolution of the grid. This emphasizes the importance of increasing spatial and temporal resolution in forecasting models to obtain accurate simulations of transport within low-pressure systems, a topic of interest in tropospheric chemistry as well as in the context of radionuclides.

3.2.5 RESUSPENSION FROM LAND

3.2.5.1 Description of resuspension

Resuspension caused major problems in the vicinity of Chernobyl requiring repeated decontamination. It is the general term used to describe the process in which previously deposited material becomes entrained in the atmosphere. Strictly, the material ought to have been previously deposited via an atmospheric process, although the term is often used to describe the entrainment of material originating from accidental spills, etc. The potential effects of resuspension may be considered as two-fold: the spread of contamination to adjoining clean areas and the occurrence of an inhalation hazard. A brief summary of resuspension is given in this section. For further information, the reader is referred to more detailed reviews (Linsley, 1978; Sehmel, 1980a, 1984; Nicholson, 1988a).

In order to assess resuspension, the resuspension factor (atmospheric concentration/ground contamination per unit area) and resuspension rate (fraction resuspended per unit time) have both been used. The resuspension factor suffers the disadvantage that resuspension from upwind sources cannot be easily taken into consideration and the resuspension rate, although preferable for use in many model calculations, is not easily measured directly. In practice, resuspension factors are usually used since they are more readily determined.

3.2.5.2 Factors affecting resuspension

Resuspension can occur due to a shearing action by the wind and by mechanical agitation of the deposited species. Such mechanical agitations might include agricultural activity, vehicle or pedestrian movements and cleaning actions. In practice, wind resuspension from surfaces such as vegetation includes the effects of a movement of individual elements (e.g. leaves) and hence, resuspension can include the effects of momentum imparted on to surface material as well as the shear of the airstream. Wind tunnel measurements have indicated resuspension rate and factors which typically increase in proportion to a given power of wind speed. This power is usually greater than or around 3 (Nicholson, 1988a). Traffic resuspension measurements have also indicated an increase in resuspension with vehicle speed, and this reflects the importance of enhanced turbulence and higher tyre shear at greater speeds (see Nicholson and Branson, 1990).

Resuspension induced by falling hydrometeors can occur (e.g. Garland, 1979) and results from the air movement created just before impaction on a surface or by a movement of the surface due to momentum being imparted on contact. Such resuspension is sometimes termed 'tap and puff' and has been well known by biologists as being instrumental in the dispersion of spores. However, after a period of rainfall surfaces are likely to become wetted and surface tension effects will reduce the resuspension rate. In addition, it is possible that precipitation may result in a downward translocation of deposited material from the vegetation to the underlying soil surface or to parts of the plants from which resuspension is unlikely.

Due to the variety of surface types and conditions found in the environment, as well as the variations likely in environmental conditions, resuspension is likely to vary significantly according to time and location.

3.2.5.3 Size of resuspended particulates

Resuspension of particles by the wind can occur because of the shear forces exerted by the airstream. Because larger particles present a greater area on which the shear forces can act, and because they may protrude beyond the viscous sub-layer associated with the underlying surface, they are more easily resuspended than small particles. Similarly, because the momentum imparted on particles due to surface movement increases greatly with particle size then the effects of an accompanying increase in adhesion force are overcome and the largest particles are most easily resuspended.

Because of gravitational effects, there is a particle size (~500 μm diameter) at which particle movement is likely to be along a surface (termed surface creep) rather than via the atmosphere. It has been noted that particles with sedimentation velocities sufficiently smaller than the vertical velocities present in the atmosphere may remain in suspension. Gillette *et al.*, (1974) equated a threshold for resuspension in which the sedimentation velocity was 12 to 68 per cent of the friction velocity. This, in practice, means that particles greater than a few tens of micrometres are unlikely to be fully resuspended in even very windy conditions. Particles larger than this rapidly fall back to earth and are said to be saltating. It is a matter of definition whether these are considered to be resuspended. They can certainly lead to a spread of contamination but are unlikely to represent any kind of respiration problem.

It is important to note that resuspended trace species are usually associated with soil particles or other host particles. Thus, the resuspended material will not have the same size distribution as the original depositing species and, consequently, an inhalation risk cannot be evaluated from such a knowledge of the size of the original material.

3.2.5.4 Magnitudes of resuspension and dependence on time

Resuspension factors have been summarised by Sehmel (1980a) and were found to lie in the range 10^{-10}–10^{-4} m^{-1} when the effects of wind were considered alone. When the effects of mechanical resuspension stresses were included, this range was extended to 10^{-10}–10^{-2} m^{-1}. Such a wide range of values presents problems in adequately assessing the effects of resuspension in dose predictions. The resuspension factor is very dependent upon the area contaminated since horizontal transport plays a major role in the measured value.

A time dependence of resuspension has been widely noted (e.g. Nicholson, 1988a). The exact nature of the time dependence relationship has been described as either inverse power (e.g. Garland, 1979) or as a negative exponential (e.g. Linsley, 1978). Both resuspension rates (and factors) have been found to rapidly fall off after an initial contamination event and this can be explained by a number of effects. Firstly, in the early stages after deposition the most easily resuspended particles are removed from the surface. Thus, there is an initial high rate of resuspension. This will rapidly fall but will remain above zero since there will remain the statistical chance that wind gusts or eddies will arrive, in time, with sufficient velocity to remove the less readily removed material. Secondly, after the initial resuspension there will be a translocation of deposited material to shielded parts of vegetation (e.g. leaf nodes) and to the soil. Finally, in the much longer term, mixing of the deposited material with surface soil will limit resuspension from the soil itself and resuspension from vegetation might be controlled by soil splash onto plants and subsequent resuspension by wind or mechanical stresses.

Considering the time dependence of resuspension, it is perhaps not surprising that such a wide range of resuspension factors have been reported. Consequently, the adoption of a resuspension factor (or rate) for modelling purposes will depend strongly on the time period over which effects are to be integrated.

3.2.6 SEA TO LAND TRANSFER

Low-level radioactive effluent is routinely discharged, under governmental authorizations, from some industrial establishments into coastal waters. In the sea the radionuclides are distributed between dissolved and particulate-associated phases, the fractions being different for different elements.

In some coastal areas, a small proportion of the radionuclides are blown back onto land in seaspray; an effect most pronounced for the particle-reactive elements such as plutonium and americium. Whilst this phenomenon is likely to be widespread, it is most readily measurable close to the points of radionuclide discharge.

The highest doses to man via this pathway (McKay and Pattenden, 1990) are believed to arise near the British Nuclear Fuels Sellafield works on the Cumbrian coast of the Irish Sea (summarized by Peirson, 1988). However the process has also been observed close to the AEA-Technology Establishment, Dounreay, Caithness,

Scotland (Pattenden *et al.*, 1989a,b) and the reprocessing works at Cap de la Hague, Normandy, France (Martin *et al.*, 1981).

Uptake by man is mainly through direct inhalation of aerosol or indirectly via inhalation of resuspended soil and consumption of foodstuffs. Whilst the doses received are currently small relative to those from contaminated sea food or external exposure, they arise mainly from the long-lived particle-reactive actinides of plutonium-239, plutonium-240 and americium-241. The transfer is thus sustained by long-lived radionuclides which may be retained for long periods by fine sediment on the seabed and it is therefore likely to persist long after discharges cease. Consequently it is important to be aware of the pathway and to be able to estimate its likely magnitude now, in the past and particularly in the future when its relative importance might increase.

3.2.6.1 Experimental observations

Sea to land transfer is observed most readily through analysis of soils. Levels generally increase with increasing proximity to the sea and the isotopic ratios become increasingly consistent with those in seawater.

In Cumbria a large number of surface (0–15 cm) soil samples were collected within a 50 km radius of Sellafield between 1977 and 1979 (Eakins *et al.*, 1981). These showed that within a coastal strip, 13 km north to 27 km south of Sellafield and stretching 10 km inland, the accumulated deposition of Pu and Am was greater than could be attributed to other processes (see Figure 3.6). This was the first clear evidence of the sea to land effect. Later atmospheric measurements (Pattenden *et al.*, 1980) showed further that (i) an aerosol with a substantial fraction of large particles (indicated by higher coast-to-land ratio in deposition compared with air) was involved, consistent with it being seaspray and (ii) the plutonium and americium in this seaspray aerosol were enriched relative to their concentrations in seawater (by up to about two orders of magnitude). The studies around Cap de la Hague, though less extensive, suggested the operation of seaspray processes similar to that seen in Cumbria. The Cumbria measurements indicate that the ratio of soil plutonium derived from sea-to-land transfer to that from weapons fallout is about 7 near Sellafield, falling to unity in other parts of west Cumbria. The total excess Pu in soils (40–70 GBq) is ~0.01 per cent of that discharged from Sellafield into the sea (440 TBq up to 1978). Autopsy measurements indicate that excess plutonium in tissues of people living near Sellafield, but not employed there, derives from aerial discharges in the early history of the works, and not from sea-to-land transfer.

Around Dounreay, the transfer appeared to be dominated by a different mechanism. Coastal soil surveys revealed a number of anomalously high accumulations of actinides within a few hundred metres of the coast. In some of these areas, the values were comparable to those found in Cumbria, despite far lower seawater levels (Pattenden *et al.*, 1989a,b). The suggested transfer medium here was spume, a stable foam commonly occurring along the northern Scottish coastline. Spume

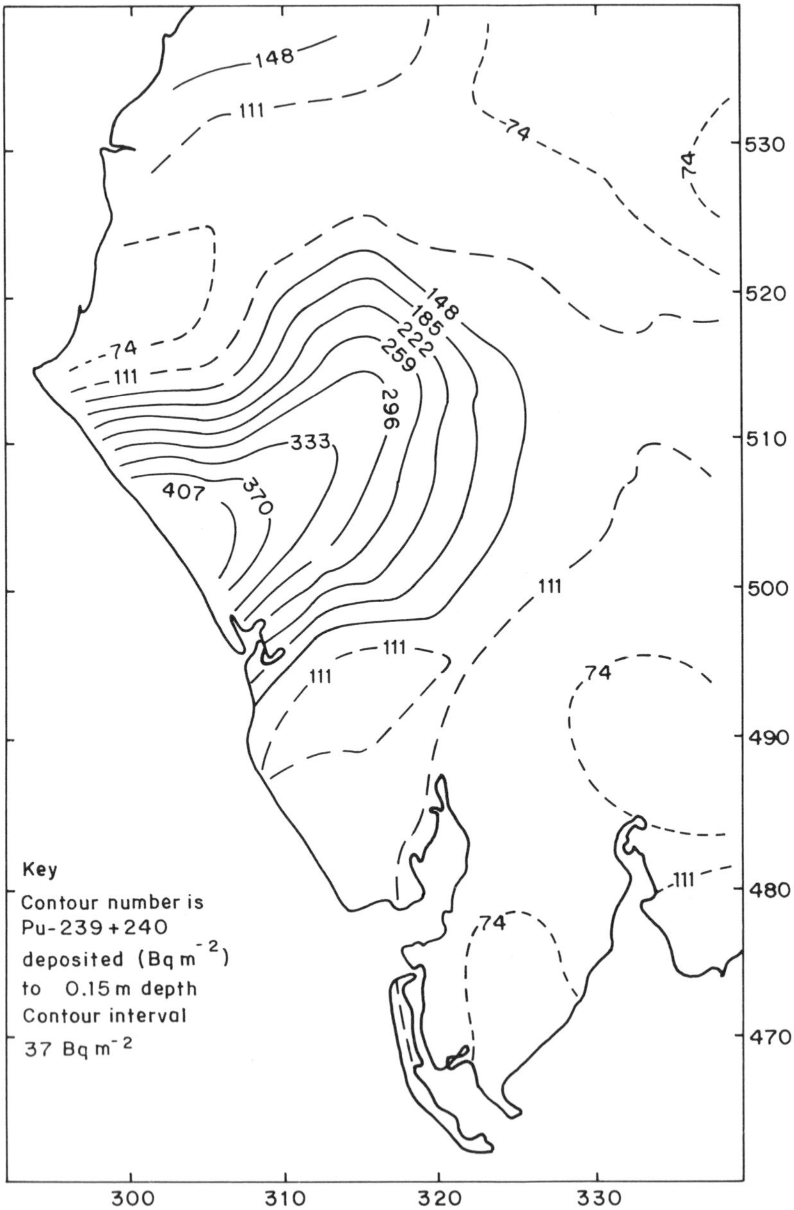

Figure 3.6 Distribution of radionuclide deposition (Bq m^{-2}) in soil in West Cumbria.

has a very high fine particulate loading and the actinides are very particle reactive, consequently the process of spume being blown onshore and washed into the soil could explain the high accumulations observed. Spume is not so readily transported inland as seaspray or as uniformly generated, and thus the sea-to-land effect observed in Caithness was more patchy and fell off more rapidly with distance from the coast.

3.2.6.2 Transfer mechanisms

In Cumbria, it has been observed that the airborne and deposited material associated with sea-to-land transfer contains radionuclides which can be related mainly to the particulate phase in nearby seawater. However, the actinide concentrations, as measured by the ratios to stable Na, are 1–2 orders of magnitude greater than those in seawater. Thus an enrichment mechanism is involved, due to scavenging of particulates by bubbles rising through the water column to the surface. At the surface the bubbles burst, generating droplets enriched in particulates, some of which are incorporated in the aerosol over the sea. The strong tendency of actinides to associate with particles means that the mechanism is particularly important for these radionuclides (Walker *et al.*, 1986). It is worth noting that the mechanism is a natural one and the actinides act as tracers for particulate material in seawater.

It has been speculated that organic coatings, arising as a result of surfactants released by plankton, may enhance scavenging through their attachment to the bubbles and particles. Such coatings, by reducing surface tension, reduce the dissolution rate of bubbles and thus increase the number of smaller bubbles of relatively high surface area reaching the surface. On inorganic particles these coatings render particle surfaces more hydrophobic and thus more attractive to rising bubbles. Recent work reported by Cloke *et al.* (1991) supports this hypothesis and suggests a linear relationship between aerosol particulate content and seawater surfactant concentration. Thus seasonal and localized increases in surfactant seawater concentration could be, in part, the cause of variations in aerosol actinide content.

3.2.6.3 Radiological consequences

Radiation doses to the population due to sea-to-land transfer have been estimated for Cumbria where a large proportion of the population is exposed through inhalation and food chain contamination. These estimates have been made using a largely empirical model (Howorth and Eggleton, 1988), and its validity outside Cumbria and its predictive capabilities thus are not yet established. However the radiological significance of sea-to-land transfer in Cumbria (and by implication, elsewhere) is certainly small, and on the basis of the empirical model probably ~2 per cent of the ICRP limit of 1 mSv per year for 1987.

3.3 ENVIRONMENTAL MEASUREMENTS

The Chernobyl accident clearly illustrates the need for measurements of radionuclide concentration in the air and on the ground to go hand-in-hand with modelling studies. Measurements on their own are rarely sufficient to answer all the questions that are inevitably raised, and this was particularly true following Chernobyl. Models complement measurements and measurements correct and strengthen models. All models require appropriate input data; in this case details of the locality, height and time profiles of the emissions. They also require suitable meteorological data. Principal of these are wind and precipitation fields in space and time, but other parameters may also be necessary. Actual requirements will change with the scale and character of the emission.

The remainder of this chapter contains three parts: the first describes in outline the various techniques and the uncertainties associated with radiological sampling methods. The second considers how meteorological information is collected, and in so doing implies the accuracy likely to be achieved in the interpolated fields. The third part of the chapter discusses the wide range of atmospheric dispersion and deposition models, and gives an example of how these models have been 'validated' against the results of the Chernobyl accident.

3.3.1 CONCENTRATION MONITORING

3.3.1.1 Air sampling methods

A variety of different methods are used for monitoring the concentrations of radionuclides in air. Most commonly, a sampling apparatus is used to collect intermittent whole air samples or to filter radionuclides of interest from the air. The collected sample is taken back to the laboratory for concentration analyses. Some instruments are available which allow *in situ* measurements to be made, either directly after sampling or even while sampling is in progress. Generally, different techniques are used for sampling gases and particles.

For the measurement of particles, the most common sampling methods are based on filtration of large volumes of air. The large volumes required provide a major constraint upon the time resolution of measurements. For applications where filters are to be counted directly for alpha particle emissions, a membrane type filter material which allows minimal penetration is desired. However, such filters tend also to restrict flow and thus limit the volume of air which can be sampled. If the radioactive material is to be chemically removed from the filter matrix before analysis, a filter medium which is easily dissolved and has low ash content and whose ash does not interfere with the chemical separation is required.

Filters cannot generally be used for collecting gaseous radionuclides although some chemically impregnated materials are available. Gaseous species are generally sampled by either collecting whole air samples or preferably by removing the gaseous nuclides from an air stream via some type of absorption process. The most widely used absorbent is activated charcoal, which is almost a universal absorbent and effectively removes nuclides such as radon, krypton, xenon, and gaseous iodine from an air stream with high efficiency. The efficiency of removal depends on flow rate and can be increased by cooling the charcoal to dry ice or even liquid nitrogen temperatures. Cold traps are frequently used to condense tritiated water vapour from the atmosphere. Similarly, cryogenic separations are used to remove noble gases such as krypton-85 from whole air streams. $^{14}CO_2$ is often sampled by bubbling air through a caustic solution of NaOH where the carbon is chemically trapped. ^{14}C can also be sampled by drawing high volumes of air through a dry molecular sieve.

3.3.1.2 Analysis methods

Once the nuclide of interest has been collected, a number of different methods are used to infer its concentration in air. The choice of method is usually governed by the nuclide half-life, the type of radiation emitted (penetrating power), the sampling matrix and the speed with which the information is required. In many cases particulate collected on filters can be analysed directly in the field either immediately after collection or even during collection. This is generally true when the radionuclide of interest emits fairly energetic gamma rays (e.g. ^{137}Cs, ^{131}I). Most commonly, filters are removed from the pump after sampling and brought to the laboratory where they are either counted directly in a fixed geometry and/or the collected particles are chemically separated from the filter for subsequent counting. In general gamma-emitting radionuclides such as ^{131}I or ^{137}Cs, and radon and thoron progeny can be measured by counting the filter directly.

For most low-level alpha and beta emitter measurements the filter is generally dissolved and the radionuclide of interest chemically separated and then electroplated onto a metal planchet or evaporated onto a thin plastic film. The resultant purified and concentrated low mass sample is then counted in a vacuum chamber using a high-quality solid-state alpha spectrometer. Further information upon radioanalytical procedures is given in Section 8.3.

All of the various analysis methods, whether based on directly counting the sampling matrix itself (filter, charcoal, etc.) or a prepared extracted low mass sample, usually rely on comparison with radioactive standards of the same geometry traceable to one of the major national or international standards laboratories (NIST, IAEA, etc.). Most reputable laboratories routinely participate in cross-calibration exercises sponsored frequently by organizations such as the IAEA.

3.3.1.3 Air monitoring networks

Routine high-volume air sampling for fission products is carried out in the environs of almost all of the world's nuclear facilities. Many nations also maintain monitoring networks to detect fallout from weapons tests or potential reactor accidents (or inadvertent releases). Many programmes in place during the 1950s and 1960s for monitoring fallout from nuclear weapons tests were terminated or scaled back after atmospheric testing was halted in 1962. In the immediate intervals following the Chernobyl reactor accident, a number of nations, particularly in Europe, increased the frequency of their routine monitoring programmes or activated previously inactive stations.

Nuclear facilities also often routinely monitor the external gamma ray exposure rate (ionization) in air. This quantity is a measure of both the concentration of radionuclides in the atmosphere and the deposition on the surrounding ground. Radon, a naturally occurring noble gas, is monitored routinely at selected sites around the world as a tracer of atmospheric transport of other pollutants. A number of studies have monitored ^{85}Kr, which is released during reprocessing of reactor fuel rods, as an atmospheric tracer in order to test local and intermediate range atmospheric diffusion. The US government, originally through its Naval Research Laboratory and since 1963 by the Department of Energy's Environmental Measurements Laboratory, has since 1957 collected monthly air samples throughout the world at a network of over two dozen sampling sites.

3.3.1.4 Sampling methods and instrumentation for the measurement of surface deposition on soil and vegetation

The deposition of radionuclides onto the Earth's surface is also monitored at many sites throughout the world. Radionuclide deposition density or concentration in soil or vegetation is monitored *in situ* as well as via routine sample collection. Generally, *in situ* monitors are designed to collect precipitation or to intercept depositing radionuclides. Samples are also collected by directly funnelling precipitation into an ion exchange column which removes most radionuclides of interest from the liquid. The ion exchange bed is removed for analysis at the end of the sampling period. Dry deposition is much more difficult to assess; it is usually collected in buckets or on other surfaces such as gummed film (sticky paper), trays of granular material, or fibrous material which simulates grass. Most such collectors do not discriminate between deposition occurring during precipitation as opposed to dry deposition. A combination wet/dry collector is now used at many monitoring sites to collect separate samples of wet and dry deposition.

Once the radionuclides have been deposited onto the soil, vegetation or other surface it is generally difficult to monitor routinely the deposition density. Techniques have been developed, however, which allow rapid qualitative and often even quantitative estimates of the deposition of gamma-emitting radionuclides *in situ* (NCRP, 1976). These *in situ* gamma spectrometric techniques utilize large-volume germanium diodes to obtain a spectrum of the incident gamma radiation above a soil half-space. This spectrum can then be unfolded using the known detector characteristics and certain assumptions regarding the radionuclide depth and surface distribution to estimate the concentration of many radionuclides on or near the soil surface. Rapid surveys of deposition of gamma-emitting radionuclides are also frequently carried out from aircraft utilizing large arrays of gamma scintillation detectors. Although these aerial measurements generally provide only estimates of total activity or gamma exposure rate, they are useful for rapidly surveying large areas and providing guidance for ground sampling.

The major problem in most monitoring programmes is obtaining samples which are representative of the average deposition over a large area. Frequently, the deposition via atmospheric transport processes is non-uniform and large numbers of samples are required to characterize adequately the average deposition over even a small area or the average concentration in any particular matrix. Sophisticated statistical sampling protocols have been devised to minimize the number of samples needed, but generally such sampling programmes are time-consuming and expensive.

3.3.1.5 Deposition monitoring networks

A number of global and national sampling programmes of radionuclide deposition density were carried out during the 1950s and 1960s to monitor fallout from nuclear weapons tests. Many of these programmes were terminated after nuclear testing in the atmosphere was halted in 1962. Deposition is still usually monitored routinely in the vicinity of nuclear facilities by many governments and most such facilities maintain their own regularly scheduled routine sample collection programmes designed to demonstrate regulatory compliance with emission limits. Many of these facilities are also surveyed periodically by airborne gamma ray spectrometry to detect previous undocumented releases and establish baseline levels in case of an accident.

3.3.2 METEOROLOGICAL MEASUREMENTS

3.3.2.1 Requirements for small releases

Radioactive material (like other pollutants) injected into the atmosphere can be carried long distances before being deposited to the underlying surface. Although a typical distance-scale is 1000 km, the scale can vary considerably according to the physical nature of the material, the current state of the atmosphere and the effective height of release. For large releases it is necessary to follow the material over such long scales and further, but much smaller releases may only be of significance out to one or two kilometres, or at most a few tens of kilometres. Transport over such relatively short distances can be assessed using one of several programs run on desk-top computers, using surface-based meteorological data as input collected either from instrumentation on-site or from a standard meteorological observing station which is ideally within about 30 km distance.

The better programs require a data-base of surface elevation expressing local topography which can result in a significantly variable windfield across the area of concern, especially in rather stable conditions. In addition the meteorological data should include, as a minimum, wind speed and direction, net incoming radiation, air temperature, relative humidity, medium and low-level cloud amounts, the precipitation rate (if any), the surface roughness length, and some knowledge of the state of the ground (principally how damp it is, as may be inferred from the 'present weather code' in a standard meteorological observation). If radiosonde or other upper-air data are available, then information on boundary layer depth, buoyancy frequency in the air above the boundary layer and the temperature jump across the capping inversion should also be used.

3.3.2.2 Requirements for large releases

Important though small releases may be, in the rest of this section emphasis will be placed on larger releases where material has to be followed through the atmosphere over hundreds or thousands of kilometres. Generally models which attempt to simulate the transport and deposition over large ranges rely on the output from numerical weather prediction (NWP) models, either simply using analysed fields resting heavily on observed data (but may use predicted values in areas of sparse data input), or on totally predicted fields when looking ahead to future movements of the plume. The transport models take from these NWP models as a minimum wind and temperature profiles, boundary layer depths and estimates of precipitation type and intensity in both space and time. These NWP models require large fast computers and depend on the vast input of a wide variety of observational data collected throughout the hemisphere or globe. The nature of these data and the means by which they are collected will now be briefly summarized.

3.3.2.3 Input data

These NWP models require, as input, data describing the quantitative state of the atmosphere. This demands regular, reliable and accurate measurements of pressure, temperature, wind, cloud and precipitation. No single observing method is capable of providing all the data required and a series of monitoring networks of different types, using different techniques, has been established.

3.3.2.4 Surface observations

Many countries have a mix of stations taking surface observations; some are 'key' stations with a spacing of about 150 km used to define the broad weather pattern over the country. These are often supplemented by other stations manned by meteorologists to give more local small-scale details, but all professionally manned stations are relatively costly to maintain. Other data are collected by synoptic automatic weather stations (SAWSs) and by auxiliary observers such as lighthouse keepers, coastguards and private individuals.

Out at sea observations are made by personnel on board dedicated weather ships, and on voluntary observing ships and from gas and oil platforms and rigs. Many ships are now equipped with automatic transmission equipment, sending data via meteorological satellites more reliably and at substantially lower cost than by other means. Other ships send their data in coded form by radio and telex to reach national meteorological centres via coastguard stations. Worldwide there are some 7500 ships from 49 countries making such observations when they are at sea. In spite of the number of observing ships, the typical density of ship reports received from some areas is still woefully inadequate. Over a one-week sampling period in a recent winter only 15 surface observations were received on average for each midday observing time from merchant ships in the entire Atlantic west of the British Isles (i.e. from 50° to 60 °N, 10° to 60 °W), since the major shipping lanes lie further to the south. It would require about 200 such observations together with satellite imagery to achieve an adequate description of the broad features of weather disturbances at the surface. Even fewer ship observations are received at night due to a lack of a duty radio operator on many vessels during the 'silent' hours. Whilst at first sight this may not seem too serious in the present context, forecasts of the movement of airborne radioactivity may be seriously degraded by such sparsity of observations in relatively distant upstream areas.

3.3.2.5 Upper-air observations

Upper-air measurements of geopotential, temperature, humidity and wind are made routinely from suitably equipped stations around the world, using balloon-borne radiosondes, up to heights of around 30 km at 00.00 and 12.00 GMT. These stations require reception, tracking and analysis systems and these as well as the sondes are periodically upgraded with the latest technologies. The measurements give information that are clearly short-time samples and that are averaged to some extent along the vertical track of the sonde as it ascends. The upper winds are obtained by tracking the sonde by radar or by using a Navaid radiosonde which receives and retransmits low-frequency radio signals of the type used for aircraft navigation.

An increasing number of boundary-layer radiosonde systems are being established which measure geopotential, temperature and humidity up to about 5000 m. These give much more detailed profiles in the region of particular interest to pollution transport modelling. Other systems, like doppler acoustic sounders, are also capable of inferring low-level winds and turbulence intensities. These however are still largely research tools and are thus not widely available.

3.3.2.6 Measurements from aircraft

In recent years an avionics system known as ASDAR (Aircraft to Satellite Data Relay) has been developed to collect meteorological data on board commercial aircraft, especially those which overfly data-sparse areas, and to transmit the data automatically via satellites to the telecommunications networks. Valuable though these are (particularly for defining the jetstream), reporting is inevitably spasmodic.

Aircraft reports, which do not originate from ASDAR, are however sometimes inconsistent, and subject to communication delays.

3.3.2.7 Weather radar

Weather radar is a very valuable source of information not only to the NWP models but also to the transport and deposition models tracking pollution. Radars provide a high-resolution picture of almost real-time precipitation. Traditionally rainfall has been measured using funnel rain-gauges but, because of the large spatial and temporal variability of precipitation, a large number of such gauges are required for accurate estimates of the total rainfall over any particular area such as the cloud of radioactive debris released from a damaged reactor. Radar, however, measures rainfall and snowfall instantaneously over a large area, thus simplifying the collection and transmission of data. This is not to say that conventional gauges are redundant. For example, a number of rain-gauges which can telemeter their data to the radar base are required to calibrate the radar returns and thereby yield optimum estimates of surface rainfall. This is particularly important in mountainous terrain where the radar beam cannot always 'see' down to ground level.

An example of the high resolution of the radar data is provided by the situation in the UK where individual radars operate on a cycle which repeats every 5 minutes and data are processed and reduced to measurements of precipitation in both 2 km × 2 km and 5 km × 5 km areas aligned with the National Grid. Although the precipitation estimates have greatest accuracy within about 75 km radius of the radar, *daily* totals of rainfall are achieved for each 5 km square within a 210 km radius. In addition, the *composite* radar images produced from the radar network every 15 minutes are also archived.

Many countries in Western Europe have come together in the last few years to exchange and harmonize their radar data to provide composite maps of rainfall. This process is bound to continue, bringing in an increasing number of countries.

3.3.2.8 Satellite data

Observations from polar-orbiting (typically at about 1000 km altitude) and geostationary satellites (at about 36 000 km above the equator), able to sense radiation from the atmosphere, provide opportunities for meteorologists to make a wide variety of very relevant measurements. By detecting the variable solar radiation reflected by clouds and the Earth's surface during daylight hours, a conventional image can be formed and transmitted back to Earth. Furthermore, by monitoring certain infra-red wavelengths, images of cloud can be obtained at all times of the day. Microwave radiometers offer the prospect of measurements of the rate of rainfall over the sea that are about as accurate as those now obtained from surface-based radar. Also by measuring infra-red or microwave radiation at a range of wavelengths, some of which are strongly absorbed and some of which are not, it is possible to discern the temperature of the atmosphere as a function of height. Although satellite soundings provide a denser network of observations in both time and space than is available from the conventional radiosonde network, there are significant problems to overcome if the desired accuracy is to be achieved. An Along Track Scanning Radiometer (ATSR) has been fitted to the European Space Agency Satellite ERS-1 and provides measurements of sea-surface temperature to an accuracy better than 0.5 °C.

3.3.2.9 Combined precipitation data

As already seen, no single system on its own can provide a totally adequate precipitation field. Considerable research effort is therefore in progress to combine information from various sources to provide the best composite pictures. One such system operated in the UK is the FRONTIERS computer-based system that receives radar and satellite data in image format and manipulates them *under forecaster control* to generate improved images of the current rainfall pattern. By 'improved' is meant enhanced, extended and quality-controlled, and interpreted by the forecasters' understanding of the meteorological situation.

3.4 MODELLING

3.4.1 INTRODUCTION

Modelling of atmospheric dispersion and deposition of radionuclides is undertaken for a variety of reasons. For example detailed modelling of some of the processes described above, such as the fate of radioactive aerosols entering a convective storm, aid in understanding the role and interaction of specific microphysical and meteorological factors. More general models, incorporating simpler parameterizations of advection, turbulent mixing, radioactive decay and removal by dry and wet deposition, are directly applicable to estimating the environmental consequences of radionuclides released to the atmosphere. These estimates are particularly important in the immediate aftermath of an accidental release in alerting and guiding monitoring teams to areas most likely to have been significantly contaminated. The type of model appropriate will depend on the spatial scale of importance, differentiated below as local within 10 to 20 km of a single source, regional out to about 100 or 200 km, continental involving a range of the order of 2000 km or more, and hemispherical or global. These different scales are principally determined on the one hand by the magnitude and duration of the release, and on the other by the differing assumptions implicit to the models. For example on the local scale, and on this scale only, the meteorological parameters may be assumed stationary in time whilst any part of the release traverses the 'scale', and the material if released in the turbulent mixing layer continues to spread vertically.

The type of model to be selected is highly dependent upon the specific purpose for which it is to be used. The selection process must consider a variety of factors such as: (1) the spatial and temporal scales to be modelled, (2) response time requirements, (3) source term characteristics, (4) the complexity of the terrain and meteorology over the domain of interest, (5) meteorological and radiological data availability, (6) model complexity relative to operator training, (7) desired modelling accuracy, and (8) computational requirements. Atmospheric dispersion models are used for emergency preparedness planning as well as for dose assessments during and after an accidental release of radioactivity. In many cases a simple model coupled with climatological data may be sufficient for basic emergency preparedness planning purposes. However, accident analysis, both during and after the accident, should be based on results obtained from a model that faithfully simulates the physical processes responsible for pollutant dispersion and utilizes the available meteorological data.

Atmospheric dispersion models may serve several different purposes during an accident in addition to estimating the expected temporally and spatially varying dose distribution pertinent to the affected population centres. These are:

1. estimating source term;
2. providing guidance for deploying field measurement teams;
3. evaluating the consequences of hypothetical accident scenarios;
4. providing a consistency check on measurements;
5. helping to interpolate and extrapolate measurements.

By integrating the radiological measurements with the model predicted concentration patterns, it is possible to estimate the source term. This, however, requires a careful optimization of the agreement between the measurements and model calculations in order to acquire realistic results. Models, in conjunction with weather forecasts, can be used to increase the effectiveness of the measurement teams by directing them to areas where the concentration patterns are expected.

The calculated patterns of time-integrated atmospheric concentration and deposition of the important nuclides provide a starting point for estimating the doses to the populations exposed; either used directly in conjunction with the appropriate dose-factors to calculate inhalation doses, or doses due to external irradiation from the cloud; or in conjunction with further models of transfer through the surface environment, and into food chains, to estimate doses from external irradiation and ingestion. This can either be for individuals exposed to more acute doses leading to direct radiation effects or requiring ameliorative action, or to population or individual doses to indicate the stochastic effects. Such analysis is extremely useful in the context of emergency situations; both in treating hypothetical releases of radioactivity for planning emergency responses and risk assessment; and in the course of a real emergency as a tool for interpreting the available information and aiding decisions on the appropriate actions and control measures. Also, if as in the case of Chernobyl, there are large uncertainties about exactly what is being released at the source, comparison of the agreement between model estimates for different assumed releases and measurements downwind can be used to estimate the source terms for the important nuclides, and their variation over time. Once the immediate emergency is over, the release brought under control, more detailed modelling reconstructions can assist in the post-accident analysis.

The level of complexity of atmospheric dispersion models varies greatly. In treating hypothetical releases for risk studies detailed accuracy in the spatial and temporal pattern is not necessary, nor is it appropriate to use a very sophisticated model requiring substantial computer time when a rapid response is required close-in to an accident; too complex a model, with a large number of variable parameters, can just be confusing in such a situation. However it is also important to recognize the simplifications introduced by a model, and the limitations induced by the indeterminant nature of the atmosphere. These are described below for the different types of model applicable over different distance scales. In essence they all represent a solution of the equation for the concentration Q:

$\partial Q/\partial t=$	$\mathrm{div}(v\,.\,Q)$	$-\lambda.Q$	$+\lambda'.Q'$	$-D$	$+S$
	advection by resolved velocity v + turbulent diffusion	decay to daughter	decay from parent	deposition dry and wet	new e.g. resuspension

(3.5)

3.4.2 TYPES OF ATMOSPHERIC DISPERSION MODEL

A variety of models are available for evaluating the consequences of an atmospheric release of radioactivity. The simplest of these models is the Gaussian plume model, which requires only the wind speed and direction at the release location along with estimates of atmospheric stability and source term. At the other end of the spectrum are complex three-dimensional models that are capable of including the effects of terrain and spatially varying meteorology. This spectrum of model capability can be divided into the following three generic categories:

3.4.2.1 Short-range Gaussian models ~10–20 km

These models can provide real-time dose assessments for distances out to 5–10 km from the source point, depending on the roughness of the terrain and the complexity of the meteorology at the time of the accident.

3.4.2.2 Mesoscale models ~200 km

Beyond 10–20 km or so topographical features and changing meteorological conditions complicate the dispersion. This requires sophisticated models for the windfields taking account of ground contours and surface characteristics. Alternatively techniques have been developed to interpolate between available wind measurements over a region, ensuring that the windfields deduced are mass consistent (Lange, 1978). It is important in such models to be able to represent such features as sea breezes, and the behaviour of flows over hills and in valleys. With these windfields, concentrations can either be estimated by integrating equation (3.5) over a three-dimensional grid of cells, or using an assembly of particles (either using statistical Monte-Carlo simulation or particle in cell).

3.4.3 LONG-RANGE TRANSPORT MODELS

Over long distances, possibly involving transport across international frontiers, the first question following an emission is where will material be transported to; or alternatively if radioactivity is unexpectedly detected, where has it come from? This requires trajectory modelling, either following material forwards from a source, or backwards from the point of observation, and depends on the availability of windfield data over the region of interest. These may be taken directly from weather forecasting models, or derived indirectly from pressure fields using a quasi-geostrophic approximation. After Chernobyl a wide variety of trajectories were produced, reflecting a plethora of different techniques and windfield data. Some of these trajectories referred to different altitudes; some were two-dimensional horizontal trajectories at a constant level, and others three-dimensional using vertical winds directly, or isentropic trajectories adjusting altitude to maintain a constant potential temperature. The diversity of results indicates the considerable uncertainty in estimating continental scale transport, which may have been somewhat obscured in the Chernobyl situation by the prolonged nature of the release. It stresses the point that the modelling of atmospheric dispersion over such distances can be no better than the windfields used, and the importance of current improvements in forecasting models with much finer spatial grids, providing data at frequent intervals (ideally hourly). The accuracy of trajectories decreases markedly with increasing travel time, and is normally considered very unreliable beyond about 72 hours, although in the case of Chernobyl the material took 8 days to cross Europe.

Some models may not adequately reflect a very important characteristic of plume behaviour at continental scale which was very evident in the Chernobyl cloud, namely that beyond about 24 hours travel the plume becomes increasingly fragmented and contorted, often with filaments and blobs. This complex behaviour results from both synoptic-scale variations in the winds associated with mobile weather fronts, depressions and anticyclones, and the variable interaction of vertical motions and marked wind-shear with height.

The simplest long-range dispersion models merely involve assuming a plume along the estimated trajectory in a similar fashion to an extended version of the Gaussian plume model used at shorter distances, and often simplified to give uniform vertical mixing over a constant mixing layer. Alternatively for uniform continuous routine releases, or risk studies, statistical models are sometimes used which ignore bending of the trajectories and changing wind-speeds, and use merely wind-rose data at the source into different directional sectors. However for more sophisticated quantitative analysis of downwind concentrations and deposition, slightly more complex models are required. These can be differentiated into two types—the Lagrangian models which essentially follow the histories of component elements of the release across the region, and the Eulerian models which simulate the dispersion of material through a three-dimensional grid of cells spanning the region of interest.

3.4.3.1 Lagrangian puff models

Lagrangian puff models have been widely used in continental scale studies, not only of radionuclides but of acidic and other atmospheric pollutants. They use horizontal windfields, and treat pollutant emissions as a series of puffs which are advected as columns of polluted air along the calculated trajectories. They vary in complexity according to the detail in which they model vertical and horizontal dispersion and other pollutant processes according to the meteorological conditions encountered. One of the earliest studies of the European scale transport from Chernobyl was undertaken with such a model, MESOS (by ApSimon *et al.*, 1989), which had already been applied to the Windscale accident in 1957 in the UK—see Section 3.5.2. As was apparent from the results, difficulties arise when there are marked changes in wind height, or the motion is very three-dimensional in nature as in the upward motion within the frontal system over Scandinavia. Another difficulty is the venting of material aloft in convective clouds. The advantage is that such Lagrangian models give a very direct relationship between emissions and the contamination they give rise to. They are well-suited to estimating source terms and their variation in time, as they differentiate between different parts of the release; also to estimating transfrontier fluxes. It is also easier to revise predictions in the light of early measurements and updated forecasting of windfields as the situation evolves following an accident release. Their computer requirements are still relatively modest, and hence they are still appropriate for use in risk studies where large numbers of potential accident situations need to be investigated (e.g. ApSimon *et al.*, 1985).

3.4.3.2 Eulerian grid models

The Eulerian grid models allow a full three-dimensional treatment in integrating the transport equation over a three-dimensional array of grid cells, but are far more demanding on computer time. They have been particularly useful in treating situations with complex atmospheric chemistry, but this is less critical when treating radioactive aerosols. Nevertheless Eulerian models were successfully applied to the Chernobyl situation; for example the GRID model of RIVM in The Netherlands (van Egmond and Kesseboom, 1983). The Eulerian models still cannot resolve sub-grid scale processes such as cloud venting and convective storms; and also suffer from some numerical difficulties such as numerical diffusion, which is particularly important close to a point source. Nor do they overcome the intrinsic errors in the windfields and advection; and it is crucial to have mass-consistent windfields which satisfy the continuity to avoid violation of conservation of the mass of pollutant material. Their main draw-back for application in an emergency situation is their inability to distinguish contributions of different elements of the release, except by repetition of the calculations.

3.4.3.3 Particle models

An approach which overcomes some of the difficulties of the Lagrangian models so far discussed whilst retaining most of their advantages involves Monte-Carlo particle simulations. The release is treated as a sequence of particles which are advected according to the evolving windfields in space and time, with random perturbations in each time-step to represent the effects of turbulent displacements with respect to the mean windfield. They are thus good for treating the complex three-dimensional nature of the windfields, but are heavy on computer resources since large numbers of particles have to be tracked. This difficulty can be reduced by the use of parallel processing techniques, and by making use of the whole trajectory of each particle through a set of surface grid-cells, instead of just the instantaneous distributions every few hours. There are no problems of numerical diffusion as in the Eulerian models, but care has to be taken to avoid artificial accumulation of particles in regions of lower turbulence by inappropriate simplification of the technique.

Quantities of the key radionuclides can be associated with each emitted particle. The airborne fractions can be reduced according to the probability of decay and of particle deposition by dry and wet processes. Wet deposition, and cloud venting to the free troposphere which can be so important with convective activity (and may well have accounted for substantial hemispherical scale transport after Chernobyl), can be treated statistically with a resolution determined by the available data on cloud distributions and precipitation. This is an area where further work on modelling is required.

3.4.4 MODEL EVALUATION STUDIES AFTER CHERNOBYL

Since Chernobyl there has been an increased interest in development of modelling capabilities to simulate continental scale dispersion. Accordingly a long-range model evaluation study, ATMES, was established jointly by IAEA, WMO and the CEC using observations after Chernobyl. Joint estimates of the emissions were used, and the REM (Radioactivity Environmental Monitoring) data bank established at the Joint Research Centre, Ispra (Girardi *et al*, 1987) provided carefully compiled and screened measurements relating to daily patterns of air concentrations of ^{131}I and ^{137}Cs and their deposition across Europe. Modellers were asked to provide corresponding estimates, many of them making use of the same meteorological data set provided by ECMWF (European Centre for Medium Range Weather Forecasting), for the Chernobyl period. Various statistical analyses have been applied to compare modelled results and observations and test spatial and temporal agreement, drawing on experience with earlier model evaluation studies in North America based on perfluorocarbon tracer experiments.

Twenty-one models from 14 countries participated in the study, which varied unduly in sophistication from very simple trajectory models to three-dimensional particle models and Eulerian grid models. It was evident that different models appeared to give better agreement according to different tests, but there was no clear indication that the most sophisticated models performed appreciably better than the simplest. However, there are major diferences between the model predictions, indicating that the results are sensitive to the manner in which the atmospheric structure and processes are parameterized and represented—for example the wide variation in treatments of mixing layer depth.

Comparing the participating models was difficult in view of uncertainties in the given source terms and observations, and also because some models may already have been adjusted or tuned to this particular accident. The results are being analysed further, and will be reported in full. Proposals have been put forward to carry the model comparisons further by conducting a tracer experiment in Europe, similar to those already undertaken in North America (ANATEX, CAPTEX) to test models in a real-time application—a very difficult but relevant test for their practical use in an emergency situation. This will emulate a release of inert gases and non-depositing pollutants.

3.5 APPLICATIONS OF MODELS TO IMPORTANT ACCIDENTS

The three major accidents at Three Mile Island, Windscale and Chernobyl have provided unique opportunities to evaluate models and test their usefulness. The latter two both released radionuclides that deposited to the surface by a variety of processes, in contrast to the inert non-depositing tracers available for experimental releases. For all three accidents models were applied to simulate the atmospheric transport, and compared with the radiological observations.

3.5.1 THREE MILE ISLAND

Several modelling approaches were taken to estimate the radiation dose received by the affected population due to the release of noble gases (mainly ^{133}Xe) from the damaged reactor. This included several independent studies based on atmospheric dispersion modelling in conjunction with the environmental radiation measurements, as well as spatial interpolation of the radiation measurements made at numerous locations surrounding the site. The various dispersion modelling efforts used a variety of models that ranged from Gaussian to complex three-dimensional models.

Meteorological data were available from the on-site tower and from several local sources. The meteorological conditions during the first five days of the accident, when the highest release rates occurred, consisted mainly of up- and down-river wind flows. From the morning on 28 March, when the release was initiated, until mid-afternoon on 29 March, the winds were primarily from the southeast–southwest direction at approximately 3 m s^{-1}. Subsequently the winds rotated to a northeast–northwest direction with an average speed of 1–2 m s^{-1}. Calm and highly variable conditions were observed during the night of 29–30 March. These calm and variable conditions continued until the evening of 30 March when strong and steady southerly winds of about 3 m s^{-1} returned. On 1 April the winds rotated into the westerly to northwesterly directions with speeds generally ranging between 1 and 3 m s^{-1}.

Atmospheric dispersion modelling, based on these meteorological conditions, permitted elucidation of the temporal evolution of the time-integrated dose pattern. This is illustrated in Figure 3.7 which shows the evolving integrated dose pattern (Knox *et al.*, 1981). These results were generated by means of a three-dimensional mass-consistent windfield model coupled with a particle-in-cell transport and diffusion model using a normalized 1 million curie release that varied in time according to that reported by the General Public Utilities (GPU), the plant operator (Woodard, 1979). The figure shows how the dose pattern was quickly and predominantly established by the generally south-to-north flow during the initial release period. Thereafter, the low levels of release resulted in only relatively minor but discernible changes in the initial dose pattern. Note particularly the southward extension of the pattern from 29 to 30 March, the east and southeast spread from 30 to 31 March, and finally the 'diffusion-like' effect of nine days of synoptic and diurnal meteorological variations with a small source term.

Integration of atmospheric dispersion patterns with the environmental radiation measurements permitted estimation of the total integrated dose to the exposed populations. Using the US Department of Energy (DoE) aerial radiation measurements made within the plume on a regular basis over a two week period, resulted in the DOE integrated dose pattern shown in Figure 3.8 (Hull, 1980). Note that the highest doses occurred in the areas immediately north of the plant with secondary nodes extending in the southeast and easterly directions. A similar pattern was generated by GPU after integrating the TLD measurements with their finite plume dispersion modelling (Woodard, 1979). These studies indicate that a total dose of 100 mrem was exceeded over an area extending several kilometres in a northeast to northwest direction from the plant, although the highest measured off-site dose by the TLD network was 75 mrem.

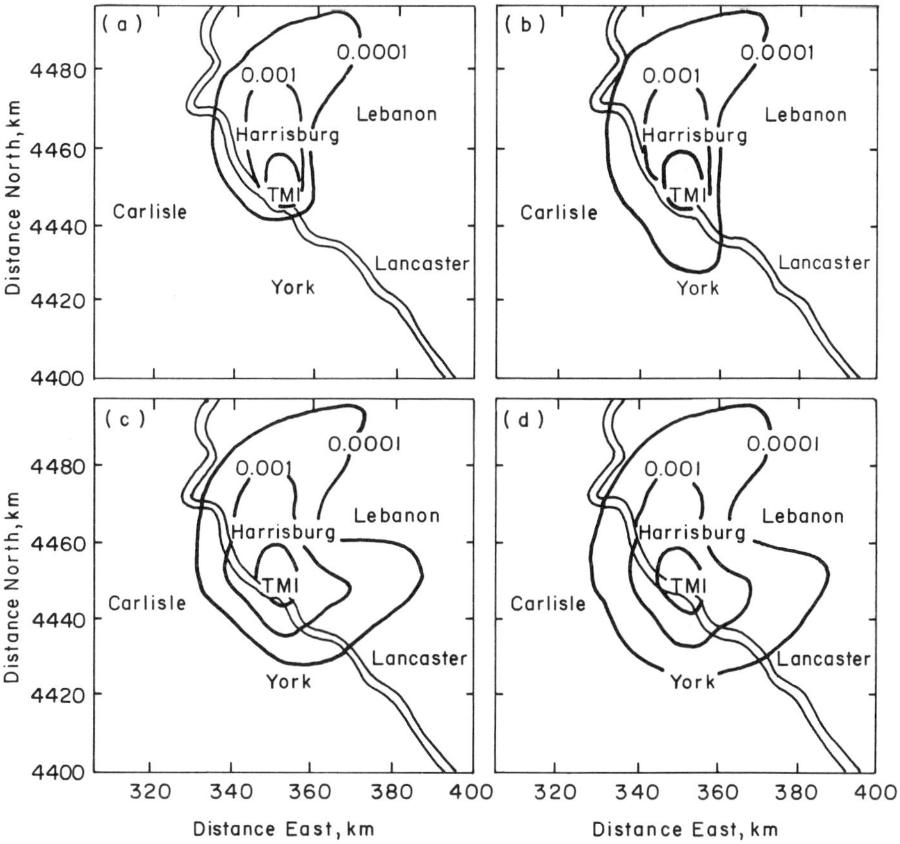

Figure 3.7 Normalized calculated integrated dose patterns in units of mSv due to TMI-2 noble gas release on (a) 29 March (24-h integration), (b) 30 March (48-h integration), (c) 31 March (72-h integration), and (d) 7 April 1979 (240-h integration). The patterns are based on the normalized release of 37 PBq of ^{133}Xe. (Reproduced by permission of the Lawrence Livermore National Laboratory. All rights reside with the US Government.)

Integration of the dose pattern with the population distribution throughout the affected region led to an assessment of the collective dose equivalents. The most credible estimates are given in Table 3.2. The estimates resulting from analysis of the DOE aerial radiation measurements was 2000 person-rem (Hull, 1980) in contrast to 3300–3400 person-rem obtained by the US Government Ad Hoc Interagency Study Group on the basis of spatial interpolation of the TLD measurements by either atmospheric dispersion modelling or by inverse distance scaling (Ad Hoc, 1979; Pasciak *et al.*, 1981). The GPU obtained 3300 person-rem by combining dispersion modelling with a 10 million curie source term and the TLD measurements. The lowest collective dose estimate, 500 person-rem, was derived on the basis of

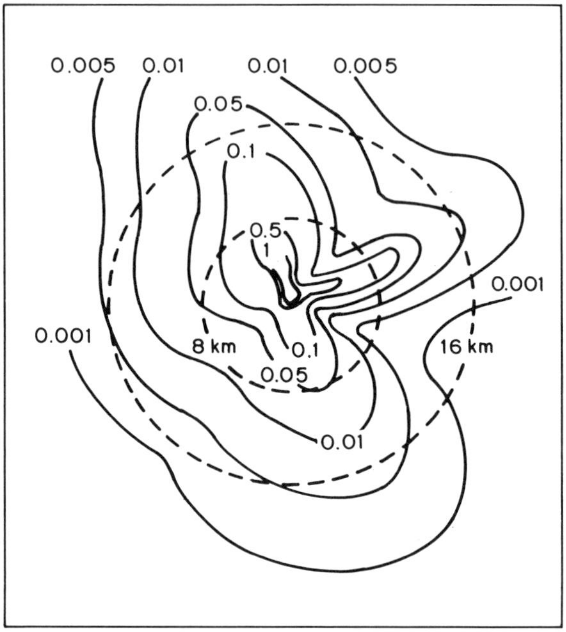

Figure 3.8 Estimated dose pattern derived from the DOE aerial measurements from 28 March–3 April 1979. The units are mSv. (Reproduced by permission of the Lawrence Livermore National Laboratory. All rights reside with the US Government.)

Table 3.2 Estimates of collective dose equivalent due to noble gas releases from TMI-2 accident

		Person-Sv	
Group	Technique	Best value	Range
DOE	Aerial measurements	20	
GPU	TLD measurements and dispersion modelling	33	16.5–66
Ad Hoc Group	TLD measurements and dispersion modelling	34	
Ad Hoc Group	Spatial interpolation of measurements	33	16–53
President's Commission	Source term and dispersion modelling	5	0.50–50

atmospheric dispersion modelling using a 2.4 million curie source term (Auxier *et al.*, 1979). Thus, the central estimate is about 3300 person-rem with a range of several orders of magnitude.

3.5.2 WINDSCALE

In the case of the much earlier accident at Windscale in 1957 relatively simple models were used at the time to estimate source terms from atmospheric concentrations measured over the United Kingdom (Crabtree, 1959). Subsequently the situation has been used to evaluate models developed for nuclear accident risk studies (e.g. ApSimon *et al.*, 1989).

The release took place mainly in two peak periods, and the situation was complicated by the passage of a weak frontal system at about the time of the first peak. During the second peak, after water was poured on the pile early on the morning of 11 October 1957, the meteorological conditions were more straightforward with a steady well-defined plume blowing towards the south. This was well represented within a few tens of kilometres by a simple Gaussian plume model (Clarke, 1974).

Over the longer distance scales the trajectory model MESOS described the general character of the dispersion fairly satisfactorily despite the complications. The initial material was transported eastwards across England, subsequently turning southwards to reach London and the continent where it was detected during the following day.

Concentrations in the southward plume from the second peak in the release were also in good agreement with observations; although as travel distances increased towards southern England the model overestimated spreading westward over Wales—possibly due to effects of the Welsh mountains in blocking the spread in that direction. Over the next two to three days the material stagnated in the anticyclone before moving off eastwards; again this was consistent with the observations. A model prediction, which could not be confirmed because of lack of measurements, was the fumigation action of the early part of the release to the surface over northeast England as the mixing layer dispersed on the morning of 11 October. Such situations can lead to temporary higher concentrations as material is mixed down to the surface.

Estimates of deposition of the principal nuclide released, ^{131}I, were consistent with model estimates using a dry deposition velocity of 3×10^{-3} m s^{-1}. Since no significant rainfall scavenged the material there could be no studies of wet deposition (although some cloud over hills may explain some observations of enhanced areas of deposition—another orographic effect not represented in current models).

3.5.3 CHERNOBYL

Radionuclides released after the accident at the Chernobyl nuclear power plant were extensively monitored and were detected throughout the northern hemisphere. Theoretical models simulating the processes that affect radioactivity transport and deposition have been used in conjunction with the observations to deduce how the material dispersed on local to global scales. This dispersion is discussed here along with selected model results. The accident at Unit No. 4 of the Chernobyl plant occurred at 1.26 a.m. local time on 26 April, 1986. It is described in Section 2.7.

3.5.3.1 The region near Chernobyl

Gravitational settling of the coarser material (particles with a radius greater than 10 μm) released over the first five days dominated the deposition pattern close to the site. This pattern emphasizes the potential importance in nuclear accidents of hot fuel particles and the need for further studies of their behaviour in the environment. During this time winds at Chernobyl blew from all directions. Material from the initial release was carried westward; subsequently, over 26 and 27 April the plume swung to the southwest and northwest with fairly light and variable winds. During 28 April material went mainly north, turning toward the east on 29 April and then to the southeast and south to complete the circle on 30 April. The calculated deposition agrees very well with the observed distribution of radiation levels for the zone within 30 km of the plant, where evacuation took place (see Figure 2.5).

3.5.3.2 The European scene

In the night following the explosion the initial plume travelled westward at a height of a few hundred metres in stably stratified air. During the day of 26 April, convective motions mixed the cloud vertically up to 2–3 km. Significant amounts were also transported to higher levels in cumulus clouds. The plume turned northward, reaching Scandinavia on 27 April. The following day radioactivity was incorporated in a developing frontal precipitation system, and localized peaks of wet deposition (in excess of 100 kBq m^{-2}) occurred in parts of central Scandinavia. A detailed analysis of this phase has been undertaken in Sweden.

Subsequently, radionuclides spread from Poland southwestward across Europe, south of a ridge of high pressure pushing across northern Europe from the Atlantic Ocean. Some of this radioactivity then turned northward to the United Kingdom behind the anticyclone that developed over Denmark. Other material travelled eastward across the Soviet Union and southward to Turkey and Greece. Much of the radioactivity eventually left Europe by moving across the North Sea and Scandinavia. Deposition of the most important long-lived nuclide, caesium 137, did not decrease smoothly with travel distance but was enhanced in regions where rain or snow intercepted the plume (see Figure 3.9).

What such a large-scale map fails to reveal is the extremely patchy nature of the deposition, which was often concentrated in a few small areas within a country and was highly correlated with local variations in rainfall amounts during passage of the radioactivity. This has been demonstrated in Scandinavia, the United Kingdom, and many other European countries. Nevertheless, except in a few particular areas, contamination from the Chernobyl accident outside the USSR was generally low compared with normal overall annual exposures from natural sources of radiation. An interesting observation is the effect of ionization on conductivity of the air over the areas of higher deposition, which in Sweden have been suggested to be associated with increased frequencies of lightning.

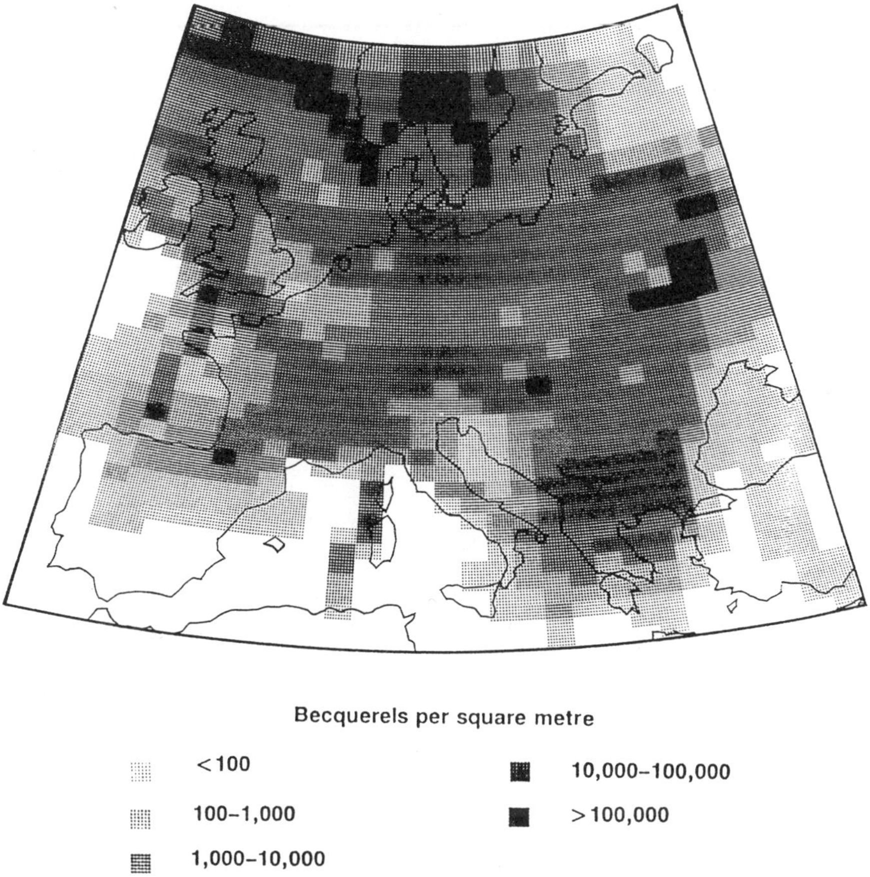

Figure 3.9 Total deposition across Europe of caesium-137 from the Chernobyl accident as estimated using the MESOS model of Imperial College, London (ApSimon and Wilson, 1987).

3.5.3.3 The northern hemisphere

On the scale of the northern hemisphere, simulations such as those undertaken by the Lawrence Livermore National Laboratory and the Japan Meteorological Institute provide good visual demonstrations of the dispersal of the radionuclides, represented by an assembly of particles. These indicate how the radioactive material became segmented during the first day, with one part heading toward Scandinavia and hence over central Europe as already indicated, while the remainder was transported in an easterly direction across Asia to Japan, the North Pacific, and the North American continent (see Figure 3.10). These simulations imply that a significant fraction of the radioactivity was transported at elevated levels in the atmosphere—up to 4 km in the Japan estimates and 10 km for the Lawrence Livermore model.

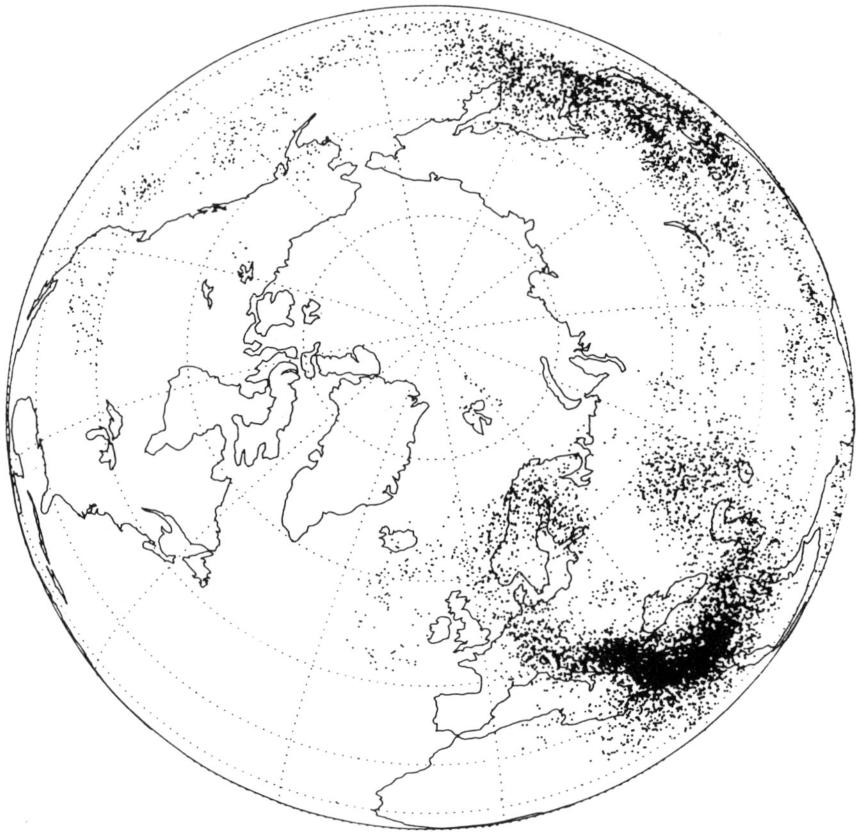

Figure 3.10 Calculated spatial distribution of radioactivity over the northern hemisphere 10 days after the Chernobyl accident as illustrated by the Lawrence Livermore National Laboratory and the US Department of Energy (Lange *et al* 1992).

3.5.3.4 Source terms

Scientists have used modelling results in conjuction with observations to deduce the quantities of various nuclides released from the Chernobyl reactor and their variation in time between 26 April and 6 May. This task is complicated by the different behaviour of different parts of the release over different distance scales. Thus, emissions of large particles could not be directly evaluated from observations at longer distances, and there are uncertainties about the height attained by the initial release on the first day and its complex transport within the frontal system. Nevertheless, estimates from European and global modelling assessments are in agreement within a factor of about 2 (see Table 3.3).

Table 3.3 Estimates of emissions from Chernobyl reactor

Source of estimate	Iodine-131 (10^{17} Bq)	Caesium-137 (10^{16} Bq)
USSR	2.8	4.7
MESOS model (European fallout)	1.7	3.9
Lawrence Livermore (northern hemisphere scale)	6.0	8.9

Note: All estimates are based on values of decay to 6 May 1986 as in the Soviet report to the International Atomic Energy Agency. Refers only to material travelling beyond the local zone where particles were deposited. Allowance for vertical dilution by convective mixing in clouds could increase this estimate by up to 30 per cent.

Analysis of the time-dependent variation from the MESOS model at the Imperial College in London accords with two peaks in the release and indicates a greater ratio of the less volatile ^{103}Ru relative to ^{137}Cs as core temperatures rose during the second peak. This is fully consistent with the Soviet accounts of the accident.

3.6 RECOMMENDATIONS AND RESEARCH NEEDS

Historically, much of the development of a scientific understanding of atmospheric transport of pollutants came from studies of radioactive tracers, either derived from bomb fallout, or in planned laboratory and field experiments. This knowledge was rapidly transferred to stable element species, as processes affecting pollutants such as heavy metals and sulphur oxides became a focus of interest. Nowadays, major atmospheric releases of man-made radioactivity are very infrequent and, as exemplified by Chernobyl, highly unpredictable. Because of the relatively short lifetimes of radioactive species in the atmosphere, experimental observations of atmospheric processes following such an accident must be made very rapidly and with a minimum of preparation. When this is considered, it is remarkable how much was learned from measurements after Chernobyl.

Many of the real research needs in this field can only be met fully by studies of nuclear accidents. However, research must nonetheless proceed at other times, and will depend upon laboratory and field experiments, theoretical studies, and in particular studies of processes using non-radioactive elements present in the atmosphere, with extrapolation to closely related radioactive elements and species. Thus, for example, research on terrestrial resuspension processes can readily progress through use of stable elements, or particles labelled in other ways, for instance, using fluorescent dyes.

Research needs therefore include the following:

(a) Studies of size distributions of radionuclides originating from different processes within a hot source, such as evaporation and mechanical disruption; the influence of aerosol processes such as coagulation and hygroscopic growth in the crucial first seconds after aerosol formation.
(b) Studies of the physico-chemical speciation of fission products and reactor fuel components at the point of emission; this is particularly important for elements such as iodine with both gaseous and particulate forms.
(c) Research into the chemical and physical processes affecting radioactive elements during transport within the atmosphere. This includes aerosol processes such as coagulation with the ambient aerosol, and factors controlling the gas/particle phase partition of elements capable of existence in both states.
(d) Studies of the water solubility and hygroscopicity of particles aged in the atmosphere. This has important implications both for incorporation in cloud water and for dry deposition to humid surfaces.
(e) Investigation of the physico-chemical changes occurring when aerosol particles are processed by cloud condensation and evaporation processes.
(f) Studies of the transport and deposition properties of large (> 10 μm diameter) aerosol within the atmosphere.
(g) Research into resuspension processes for radionuclides from both terrestrial and marine environments, to provide a better predictive capability for both short-term (days) and long-term (months, years) problems of resuspension.
(h) Further study is needed of the dispersion of gases and particles in complex terrain, and in air involved in frontal systems.
(i) Research to improve understanding of wet deposition of pollutants in frontal systems.
(j) Inclusion of 'occult' deposition of radionuclides in cloud and mist in numerical models as such processes can lead to considerably enhanced local deposition, which is not otherwise accounted for.

4 Terrestrial Pathways

Co-ordinator: **R. Kirchman**

Contributors: **J. N. B. Bell, P. J. Coughtrey, M. Frissel, T. E. Hakonson, W. C. Hanson, D. Horrill, B. J. Howard, L. J. Lane, C. Myttenaere, W. L. Robison, C. Ronneau, G. Shaw, W. R. Schell, J. Van Den Hoek, A. Konoplyov, and N. Zezina**

4.1 INTRODUCTION

Artificial radionuclides enter terrestrial ecosystems from atmosphere (e.g. as a result of stack discharges or worldwide contamination following weapons testing), via surface waters (e.g. as a result of discharges to inland waters or leaching from shallow land disposal sites), or from ground water (e.g. as a result of releases from waste repositories). The emphasis in this chapter is on behaviour after atmospheric input, though many of the pathways and processes discussed are relevant to other modes of input.

Behaviour of radionuclides in terrestrial ecosystems is determined by the initial chemical form of the radionuclide entering the ecosystem. The global fallout after nuclear weapons testing was largely made up of water-soluble and exchangeable forms of ^{137}Cs and ^{90}Sr. In contrast, deposition of ^{137}Cs and ^{90}Sr after the Chernobyl accident was often in non-exchangeable forms (Bobovnicova *et al.*, 1990). Also, close to Chernobyl, a considerable fraction of the deposited radionuclides was present as fuel particles insoluble in water or neutral solutions (Konoplyov *et al.*, 1988).

4.1.1 DEPOSITION, INTERCEPTION AND RETENTION

Deposition processes to ground are discussed in Chapter 3. Transfer of radioactivity from atmosphere to vegetation involves: deposition (usually treated by means of a deposition velocity, V_g); interception (usually treated as an interception fraction, r, defined as the proportion of the total deposit to ground which is initially intercepted by vegetation); and retention (usually treated with a weathering half-life, T_w, defined as the half-life for decline in vegetation concentration on a mass basis, or content on an area basis, following initial deposition). All three parameters are sensitive to the particle size and chemical characteristics of the depositing material, the vegetation or ground surface characteristics, prevailing weather conditions, and state of growth of the ground cover.

Radioecology after Chernobyl
Edited by F.E. Warner and R.M. Harrison
 Published by John Wiley & Sons Ltd

4.1.1.1 Deposition

There is substantial published information on the interaction of particulates with vegetation surfaces both from experimental (e.g. wind tunnel) and field studies. Much of this work was pioneered by Chamberlain and co-workers who identified the three important processes as: gravitational sedimentation, inertial impaction, and eddy diffusion deposition. Chamberlain (1970, 1975) provided mathematical treatments for these processes on the basis of experimental investigations. Many of the field data are for ^{90}Sr and ^{137}Cs deposited in weapons-testing fallout, though information has also been obtained for particulates contaminated with heavy metals (Little, 1979; Martin and Coughtrey, 1982).

It is generally very difficult to obtain reliable estimates of V_g from field studies, especially if deposition occurs over prolonged periods and by both wet and dry mechanisms. As a result, there is often a significant discrepancy between values obtained from carefully controlled experiments and those obtained from field measurements. Typically, values of 0.001 to 0.005 m/s have been used for V_g in modelling studies as appropriate to small particles with higher values (0.01 m/s) for reactive gases such as iodine and lower values (0.0001 m/s) for unreactive gases and organic forms of iodine (Coughtrey *et al.*, 1983–1985).

4.1.1.2 Interception

Interception varies according to the leaf area index, botanical composition, morphology and surface roughness characteristics of the underlying vegetation. Compared to studies on deposition, there are relatively few published investigations on interception. A well-cited series of studies is that of Milbourn and Taylor (1965) involving application of radioisotopes in solution as sprays to different types of pasture. Chadwick and Chamberlain (1970) proposed that initial retention by herbage could be modelled according to:

$$r = 1 - e^{-\mu W} \tag{4.1}$$

where μ = the 'absorption coefficient' (m^2/kg);
and W = the herbage density (kg/m^2).

Chadwick and Chamberlain calculated a mean value for μ of 2.3 m^2/kg. The data of Milbourn and Taylor (1965) indicate a range of 1.2 to 6.2 m^2/kg for soluble deposits applied to a variety of pasture types. Miller (1980) advocated the use of a normalized deposition velocity (V_d) to take account of the effects of herbage density. Relative to pastures, very few data exist for interception by crops. For cereals, Aarkrog (1983) showed a strong effect of stage of development on interception and retention of radionuclides applied in solution. Eriksson (1977) showed large differences in interception between particulate-bound and soluble radionuclides in wet or dry conditions.

For prolonged deposition to vegetation (as occurred after weapons testing), Chamberlain introduced the concept of normalized specific activity (NSA), defined as the ratio of radionuclide concentration in foliage (Bq/kg dry weight) to the rate of deposition to ground ($Bq/m^2/d$). A value of 40 m^2d/kg was considered appropriate for herbage in good growing conditions. NSA values are much higher for poor growing conditions and some types of native vegetation (Martin and Coughtrey, 1982).

Radioactive materials can be removed from atmosphere by rain-out (i.e. incorporation in raindrops) and by wash-out (i.e. removal of particulates from below the cloud base by falling raindrops). Generally, both wash-out and rain-out are incorporated into a dimensionless wash-out ratio (WR) (Chapter 3). Very little information exists on the interception of radionuclides after wet deposition, though the data base on this subject has been extended recently by Hoffman *et al.* (1989) and Caput *et al.* (1989).

4.1.1.3 Retention

Weathering half-lives in the range of 12 to 17 d are commonly encountered in the literature. Such values are probably appropriate to initial retention of fission products after weapons testing for which losses occur as a result of re-entrainment of carrier particles, sloughing of leaf surface waxes, and leaching by rainfall. Much longer half-lives or more complex patterns in retention can be associated with radionuclides applied in solution, sub-micron particles, and gaseous or volatile radionuclides such as those of sulphur and iodine. Losses of tin and iodine by volatilization have been reviewed by Arnold (1990).

To obtain reliable estimates of T_w from field data, supporting data are required on the pattern of growth of the vegetation and the timing and type of precipitation after initial interception. It is difficult to evaluate models for retention from field data because of the confounding effects of foliar absorption and translocation, or root uptake.

4.1.2 MOVEMENT IN SOILS

Most published information on behaviour and movement of radionuclides in soils reflects patterns observed after initial deposition to soil surface, either in field conditions (from sites surrounding nuclear installations or affected by fallout from weapons testing), or from laboratory experiments using soluble tracers. Behaviour in soil and subsequent removal reflect the physico-chemical characteristics of the radionuclide, the properties of the soil (including changes with depth from the surface), the type of vegetation, hydrology, and underlying geology. There have been several studies involving sampling of soils at different depths and comparison of the observed distribution and total content as a function of time. Residence times for turnover of radionuclides in a defined soil layer have been derived but generally on the basis of the total contents of radionuclides rather than on the basis of

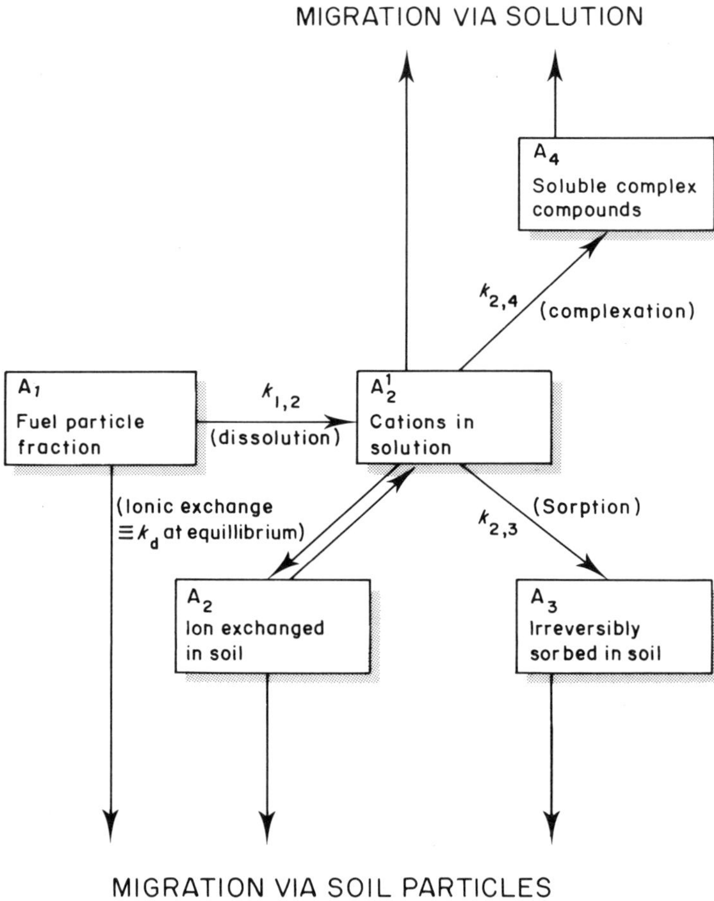

Figure 4.1 Schematic model for radionuclide transformation processes in soil.

Table 4.1 Rate constants (d^{-1}) for dissolution of ^{90}Sr and ^{137}Cs from fuel particles in the 30 km zone surrounding Chernobyl

Site	Soil type	^{90}Sr	^{137}Cs
Chernobyl	Soddy-podzolic loamy sand	1.5×10^{-3}	0.87×10^{-3}
Benyouka	Alluvial soddy acid	1.4×10^{-3}	1.0×10^{-3}
Kopachi	Slightly podzolic loamy sand	4.0×10^{-4}	—
Korogod	Arable soddy-podzolic loamy sand	3.6×10^{-3}	1.5×10^{-3}

From Konoplyov and Bulgakov, 1991.

specific fractions extractable with defined extracting agents. As a result, models for radionuclide movement in soils are generally rather simple and usually empirical (Coughtrey, 1988). For some radionuclides e.g. ^{90}Sr and ^{137}Cs, more detailed observations have been obtained and movement simulated on the basis of diffusion theory (Prokhorov, 1973; Frissel and Pennders, 1983; Kirk and Staunton, 1989; Voitsekhovitch *et al.*, 1991).

The simplest equation to describe migration is the mass conservation equation, i.e.

$$\frac{\partial S}{\partial t} + \frac{\partial C}{\partial t} = D\frac{\partial^2 C}{\partial x} - V\frac{\partial C}{\partial x} + S \qquad (4.2)$$

where C is the concentration in solution;
S is the concentration in solid phase;
t is time;
x is the vertical distance;
V is the interstitial solution velocity;
D is the apparent diffusion coefficient; and
S is the source term that describes processes other than adsorption and desorption.

The application of this equation is not limited to soils. It can also be applied to migration in rivers or to migration near a waste repository.

The whole collection of radionuclide transformation processes in soil can be described schematically as shown in Figure 4.1. Radionuclides present as original deposited particles, irreversibly sorbed to soil particles or present in ion-exchange sites on soil particles are subject to processes involved in particle migration. Radionuclides present in soil solution or as organic complexes are subject to processes involved in solution transport.

Processes involved in the movement of radionuclides in soil have been reviewed by Morgan (1990b). Important processes include: sorption, resuspension, mass transport and leaching.

With respect to dissolution of fuel particles deposited after the Chernobyl accident, rate constants derived from the 30 km zone surrounding the source are given in Table 4.1. Similarity between the values obtained for both ^{90}Sr and ^{137}Cs indicate that the processes involved are not determined to a large extent by the chemical characteristics of the different radionuclides.

4.1.2.1 Sorption

The potential for transport of a radionuclide through the soil profile is determined by chemical and biological 'availability'. Information on this subject derives from many sources, the most important of which are studies involving the extraction of soil with a sequence of chemicals thought to remove different chemical fractions. Results are then compared with studies on plant uptake or vertical transport to provide chemical explanations for the observed behaviour. In the context of environmental radioactivity this is a relatively new field, however it is well-supported by many years of work in the soil and plant sciences.

In the absence of information on chemical extractability, the most common approach has been to use a distribution coefficient (K_d), defined as the ratio of the concentration of a radionuclide in the solid phase (Bq/g dw) to that in the solution phase (Bq/ml). K_d is sensitive to a number of factors including: chemical characteristics of the radionuclide, chemical and physical properties of the soil, methods used to obtain the K_d, and the solution concentration range.

Application of K_d in terrestrial radioactivity studies is confounded by the fact that the chemically soluble fraction of a radionuclide in soil that is available for transport via mass flow is not necessarily that fraction which is biologically available. Nevertheless, determination of K_d values for a range of radionuclides under comparable conditions does allow general conclusions to be drawn about the likely behaviour of a radionuclide in the environment. Radionuclides with high K_ds under specified conditions (e.g. plutonium, caesium) generally show little movement in soils and have a high potential for resuspension, whereas those with low K_ds (e.g. technetium in lower valency states, strontium) show greater movement and have a low potential for resuspension.

4.1.2.2 Resuspension

Resuspension of radionuclides attached to soil particles provides a mechanism for loss from the system by water- or wind-mediated erosion, and for contamination of vegetation surfaces. Resuspension has generally been treated by means of an empirical resuspension factor (RF) defined as the ratio of a radionuclide concentration in air (Bq/m^3) to the ground surface content (Bq/m^2).

4.1.2.3 Mass transport and bioturbation

Surface soils are subject to considerable turnover as a result of the action of burrowing organisms. Bishop (1989) reviewed information on this subject and concluded that biotic transport of radionuclides by, or as a result of the actions of, burrowing animals, could represent a significant and rapid pathway for radionuclide migration in surface soils. Bishop noted that the maximum amount of soil moved by earthworms in established pastures is in the order of 10 $kg/m^2/y$ (equivalent to a soil depth of around 14 mm). For radionuclides which become fixed to soil particles and therefore unavailable for leaching, redistribution within soil profiles will be dominated by mass transport. Mass transport may also occur by physical mechanisms, e.g. the movement of soil particles through macropores.

4.1.2.4 Leaching

Net losses from a soil profile via leaching are difficult to quantify, even with complex models. For well-mixed soils the generally accepted approach is to make use of a soil leaching coefficient, S_1, which is defined as follows:

$$S_1 = \frac{V_w}{d_s}\left[1 + \frac{(\rho K_d)}{\theta}\right] \tag{4.3}$$

where: V_w = velocity of water percolation;
d_s = depth of soil;
ρ = soil bulk density; and
θ = soil water content

To apply the above, both V_w and K_d must be defined. For V_w, a complete hydrological model is required. Such models are beyond the scope of this work. Since radionuclides often exist in soil as more than one chemical species, a separate K_d is required for each. Also, although ion exchange is often the major absorption mechanism, for some radionuclides and soil types, other mechanisms of sorption or fixation apply.

For unmixed soils which have distinct horizons of contrasting physical, chemical and biological characteristics, more complex models are required. These are generally multi-compartment models with transfer from one compartment to another controlled by first order kinetics (Haywood *et al.*, 1980; Thorne and Coughtrey, 1983).

4.1.3 PLANT UPTAKE AND DISTRIBUTION

The two routes for uptake of radionuclides by plants are foliar absorption and root absorption. Scott Russell (1966) used the term 'plant base' absorption for certain specific conditions but the physiological meaning of this term is unclear. Translocation has generally been considered only in the context of foliar contamination with the 'translocation factor' defined as the proportion of a surface deposit subsequently transferred to edible fractions.

4.1.3.1 Foliar absorption

Foliar absorption of soluble radionuclides has been given little attention in the literature; the processes involved and the patterns to be expected for different radionuclides are not clear. Foliar absorption of gaseous materials is better described and quantified, primarily from analogy with non-nuclear contaminants. Thus, for ^{35}S, studies on sulphur dioxide can be used to determine the resistance to passage at various stages of the transfer. For ^{14}C and ^{3}H a substantial literature exists on photosynthesis, respiration and transpiration, and models of varying complexity exist to simulate absorption and subsequent redistribution. The degree to which surface-deposited radionuclides are absorbed by plant surfaces is very important in interpreting information on deposition, interception and retention.

Watkins (1990) reviewed information on absorption and translocation of radionuclides applied in solution. Foliar absorption may occur via the stomata or across the cuticle. Stomatal penetration by aqueous solutions involves many mechanisms and the main determining factors are: surface tension of the solution; contact angle (wettability); and morphology of the pore. Cuticular penetration is determined by the morphology of the surface and the chemical composition of the cuticle. The longer a contaminant remains on the plant surface the more likely it is to be absorbed. Thus rainfall is an important factor. An additional factor is the role of phyllosphere micro-organisms which may utilize surface contaminants rendering them either more or less available for foliar absorption. Foliar absorption and translocation of ^{137}Cs proved to be of some significance after the Chernobyl accident for fruits and pulses. A notable example was the observed transfer to hazelnuts (Monte *et al.*, 1990).

4.1.3.2 Root uptake

Both soil and plant factors are involved in root uptake. Some information is available on the capacity of roots of different species to absorb different radionuclides from solution. This has generally been obtained from solution culture experiments using carefully controlled conditions. Unfortunately, such information is rarely directly applicable to field conditions due to the lack of knowledge on the partitioning of radionuclides between 'non-available' and 'plant-available' fractions in soils. Additionally, root absorption is not explained in all cases by physical and chemical theory. In many cases absorption relies on biological factors such as the presence or absence of micro-organisms intimately involved with the root (the mycorrhiza), and the release and uptake of carrier molecules at the root surface.

In general, the majority of studies on plant uptake from soil have evaluated the available data in the context of a soil-to-plant transfer factor (SP) defined as the ratio of a radioisotope concentration in plant (Bq/kg dry, wet or ash basis) to that in soil to a defined depth (Bq/kg dry, wet or ash basis). The main soil characteristics affecting SP are: pH; cation exchange capacity; organic matter content and composition; and clay content and composition. Extensive data exist in the literature for a range of plant and soil combinations, climate and growth conditions, different radionuclides and forms of contamination. Unfortunately, many of the available data are difficult to interpret without supporting data for the state of growth and biomass distribution of the plant concerned, relevant physical and chemical characteristics of the soil, and details on depth distribution of the radioisotope.

To overcome these difficulties, a working group of the International Union of Radioecologists (IUR) defined standard conditions for experimental studies and evaluated data from these experiments in relation to measured soil parameters (IUR, 1982). The main difficulty in the derivation, understanding, and application of SP is that it is defined on the basis of total concentration of a radionuclide in soil. For many radionuclides the total concentration in soil bears little or no resemblance to subsequent plant uptake. Additionally, SP (like K_d) is defined for equilibrium conditions.

For radiostrontium and radiocaesium an alternative approach is based on the similarity of the ecological behaviour of Cs and K and of Sr and Ca. The hypothesis is that the mineral composition of a crop is rather constant and that the Cs and Sr content of the crop can be determined from the corresponding Cs/K and Sr/Ca ratios for the soil (Wirth *et al.*, 1985).

A further approach involves analysis of the kinetics of solution to root transport (Nye and Tinker, 1977; Shaw and Bell, 1989). This requires substantial information on the chemical composition of solutions and roots.

The study of soil solution and its interaction with soils and plant roots is only possible if suitable methods are available for the isolation of soil solution. A review of these methods is given by Lembrechts (1991).

Once in the plant, retention of a radionuclide depends on relative distribution and a variety of plant physiological parameters. 'Availability' of radionuclides for leaching has been given little attention in the literature, though early studies reported by Tukey *et al.* (1958) indicated substantial differences between those that are readily leached (isotopes of Na and Mn) and those that are effectively retained (isotopes of Fe, Zn, P and Cl).

4.1.3.3 Translocation

Following root or foliar absorption, radionuclides can be translocated either above or below the point of entry. In many published studies, emphasis has been placed on the above-ground fractions of the plant and particularly the edible fractions. This neglects the fact that radionuclides may have been taken up by the plant but not translocated to the organ(s) of interest and, as a result, may have undergone chemical transformations which could have a profound effect on subsequent behaviour. In these circumstances, root uptake and downwards translocation may even affect the observed distribution of the radionuclide in soil as has been suggested for plutonium (Coughtrey *et al.*, 1983–1985).

4.1.4 TRANSFER TO, AND METABOLISM IN, ANIMALS

There is a substantial literature on the uptake, distribution and retention of radionuclides in animals. Much of this literature concerns laboratory or non-domestic mammals and has been obtained and evaluated for the specific purpose of developing models for assessing dose to man following inhalation or ingestion of radioactive materials.

Coughtrey *et al.* (1983–1985) reviewed data on transfer of radionuclides to animals and specified mathematical models for use in radiological assessments. More recent information was reviewed by Van den Hoek (1989) and by Coughtrey (1990). The important parameters involved in transfer to and metabolism in animals are:

1. the fraction of an orally ingested radionuclide which is absorbed by the gastrointestinal tract and subsequently enters systemic circulation (f_1);
2. the fractions of a radionuclide entering systemic circulation which are deposited in different organs and tissues of the body;
3. fractions and half-lives of a radionuclide associated with specific pools in the animal or with whole-body retention;
4. fractions of absorbed activity excreted in urine, faeces, milk and sweat; and
5. the longevity of retention of the radionuclide in the lungs, taking into account chemical and physical form of input.
   The last parameter is predominantly of interest for actinides.

Unlike studies on the behaviour of radionuclides in soils and plants, many published studies on animals were undertaken in the context of mathematical analysis of time-dependent trends. Nevertheless, there are a number of radionuclides and animals for which data are less extensive and for which more empirical approaches were required. In particular, transfer of radionuclides to eggs has been studied in only a few cases, as has transfer to milk of animals other than cows. An area of particular interest is transfer and dynamics of radionuclides in young animals compared to mature animals. In this context, changes in f_1 with increasing age, possibly as a result of changes in gut structure, changes in distribution between different tissues, and changes in retention during development of the animal all need to be considered.

The traditional analysis of transfer from pasture to animal is the use of a plant-to-animal transfer factor (PA) defined as the ratio of the radionuclide concentration in meat or milk (Bq/kg or Bq/l) to the daily intake (concentration in diet (Bq/kg) × dietary intake (kg/d)). Values of PA have been derived from field data for a number of animal and radionuclide combinations, but these values are usually subject to uncertainty due to difficulties in obtaining accurate estimates of radionuclide concentration in diet, dietary intake, and other factors. These latter factors include: plant species composition; form of contamination; the extent to which it includes soil ingested directly or indirectly; dry weight and ash weight content (including supplements); and stable element content. Other factors include animal husbandry practices and the metabolic state of the animal during the period of observation. Additionally, the PA obtained from a field study is specific to the chemical form of the radionuclide and other conditions in that study since estimates of f_1 are not included explicitly in the calculation of PA. Finally, PA was defined originally by Ward and co-workers in the 1960s in the context that equilibrium should have been achieved between intake and meat or milk concentration (Ward and Johnson, 1986).

In the absence of equilibrium conditions, estimates of PA can be based on integrated intake over part or the whole feeding period, though even this can be unsatisfactory when there are changes in the level of dietary contamination or dietary intake, or the approach to equilibrium is slow.

Most interest in PA has been in respect of transfers to milk, with transfers to meat receiving less attention. A collation of transfer factor data was provided by Ng (1982).

4.1.5 SEMI-NATURAL AND NATURAL ECOSYSTEMS

Interest in the behaviour of radionuclides in natural ecosystems first developed during the 1960s because of the observed transfer of ^{137}Cs along the lichen–reindeer–man food chain, though a wider interest also developed at the same time because of the recognition of the importance of artificial radionuclides as a tool for tracing and understanding nutrient turnover in ecosystems. A survey of the literature from the 1960s and 1970s shows a strong interest in the then new subject of 'radioecology' with several symposia devoted to its various aspects (e.g. Schultz and Klement, 1963; Cushing, 1976). An introduction and historical perspective of radioecology is provided by Whicker and Schultz (1982) and introductions to the Russian literature by Klechkovskii *et al.* (1971) and by Kulikov and Molchanova (1982).

Radioecological studies in semi-natural and natural ecosystems have not been limited to radioisotopes associated with weapons testing or to those providing tracers for nutrients. Auerbach (1987), for example, cited information on the long-term contamination of forest ecosystems with plutonium and technetium in the vicinity of Oak Ridge National Laboratory. He also emphasized the importance of such studies in understanding and predicting the behaviour of long-lived radionuclides. Grasslands, forests, and tundra environments have all been considered in the context of actinide distribution and transport (Hanson, 1980) and commonly provide conditions in which enhanced transfers from soil to plant or plant to animal can occur.

For radiocaesium, extensive studies in the 1960s during the period of weapons testing indicated that arctic and alpine ecosystems were particularly important in relation to impact on man. Subsequent studies (e.g. Hutchinson-Benson *et al.*, 1985) demonstrated the persistence of ^{137}Cs in arctic environments many years after the peak of fallout, and a similar situation was recognized by Bunzl and Kracke (1984) for heathlands in Germany.

Decomposition provides a mechanism for transfer of radionuclides from dead vegetation to the soil system. Many semi-natural ecosystems have an accumulated organic layer at the soil surface which may provide a source of available radionuclides for subsequent transfer to other ecosystems or man.

The significance of semi-natural systems was further emphasized in many studies following the Chernobyl accident as summarized in the work by Desmet *et al.* (1990).

4.1.6 SEASONAL EFFECTS

The consequences of radionuclide contamination of terrestrial ecosystems are strongly dependent upon a number of seasonal parameters (e.g. state of growth of the vegetation, agricultural practice, weather conditions) which control many of the processes involved in radionuclide transport summarized above. Experience after the Chernobyl accident highlights the need to take account of seasonal factors when describing the behaviour of radionuclides in terrestrial ecosystems after acute or chronic contamination. Seasonality is also highly relevant when making decisions on the application or relative effectiveness of countermeasures (NEA, 1991).

The following sections of this chapter provide state-of-the-art overviews of the behaviour of radionuclides in different types of ecosystem. Conclusions and recommendations are presented in Section 4.5.

4.2 AGRICULTURAL ECOSYSTEMS

4.2.1 TEMPERATE ZONES

The temperate zone has received relatively more attention in radioecological studies than other climate zones. Reasons are as follows.

1. The northern temperate zone, roughly the area between latitudes 50 and 60 degrees, received considerably more nuclear weapons fallout in the early sixties than did other latitudes. This prompted authorities to initiate fallout survey programmes, which later developed into radioecology. Also, the first generation of nuclear power stations was mainly built in the northern temperate zone and this stimulated radioecology.
2. Rainfall in the temperate zone is comparatively low. This leads to limited removal of fallout radionuclides by leaching and consequent accumulation in the upper layer of soil.
3. Food production in the temperate zone is high for all human food crops, animal feed and animal products. The population of the temperate zone consumes mainly products of this zone. In particular the possible contamination of milk stimulated the application of radioecology to food chain studies. Environmental awareness is high and most radioecological studies refer therefore to agricultural production zones.

The main areas studied were as follows.

1. The direct contamination of vegetation by radioactive aerosols. In the case of an accident this is by far the most important contamination process. Most of the crops are annual ones and thus effects of direct contamination are limited to a maximum of one year. For perennials the effect of external contamination

may last for many years. Radionuclides are transferred from older parts of the plant to newer ones, so that crop products which were not previously in contact with radioactive aerosols may become contaminated.

2. Accumulation and leaching in soils. There are no radionuclides of which all chemical forms are either insoluble or soluble. The major reason for accumulation of radionuclides in the upper zone of soils is, however, their adsorption on soil constituents. Studies on accumulation and leaching mechanisms generally focused on the chemical behaviour of the radionuclides (speciation) and on adsorption and desorption.
3. Indirect contamination of plants. Because of accumulation in the upper soil layer, the uptake of radionuclides by food and animal feed crops received substantial attention in the past. For routine releases of radionuclides, uptake by the roots is generally the main contamination process. Although the uptake of radionuclides is often determined without consideration of the composition of the soil solution, knowledge of the latter is essential for a good understanding of the uptake mechanisms. Besides focusing on straightforward determination of soil-to-plant transfer, studies also focus on speciation, adsorption and desorption, and provide a link to studies on accumulation and leaching.
4. Transfer to and metabolism in animals after a release of radioactivity into the environment. Milk is one of the food products for human consumption which becomes contaminated very rapidly. Moreover grazing animals may ingest relatively large quantities of radionuclides which may be secreted into milk to a greater or lesser degree. It is not surprising that secretion of radionuclides into milk has received quite a lot of attention in studies on radionuclide metabolism in animals. Since the Chernobyl accident, the contamination of meat with caesium isotopes in free-ranging domestic and wild animals has also been studied quite intensively.

4.2.1.1 Soils and plants

Direct contamination

The general theories for deposition, interception and retention (Section 4.1.1) also apply to agricultural products. In fact most models and parameters were derived for agricultural products from temperate zones. A distinction is made between dry deposition and wet deposition. In both cases only a part of the radionuclides is intercepted and/or retained by the vegetation. Interception factors are given in Table 4.2. Calculations of deposition are usually rather simple (IAEA, 1991a) and three types of air contamination are distinguished. First noble gases; these are very inert and not deposited. Secondly aerosols of chemically non-reactive compounds, e.g. aerosols of Sr, Cs and transuranic elements. Rates of deposition and wash-out are dependent on particle size and not on the type of radionuclide (Table 4.3). Therefore general aerosol theories apply. Thirdly chemically reactive gases or aerosols, e.g. iodine.

Table 4.2 Mass standardized interception factors for pasture grass as a function of the amount of rain and crop mass (IAEA, 1991b)

Crop mass (kg/m^2 fresh weight)	Amount of rainfall (mm)			
	0.1	1.0	5.0	10.0
0.1	1.5	1.2	0.6	0.3
0.5	1.2	0.72	0.24	0.12
1.0	0.97	0.57	0.18	0.09
2.5	0.4	0.33	0.10	0.05

Table 4.3 Dry deposition velocities (cm/s) for aerosols of chemically non-reactive compounds (IAEA, 1991a)

Type of surface	Chemically non-reactive compounds			Reactive gases
	0.01–0.1μm	0.1–1μm	1–5μm	
Smooth surface	0.02	0.01	0.02	0.2
Grassland canopy	0.2	0.1	0.2	1
Woodland canopy	2	1	2	—

Radionuclides react with vegetation or soil and scavenging is very intense, consequently deposition is relatively high. An important factor is the type of the surface and/or vegetation. The rougher the surface, the higher the scavenging and thus the deposition. Deposition increases in the order: smooth surface, rough surface, small crops, large crops, forests. The difference between the lowest and highest deposition rates is a factor of 100 (Frissel *et al.*, 1989). For larger crops sometimes the state of growth of the crop or leaf area index (LAI) is proposed as a measure for interception (IAEA, 1991b).

For accidental releases, rainfall can contribute substantially to total deposition. This is an important lesson which radioecologists have learnt from the accident at Chernobyl. With the exception of the immediate vicinity of Chernobyl, all severely contaminated areas were contaminated via wash-out.

Accumulation and leaching in soils

The solubility of most radionuclides which are formed in nuclear reactors or other nuclear devices is low, consequently the major part of radionuclides released into the environment will finally accumulate in either the upper layer of soils in terrestrial systems or of sediments in aquatic systems. In general, Pu and Am hardly migrate in soil. Neptunium and Sr migrate over short distances, i.e. only 1 cm in a few years. The migration of Cs depends very much on soil characteristics. Usually it is very strongly adsorbed and migrates less than Sr. Exceptions to these rules are discussed later.

Migration can be described by a process which in chemistry is known as chromatography. The soil is considered to be divided into a number of layers. Radionuclides are assumed to be present in the top layer and partly in solution. Rain or irrigation water causes the solution to flow from the upper layer to the second layer. Part of the radionuclides adsorb in the second layer, depending on K_d value (Section 4.1.2). The studies of Frissel and Pennders (e.g. Frissel and Pennders, 1983) using the mass conservation equation cover a period of 20 years and indicate vertical migration velocities for Sr, Cs, Pu and Am of 0.3 to 0.5 cm per year.

In illitic soils, K, NH_4 and Cs ions are adsorbed within the clay mineral platelets of the illite. This adsorption is so strong that release occurs very slowly, only large amounts of K or NH_4 fertilizers promote growth of crops. However, desorption may counteract any reduction. This problem is associated with the composition of the soil solution and is discussed later. A very detailed analysis of Cs fixation is given by Cremers *et al.* (1988). Despite limitations, K_d is one of the best variables to indicate differences in leaching behaviour in soils of the temperate region. Table 4.4 provides default values from some radionuclides.

Table 4.4 Representative K_d values for agricultural soils (IAEA, 1991b)

Radionuclide	Sand	Loam	Clay	Organic soil
Sr	1.4×10^1	2.0×10^1	1.1×10^2	1.5×10^2
Cs	2.7×10^2	4.5×10^3	1.8×10^3	2.7×10^2
Np	4.1×10^0	2.5×10^1	5.5×10^1	1.2×10^3
Pu	5.5×10^2	1.2×10^3	4.9×10^3	1.8×10^3
Am	2.0×10^3	9.9×10^3	8.1×10^3	1.1×10^5

The chromatography concept is rather complicated, and sometimes simpler concepts are preferred. A very simple approach is the apparent half-life concept. The assumption is that in a certain period half of the radionuclides disappear from a certain layer; the processes are not considered. A refinement of this concept is that of the residence time. The soil is assumed to be distributed into different layers. The radionuclide moves through subsequent layers, staying in each layer with a certain residence time. The advantage of the approach is that the processes which cause migration do not have to be considered; the disadvantage is that diffusion, dispersion and vertical transport cannot be distinguished.

Migration theories have two shortcomings. The first is that mechanical mixing is not considered. The effect of ploughing is obvious, but also the mixing effect of earthworms and trampling by cattle can be important in the upper soil layers. The second is incorporation of radionuclides in clay or other soil minerals which causes migration to decrease with time. Radionuclides produced by a natural chain reaction in Oklo (Gabon), some 1.7 billions of years ago, and running for about 100 000 years, migrated over much shorter distances than calculated with commonly used K_d values. The accident at Chernobyl occurred only 6 years ago, a rather short period to measure migration in soil. In almost all reports it is concluded that radionuclides are adsorbed in the upper few centimetres of the soil and migrate very slowly (Bokori *et al.*, 1989). In areas with high rainfall a small part of the radioactivity may immediately have penetrated deeper, thereafter it did not migrate further (CEC, 1991). Because of increasing fixation with time; the velocity decreases with time, an example is described by Bunzl *et al.* (1988). A remarkable, so far unexplained, exception was reported by Pavlotskaya *et al.* (1991), who noted shortly after the accident at Chernobyl that in hydromorphic soils the major part of Pu was already present at a depth of 10 cm. The accident at Kyshtym in 1957 caused severe contamination of soils with ^{90}Sr. Most soils have been ploughed and no migration pattern was determined. Recent investigations on some non-ploughed areas showed that ^{90}Sr had reached a depth of about 15 cm, indicating a migration rate of 0.5 cm per year. However, the ^{90}Sr content in the upper layers of soil is lower than expected. It is possible that part of the ^{90}Sr leached to deeper soil layers. Detailed information is lacking and cannot be traced back (Molchanova, J. V., personal communication, 1990).

A special problem is that caused by hot particles which may constitute a reason for an increase of the available fraction of radionuclides with time (Alexakhin, in press).

Soil to plant transfer of radionuclides

Application of transfer factors is complicated by the thickness of the layer in which radionuclides are assumed to be present. A good choice should be the thickness of the rooting layer, but that layer is variable. The IUR recommends as standards for TF calculations average concentrations over 0–10 cm for 'grass' and 0–20 cm for all other crops. If the contamination is very superficial, and for instance in a litter layer or sward, the IUR convention is not practical. Instead Bq per kg crop/Bq per m^2 has been used, particularly for Chernobyl data.

Table 4.5 shows some representative TF values. They are suitable for screening purposes, but the number of conditions which influence uptake is so large that for realistic assessment calculations, site and condition specific values have to be used. For Cs, time lag, i.e. the time elapsed since the contamination, is most important. A decrease of availability of Cs by 50 per cent in the first year after contamination is the most frequently reported value. Values range, however, between 30 and 90 per cent. Another problem is the large variation, partly caused by factors which vary from year to year depending on weather conditions, and local conditions. After three to five years the uptake drops a further 50 per cent. After 10 years usually only 10 per cent of the original availability remains; but this value certainly does not apply to all soils. An example of the reduction of Cs content of crops in the Gomel area of Belarus is shown in Table 4.6 (Grebenshchikova *et al.*, 1991).

Table 4.5 Some representative soil-to-plant transfer factors (Bq/kg dry crop per Bq/kg dry soil in 0–20 cm layer). Values for grass refer to the 0–10 cm layer (Frissel and Van Bergeijik, 1989)

Nuclide	Crop	Crop part	TF value
Sr	Cereals	Grain	0.13
Sr	Fodder		0.95
Sr	Grass		1.3
Sr	Bean	Pod	1.2
Sr	Carrot	Root	0.46
Sr	Potato	Tuber, flesh	0.17
Sr	Green vegetables		2.3
Cs	Barley	Grain	0.03
Cs	Wheat	Grain	0.018
Cs	Alfalfa	Fodder	0.26
Cs	Grass		0.21
Cs	Bean	Pod	0.023
Cs	Potato	Tuber	0.1
Cs	Spinach		0.24
Np	Wheat	Grain	0.0023
Np	Potato	Tuber	0.0067
Np	Green vegetables		0.037
Pu	Wheat	Grain	0.00001
Pu	Potato	Tuber	0.00018
Pu	Green vegetables		0.00007
Am	Wheat	Grain	0.00002
Am	Potato	Tuber, flesh	0.00039
Am	Green vegetables		0.00066

Table 4.6 The reduction of the Cs content of crops with time observed near Gomel, Belarus (Grebenschikova *et al.*, 1991)

Crop	Part	Bq/kg per kBq/m^2		
		1987	1988	1989
Winter rye	Grain	0.24	0.12	0.05
Barley	Grain	0.19	0.06	0.04
Oats	Grain	0.65	0.30	0.22
Potato	Tuber	0.16	0.11	0.04

The bond between Cs and illitic clays is very strong and can only be broken by an excess of K or NH_4 ions. This explains why an application of K-fertilizers can have different effects on contamination of crops. An excess of K in soil dilutes Cs ions which decreases uptake, the excess may cause desorption of fixed Cs which increases the uptake. Application of ammonium fertilizers causes the same effect.

A complication is that ammonium ions may be formed by microbiological decomposition of organic soil material. Most arable soils are well drained, in which case ammonium is oxidized to nitrate which hardly influences Cs uptake. In wet soils, particularly if they are rich in organic material, ammonium is not oxidized and consequently, Cs remains more available to the solution phase. This situation applies to all soils with a sward or litter layer, and is important for forests and pastures of upland and alpine soils.

The availability of Sr also decreases with time, but the decrease is small and limited to a few per cent per year. In short-term experiments it may be overshadowed completely by annual variations. The reason for the decrease is that Sr ions are incorporated into the crystal lattice of soil minerals.

Another important factor which influences uptake is the pH. For important fission products, uptake increases with decreasing pH values. A factor of 10 can be obtained easily (Frissel and Pennders, 1991). A complication is that soil type, pH and organic matter content are related. The uptake of Cs from a sandy soil is about five times higher than from a clay soil; for a loam soil this factor is about 2. For well-cared agricultural soils, soil type and organic matter content describe the uptake sufficiently and the pH does not have to be considered separately (CEC, 1991). For less well-cared soils and semi-natural soils the pH will be the first parameter which has to be assessed to estimate uptake. After the accident at Chernobyl Soviet scientists applied three pH classes for their rehabilitation programme: 4.5–5.5, 5.5–6.5 and 6.5–7; the ratios of the Cs uptake of the classes are about 4:2:1 (Table 4.7; Prister, 1991). The interaction between soil type and pH is also important for the uptake of Sr. Jouve (1990) found that the uptake from sand soils is 5 times as high as from clay soils. In his prognosis model he included pH, K_d and exchangeable Ca content of a soil.

Depth of contamination, rooting depth and soil moisture regime form another set of related factors. It is obvious that uptake will increase if root activity increases in the contaminated layer. The presence of soil moisture activates root activity. In

Table 4.7 Some soil-to-plant transfer values determined for ^{137}Cs in the Kiev region of the Ukraine (Prister, 1991)

	Bq/kg per kBq/m^2		
Crop	pH range 4.5–5.5	pH range 5.5–6.5	pH range 6.5–7.5
Wheat	0.5	0.2	0.5
Barley	0.3	0.1	<0.01
Lucerne	0.8	0.4	0.2
Clover	0.8–2.3	0.3	0.3
Potato	0.3	0.1	0.04
Onions	0.6	0.2	0.11
Cabbage	0.3	0.1	0.04

particular, in superficially contaminated dry soils this behaviour offers a possibility to influence the uptake of radionuclides. Subsoil irrigation will activate the root below the contaminated layer and decrease uptake. On the contrary, surface irrigation promotes uptake. N. Zezina (personal communication, 1990) succeeded in increasing Cs uptake by maize by a factor of 100 after applying sprinkler irrigation to a coarse sandy soil in an experimental field 10 km west of Chernobyl.

Flooding is responsible for changes in physical, chemical and biological properties of soils. It also results in modification of growth and development, particularly of the root system. An important example of the effects of flooding on radionuclide uptake is rice growing (Bourdeau *et al.*, 1965; Myttenaere, 1972; Myttenaere *et al.*, 1975), where besides direct contamination from deposition of fallout, the plants can absorb radionuclides from the soil or from the water in which they stand. Investigations have been carried out in the field using fallout radioisotopes as tracers, and under controlled conditions in lysimeters, glasshouses and growth chambers. In the field, covering had little or no effect on the levels of ^{144}Ce, ^{106}Ru, ^{137}Cs, ^{54}Mn or ^{90}Sr in the roots of flooded rice plants. In contrast, covering reduced ^{144}Ce in shoots to 28 per cent of the control, and ^{106}Ru, ^{137}Cs and ^{90}Sr to 75–77 per cent. ^{54}Mn was affected only slightly (down 4 per cent). The ^{137}Cs/^{90}Sr ratio in the grain hull was 1.6 whereas that for hulled rice was 10. For covered plants, the ^{137}Cs/^{90}Sr ratio in the hull was 3.1 whereas that in the hulled grain was 8.2, indicating a lower mobility of ^{90}Sr compared to ^{137}Cs within the plant. The major fraction of ^{137}Cs and ^{90}Sr found in the shoots and grain of flooded rice was due to indirect contamination, the two radioisotopes being equally available via this route.

The relative importance of soil and water in the indirect contamination of flooded rice with radiocaesium has also been investigated in model rice fields using a dual tracing technique. Concentration factors between plant and water and plant and soil (Table 4.8) varied with the plant part and the origin of the contamination, but were much higher for ^{137}Cs coming from water than for ^{134}Cs coming from soil.

Table 4.8 Concentration factors between plants and water (^{137}Cs) and between plants and soil (^{134}Cs)

Plant part	CF_1 Plant/water[a]	CF_2 Plant/soil[b]	$\frac{CF_1}{CF_2}$	CF in nutrient solution
Roots	63	0.572	110	225
Leafy shoots	111	0.052	2135	55
Panicle minus grains	70	0.019	3684	—
Hull	45	0.010	4500	94
Caryopsis	20	0.005	4000	32

[a] $\frac{\text{Activity per g dry matter}}{\text{Activity per cc irrigation water}}$.

[b] $\frac{\text{Activity per g dry matter}}{\text{Activity per g dry soil}}$.

Another important factor is the adhesion of soil to crops. Even the presence of 10 g soil per kg dry crop equivalent to 1 g soil per kg fresh weight (for many green crops) causes a TF of 0.01 Bq/kg crop per Bq/kg soil. Strontium TF values range from 0.1 to 3, thus the increase of the TF due to soil adhesion is unimportant. Apparent TF values for Cs range from 0.01 to 0.3; soil adhesion therefore contributes significantly. Soil adhesion is mainly caused by rainsplash and in that case is limited to heights of about 40 cm. Rainsplash, together with wind erosion may, however, also be the cause of resuspension. It is difficult to give a default value for soil adhesion, as the value depends so much on local conditions. The IUR (1990) reported typical values of 10 and 4 g soil per kg dry crop for crops lower than and above 40 cm respectively. An extreme value is perhaps 250 g per kg soil.

The determination of soil adhesion values is rather difficult. One way is to determine the load of an element which is not or hardly taken up by crops (e.g. Ti, Sc). Plutonium concentration is sometimes used for this purpose. Alternatively experiments can be done with and without conditions which prevent resuspension. Resuspension must be prevented if reliable TF values are required for Pu, Am and Np, otherwise soil adhesion dominates the root uptake contamination. The TF values for transuranics shown in Table 4.5 and reviewed by Frissel (1990) are based on experiments which avoided soil adhesion. The same review shows that for field experiments higher TF values for Pu and Am were observed, suggesting that soil adhesion played a dominant role.

Only generic values for soil to plant transfer are needed to answer assessment questions in which only simple steady-state screening models are applied. In this case all processes leading to plant contamination can be included in one parameter which describes the maximum transfer from soil. Greater accuracy is required for models intended to assess dose or environmental concentrations. Such models should consider more transfer processes and should use best estimate values instead of maximum values. In particular, for transuranics, one should consider the direct and indirect uptake separately. The biological availability of both contaminants is expected to be quite different.

Countermeasures

Once an area has been contaminated two major remedies exist, i.e. removal of radionuclides by removing contaminated soil or reduction of uptake. Removal of bare soil is not discussed here. An alternative method starts with temporary fixation of Cs and Sr by applying polyacrylate or carboxy methyl cellulose derivatives. Sowing of clover or grass produces a sod which contains the major fraction of the radioactivity and which can easily be removed by conventional equipment. Another effective way to reduce uptake is deep ploughing to bury radioactivity below the rooting zone. In districts of the Southern Urals which were contaminated with ^{90}Sr after an accident at Kyshtym, good results were obtained with ploughing to a depth of 50 cm.

Alternative methods consist of the application of fertilizers, lime, peat, zeolites and other compounds which may influence the uptake of radionuclides. The reduction factors which can be obtained are usually not greater than a factor of four. On poor soils, fertilization will boost crop production so much that contamination expressed per unit weight crop decreases considerably. However, total uptake of radiocaesium may be higher. For ^{90}Sr countermeasures which increase pH are useful. Table 4.9 shows the impact of applications of various products; lime is the most successful immediate measure. Ploughing and waiting appear to be the best measure.

Table 4.9 Impact of agricultural countermeasures on the uptake of ^{90}Sr on meadows observed near Gomel, Belarus (Grebenschchikova *et al.*, 1991)

	kBq/kg grass fresh weight		
	1st cut 1988	1st cut 1989	1st cut 1990
Control	9.25	3.33	2.33
Ploughed, N,P,K applied	2.41	1.66	0.33
Ploughed, N,P,K + zeolite (10 000 kg/ha applied)	1.37	0.70	0.33
Ploughed, N,P,K + phosphorgypsum (5000 kg/ha) applied	2.55	1.48	1.04
Ploughed, N,P,K + lime (5000 kg/ha) applied	1.30	0.67	0.56

4.2.1.2 Transfer to and metabolism in animals

The importance of animals, and more particularly of domestic mammals, with regard to the radioactive contamination of the food chain resides in the fact that edible products such as milk, meat and eggs are important items of the human diet. The general principles of metabolic behaviour of radionuclides in domestic animals and in free wild animals are similar and they need not be discussed separately.

In this section, the factors influencing the transfer of the most important radionuclides to the major food-producing animals are discussed as well as the metabolism and distribution in tissues, and secretion into milk and other products. As a result of the importance of milk in the human diet, particularly for young children, the transfer of radionuclides to milk of lactating cows has occupied a central place in many studies.

Absorption of radionuclides

The important parameters for the transfer of radionuclides to and their metabolism in animals are listed in Section 4.1.4. In order to be metabolized, a radionuclide will have to enter the animal organism. Three possibilities need to be considered:

absorption through the skin, inhalation, and oral ingestion. The superficial layer of the skin in domestic animals is impermeable to many substances even when they are applied in aqueous solution. An exception must be made for tritium which may penetrate through the skin when it occurs as tritiated water.

A damaged skin may lose its protective function to a certain extent. Appreciable absorption of radionuclides through the intact skin of domestic animals under field conditions is very unlikely to occur. Inhalation is the process of air movement in the respiratory system which consists of air passages leading to the lungs where the actual gas exchange of oxygen and carbon dioxide takes place. Radionuclides may enter the respiratory system either as a gas or adhered to particles of variable size. Particulate material will only reach the lungs if the particles are small enough, since larger particles will be trapped by the cilia of the epithelial cells lining the mucous membrane of the air passages, e.g. trachea, bronchi, bronchioli. Such trapped particles will be removed quite efficiently by the upward movement of the cilia. Particles larger than 5–10 μm in diameter are generally considered to adhere to the mucous membrane of the air passages and do not reach the lungs. When the particle size decreases, more particles will reach the lower respiratory tract and the lungs where retention may occur. Translocation of radionuclides from retained material or from inhaled gases to other tissues depends predominantly on their solubility. Rapid clearance with retention half-times of a few days or less have been observed for certain chemical forms of iodine and strontium. Much longer retention half-times have been found for plutonium oxide (Bair, 1960).

Absorption of radionuclides by animals via inhalation is of limited importance. Hvinden *et al.* (1964) showed clearly that the intake of radionuclides through inhalation was at least three orders of magnitude less than that from pasture. Since the respiratory tract is not usually an edible product for man, the significance of the inhalation route for food chain contamination is further reduced.

Oral ingestion of radionuclides with contaminated feed is the predominant route for uptake by animals. However, contamination of animal produce will occur only if the radionuclide under consideration can be absorbed from the gastro intestinal tract. Whether or not it will then be retained in edible tissues or secretions of an animal, depends on its physiological properties.

Differences in length, size and number of compartments of the digestive tract exist between carnivorous, omnivorous and herbivorous animals (Swenson, 1975). The digestive tract is much longer in herbivores which are very important species from the point of view of transferring radionuclides to man. Ruminants have a complex stomach system in which the very large first stomach, the rumen, plays a significant role in the digestion of fibrous cellulose-containing plant material. This function is performed in the large intestine, particularly the colon, of non-ruminant herbivorous animals such as the horse and rabbit. Although some absorption of ingested material may take place in the rumen, e.g. part of the ingested ^{131}I, and also in the large intestine, it is generally agreed that the bulk of the digested food is absorbed from the small intestine. Absorption rates are different for different radionuclides. This is because the epithelial lining of the mucous membrane in principle is impermeable to the large majority of electrolytes and organic compounds in the lumen of the gut. Absorption may be a passive process such as simple diffusion, or a result of special mechanisms requiring energy which ensure rapid and efficient transport across the epithelial cells from the lumen of the gut into the blood capillaries and small lymph vessels in the wall of the alimentary tract. These different transport mechanisms may lead to different absorption rates for certain elements relative to their physiological significance. For example, calcium is absorbed by both active transport and simple diffusion, and strontium by simple diffusion only. This partly explains the much slower absorption of strontium compared with that of calcium. For many elements the epithelial cells of the gut wall are completely impermeable such as is the case for the rare earths. Near impermeability is found for barium and the actinides resulting in their very poor absorption.

Another point of interest is the movement of substances in the opposite direction, that is excretion from the blood into the lumen of the gut. This so-called 'endogenous excretion' has been observed for many radionuclides and it may be quite significant when the actual absorption is high, e.g. for caesium.

Following the entry of absorbed radionuclides into the systemic circulation, their physiological characteristics will eventually determine their fate. They may be concentrated in a particular organ, e.g. in the thyroid for iodine, the bone for strontium and the bone or soft tissues for caesium. Radionuclides may also be secreted into milk of a lactating animal, or excreted into urine or faeces. The selective capacity of the epithelial cells of the mammary gland for secretion into milk is responsible for the different amounts of various radionuclides which can be recovered from milk.

Transfer of a radionuclide from an animal's diet into milk or meat is usually described by a transfer coefficient (Section 4.1.4). For milk, the symbol is F_m, and for meat it is F_f. The use of a transfer coefficient for predictive and other purposes has obvious advantages but caution is necessary because of its shortcomings. These

Table 4.10 Transfer fators (d/kg or d/l) for caesium, strontium and the actinides in various animal products

		Nuclide		
Animal	Product	$^{134,137}Cs$	$^{80,90}Sr$	Actinides (^{241}Am)
Cow	Milk	0.01	0.002	4×10^{-7}
	Meat	0.06	3×10^{-4}	2×10^{-5}
Calf	Meat	0.6	0.003	2×10^{-4}
Sheep	Milk	0.06	0.02	—
	Meat	0.3	0.002	2×10^{-4}
Lamb	Meat	0.8	0.02	0.002
Hen	Meat	4.0	0.02	0.005
	Eggs	3.0	0.3	0.008

are summarized in Section 4.1.4, and are discussed in detail by Ward and Johnson (1989). Table 4.10 (CEC, 1987) presents F_m and F_f values in milk and meat for radionuclides of greatest radiological importance for man under conditions of continuous intake.

The metabolism of iodine

After accidental releases, such as was the case at Windscale and Chernobyl, ^{131}I (half-life about 8 days) is the radioisotope of immediate concern. In soluble form, it is rapidly and almost completely absorbed from the gastrointestinal tract. The absorption of iodine may begin in the first stomach of ruminants, and this explains the rapid appearance of ^{131}I in milk when animals are out on pasture at the time of release. The metabolism of iodine is characterized by the ability of certain tissues to concentrate it. This is particularly true for the thyroid gland which incorporates iodine in its hormones. The mammary gland also has this capacity to concentrate iodine from plasma but differences in efficiency between species do occur. The mammary gland of the lactating cow is rather inefficient as compared to that in goat and sheep. A release of radioactivity into the environment usually is a single contamination event, at least initially. This means that animals will ingest decreasing amounts of radioisotopes each day. Maximum concentrations of ^{131}I in milk will be found after 2 to 4 days and these will generally represent an F_m of 0.002–0.004 d/l (Bustad *et al.*, 1964; Handl and Pfau, 1987). The decrease in amounts of ^{131}I ingested result not only from the physical decay of the radioisotope but also from its removal from pasture due to meteorological and other conditions. Experimental data on the amounts of iodine transferred to milk from goats after a single contamination event have not been recorded. When extrapolating data for continuous daily dosing experiments, 5 to 10 times as much ^{131}I may be expected in goat's and ewe's milk than in the milk of dairy cows.

The metabolism of strontium

Strontium belongs to the alkaline earth metal series, of which calcium is the most important member from a physiological point of view. It is convenient to consider strontium metabolism in relation to that of calcium because dietary calcium levels have been observed to influence decisively the absorption of strontium from the digestive tract (Comar *et al.*, 1961). Another reason is the homeostatic control of calcium which results in a high degree of constancy of calcium levels such as in blood plasma and milk. The transfer of calcium across biological membranes is usually more efficient than that of strontium which leads to discrimination against strontium. The value of this discrimination factor at the level of the digestive tract in the ruminant and also in other animal species is about 0.2, and that at the level of the mammary gland is 0.5. This means that strontium is transferred to milk about 10 times less efficiently than is calcium. In practice, most values of the observed ratio between strontium and calcium levels in milk and diet vary between 0.09 and 0.16, although quite important fluctuations have been observed, e.g. from 0.04 to 0.20 (Van den Hoek, 1989). Satisfactory explanations to account for these differences could not be given in all instances.

After absorption from the digestive tract, strontium, like calcium, may be incorporated into bone from which it is released very slowly. There is a wealth of information on strontium metabolism in bone because of the radiation hazard to bone marrow in the human being. The significance of the contribution from deposits in bone to contamination of other products has not yet been quantified.

In an experiment simulating an accidental contamination event while cows were out on pasture, peak levels of strontium in milk were found 3–4 days after contamination, representing an F_m of 0.0005–0.0007 d/l (Van den Hoek *et al.*, 1969). After reaching their maximum these levels decreased with half-lives of about 10 days. It may be expected that administration of excess calcium will lead to a decrease in strontium absorption. This was reported to be the case in several experiments.

The metabolism of caesium

Caesium, like potassium and rubidium, belongs to the alkali metals. Up until the accident at Chernobyl, the contamination of animals with caesium was considered generally of lesser importance for contamination of the agricultural food chain than that of iodine and strontium, and caesium metabolism received proportionally less attention. The high caesium levels in sheep in some upland areas in the United Kingdom and in Ireland, and also those in free-ranging wild animals such as reindeer, moose and deer in Scandinavia and in some alpine regions as well as the persistence of these high caesium levels have prompted a great deal of experimental work. Our knowledge of caesium metabolism in animals has extended considerably since 1986. Caesium is taken up and metabolized very easily by mammals. Like potassium it is distributed in the soft tissues of the body (muscles and organs) and it occurs intracellularly. However, there is not the same degree of interdependence between caesium and potassium, as there is between calcium and strontium.

Absorption of caesium from the gastrointestinal tract varies quite significantly between animal species, and is influenced by a number of other parameters, the effects of which are not always understood. Apparent absorption of caesium in monogastric animals, such as the pig, may be virtually complete whereas in polygastric animals usually it is not much higher than 60 per cent. Absorption will be greater when caesium is ingested in soluble form, e.g. as CsCl in tracer experiments or in milk. Other factors such as the crude fibre content of the feed and the amount of clay particles ingested have been reported to reduce the absorption of caesium. After absorption, caesium is distributed readily and fairly uniformly into the soft tissues. At equilibrium it occurs in approximately similar concentrations in muscular tissue and in several organs, notably the kidney. Its levels in liver and spleen are lower. Caesium concentrations in muscles and organs of young animals are generally higher than in those of adult animals for the same intake. The higher availability of caesium in the milk that young animals drink, a greater capacity of their digestive tract for the absorption of caesium (and of some other radionuclides as well), a higher tissue retention of caesium and perhaps other factors may be responsible for the higher caesium levels observed in many young animals. This is illustrated by the different F_f values for lamb and ewe muscle reported for different experimental conditions of caesium intake (Howard, 1989).

Caesium is also readily secreted into milk. Important variations in milk levels for the same intake of caesium were reported after Chernobyl (i.e. $F_m = 0.002–0.02$ d/l). Typical values in cow's milk lie around 0.008 d/l for continuous intake of caesium. These values are 0.1 d/l for the goat and 0.03 d/l for sheep. After accidental contamination, the milk levels will be lower because animals are ingesting decreasing amounts of caesium each day. A value of 0.0026 d/l was reported in cow's milk in May 1986 (Handl and Pfau, 1987). This compares quite well with values of 0.0031–0.0045 d/l which had been obtained in a simulation experiment under natural conditions (Van den Hoek *et al.*, 1969).

The metabolism of other radionuclides

Although a number of radionuclides may be released under accidental conditions, transfer to man via animals is usually of limited significance either because of short physical half-life, e.g. ^{86}Rb or as a result of very low absorption from the digestive tract, e.g. ^{95}Zr, isotopes of ruthenium, cerium and other elements of the lanthanide series.

After the Chernobyl accident, ^{110m}Ag was reported to be present in the livers of cows and sheep. Noble metals are absorbed very poorly from the digestive tract but their absorption may be enhanced by the formation of complexes, which is more likely to occur in polygastric animals. Silver-110m is not metabolized actively after absorption. Most of the silver in the plasma is filtered out by the liver. This explains the relatively high levels of ^{110m}Ag in liver tissue and its virtual absence in muscular tissue and milk (Beresford, 1989).

The transuranic elements (plutonium, americium and neptunium) occupy a special place because of their great radiological significance for man. This is particularly true for plutonium which has been studied in greater detail than the other elements. Absorption of transuranics from the digestive tract, although generally very poor, may vary according to the chemical state of the element, the age of the animal and other factors. In adult animals, typical absorption values lie between 0.01 and 0.001 per cent; these values may be higher in young animals by one to two orders of magnitude. After absorption, actinides are preferentially deposited in bone and liver in variable fractions although the smaller fraction is usually found in liver. Their transfer into milk is very poor and variable, with F_m in the range 1×10^{-8} to 1×10^{-5} d/l.

The influence of selective grazing and of management practices

Evidence was presented in several studies carried out during the 1960s that farm management practices might influence quantitatively the transfer of deposited radionuclides into animal produce (Hansen *et al.*, 1964). As a result of better pasture management and adequate use of fertilizers, a higher grass yield could be obtained. This would lead to a greater dilution of the radionuclides deposited per unit area, although the greater interception of radionuclides by the pasture would presumably counteract partly the diluting effect. The net effect, however, was a reduced intake of radionuclides by grazing animals.

The contamination of large areas of land with caesium isotopes after the Chernobyl accident has drawn attention to the importance of plant selection by free-ranging animals (Howard, 1989). Seasonal changes in dietary composition led to unexpectedly high caesium levels in sheep and also in wild animals. For example, the increased intake of fungi by roe deer and other species in autumn, led to peak values of caesium in these animals during the hunting season (Johanson *et al.*, 1990). Other dietary changes such as higher intake of lichen, heather and other plants have had similar effects.

The ingestion of soil by animals may be a significant source of radionuclides in certain situations. Whether or not the ingested radionuclides will also be available for absorption by the animal will depend primarily on how firmly they are bound to soil particles. Specific soil characteristics such as clay content are probably decisive for the availability of caesium for animals.

Countermeasures

Countermeasures are used to prevent animal produce from becoming unfit for human consumption. Milk, contaminated with ^{131}I, may be converted into milk powder, and the milk powder can be stored until the radionuclide has decayed to acceptable concentrations. For radionuclides other than ^{131}I attempts are made to prevent absorption from the digestive tract. The secretion of strontium into milk may be reduced by increasing the calcium intake of the animal's diet or by the application of fertilizers containing calcium to pasture (Van den Hoek and Binnerts, 1967). The absorption of caesium isotopes can be efficiently reduced by including ammonium ferric cyanoferrate (a Prussian Blue derivative) in the diet (Unsworth *et al.*, 1989). Inclusion of bentonite and other clay minerals may also reduce significantly the absorption of caesium from the alimentary canal.

4.2.2 TROPICAL ZONES

Only very scattered information is available on the behaviour of radionuclides within tropical agricultural ecosystems, presumably reflecting the more limited and relatively recent development of a nuclear industry in such areas, compared with temperate regions. In view of the paucity of data, pathways of radionuclide transfer in tropical agricultural systems are considered here in generic terms, taking into account soil characteristics and relating these to knowledge of transfer processes in the temperate world.

Ecosystems in the tropics have an extremely wide range of soils and climatic conditions. Thus, depending on altitude and location, ecosystems may range from desert to rain forest to snowline communities. However, this section considers only those agricultural ecosystems which are restricted to tropical zones. The classical notion of a tropical soil involves the concept of a highly leached, well-weathered mineral substratum which has evolved over a period of some millions of years. While there are some specific tropical soil weathering processes, such as laterization, differences from temperate soils in many cases lie in the degree and intensity of these. However, the main difference from temperate zones is the major role of termites in soil development in many areas, which may be considered as comparable to the bioturbation induced by earthworms (Russell, 1973). Lack of disturbance during the soil development period may result in soil profiles being extremely deep. These profiles are commonly characterized by the development of deep porous oxic horizons, in which chemical alteration is so complete that the original parent material cannot be identified (Birkeland, 1974). Long and intense periods of weathering result in the almost complete depletion of cations and silica, commonly rendering the principal fabric of the soil as quartz grains, strongly cemented by oxides and hydrous oxides of iron and aluminium. The silt to clay ratio is commonly below 0.15, although the clay fraction is dominated by 1:1 kaolinitic minerals (Russell, 1973). The dominance of these minerals renders tropical soils relatively unfertile (Pitty, 1979). However, the cation exchange capacities of the soils are typically over 10 meq/100 g of clay and it is postulated from this that other 2:1 clay minerals (including illite) must usually be present.

Biological productivity in tropical soils is high, with a rapid turnover of organic matter, resulting in very rapid nutrient cycling. Infiltration of water is often at a high rate in tropical soils, resulting in the intense leaching which is associated with the development of the oxic horizons noted above. Loss of nutrient anions, such as nitrate, can therefore be considerable, despite the domination by hydrous oxides which result in a much higher anion retention capacity than in temperate soils, whereas for cations the converse occurs. Tropical soils tend to be acid, with penetration of low pH to deeper horizons than for soils in temperate regions, due to high cation uptake from depth by roots of the natural vegetation.

Soil loss due to rapid runoff is a problem in many tropical areas, but this can be mitigated by modifying agricultural practices, to avoid the development of rills and gulleys. A large proportion of agricultural land in the tropics is derived from rain forest, which is often cleared with minimal attention given to subsequent adverse environmental consequences in the form of erosion by wind and rain. Thus development of agriculture in tropical areas could potentially give rise to large-scale resuspension and mobility of radionuclides if contaminated forest lands were cleared. On the other hand, removal of timber could result in transport of large amounts of radioactivity out of a contaminated ecosystem; a large proportion of nutrient elements is locked up in standing biomass, because of highly efficient nutrient cycling.

While many radionuclides may be released into tropical ecosystems under different circumstances, a limited number of these are of particular interest in view of their radioecological significance over human life-span and their importance with respect to discharges by the nuclear industry. Radionuclides of particular interest include isotopes of caesium, strontium, technetium, actinides, and miscellaneous fission and activation products of the transition metal and lanthanide series, such as ruthenium, cerium and cobalt.

It is well known that caesium can become strongly fixed into specific clay minerals, the degree of fixation depending on the nature of the clay mineral concerned. Wahlberg and Fishman (1962) reported that caesium adsorption onto clay minerals could be ranked in the order montmorillonite > illite > halloysite > kaolinite. As kaolinitic clays dominate in tropical soils, it can be assumed that fixation of caesium will be generally lower than occurs in most temperate soils, in which illitic minerals are more prevalent. Caesium may remain more available for cycling through soil–plant systems in the tropics than in other areas, as unfixed radionuclides are readily absorbed by roots. In view of the heavily leached nature of many tropical soils, reduced concentration of potassium may lead to increased caesium availability (Shaw and Bell, 1991). While deep penetration of caesium may occur within the profiles of lateritic soils, there are reports of fallout ^{137}Cs being restricted mainly to the top 15 cm of soils in different parts of India, over the 1966–70 period (Mishra and Sadavisan, 1972). Similarly Lowe (1978), reported strong ^{137}Cs fixation in the top 10 cm of 'claylike' soils in the Malayan Peninsula.

Strong eluviation of clay minerals has been reported from surface horizons of monsoon-affected soils on the Ganges Plain (Tomar, 1987). This depletion of the surface horizons may lead to Cs remaining in a relatively unfixed state and hence more available for biological absorption. There is evidence of the monsoon in itself influencing ^{137}Cs uptake, with D'Souza and Mistry (1980) demonstrating increased concentrations in foliage of grasses growing on a high cation exchange capacity soil in Bombay in the monsoon season. D'Souza and Mistry stated that the observed greater uptake may, in all probability, be due to the greater availability of radionuclides from soil under conditions of high moisture status.

Strontium, a divalent element which can undergo simple cation exchange, shows less fixation within soils than does caesium. As a result, it has a marked degree of environmental mobility, exacerbated by the depletion of other basic cations in the surface layers of tropical soils. Evidence of this is provided by Mishra and Sadavisan (1972) who showed fallout ^{90}Sr penetrating to greater depths than ^{137}Cs in a range of soils in different regions of India. Technetium as pertechnetate may be more readily retained in leached tropical soils as a result of their relatively high anion exchange capacity.

Soils are known to form the major environmental reservoir for Pu, whose major chemical sinks are organic and oxide fractions (Livens *et al.*, 1986). Further, the chemistry of Pu and other actinides is strongly controlled by redox conditions in the soil. The preponderance of metal hydrous oxides in highly leached tropical soils may be expected to provide a major chemical sink for Pu, although in soils affected by monsoon rainfall the chemistry may be made significantly more complex in a manner not clearly understood. The degree of overall biological incorporation of the actinides is generally much less than for Cs, Sr and Tc, but any accumulating layers of dead organic material may again provide a significant environmental sink. The influence of rapid organic matter turnover on biological availability of organically bound actinides remains to be elucidated.

Very few studies have been performed on soil to plant transfer with crops which are only grown in tropical regions. One such study was reported by Delmas *et al.* (1987), in which uptake of ^{90}Sr and ^{137}Cs into peanut, banana and pineapple from three African soils was examined in an experiment in Senegal. The substrata concerned were a hydromorphic, a tropical ferruginous (lateritic) and a volcanic soil, respectively. However, Delmas *et al.* (1987) only gave results for peanut growing on the lateritic soil.

Scarcely anything is understood about the behaviour of radionuclides deposited directly onto foliage of tropical plants, and it is more difficult to make generalizations here than for contamination of soils. D'Souza and Mistry (1980) reported the pattern of field loss of ^{106}Ru, ^{125}Sb, ^{144}Ce and ^{134}Cs applied to hybrid napier grass (*Pennisetum purpureum*) which was then exposed in the field in Bombay. The rates of loss of the four radioisotopes were fairly similar, with a retention half-life of 12–16 days, on a concentration basis which is comparable to that reported in temperate regions. However, as is pointed out by the authors, there may be a marked seasonal influence on field loss in tropical regions, depending on when contamination occurs in relation to the monsoon and other rainy seasons. Rainfall of monsoon intensity is rarely experienced in temperate zones and its influence in contaminating crops via wet deposition, in removing foliar contamination, and in soil splash remain areas of ignorance.

The leaves of trees in moist tropical regions often bear large numbers of epiphyllae, such as algae, lichens and bacteria, on their surfaces. These epiphyllae are adapted to receive their nutrition from the atmosphere and consequently have been shown to accumulate large quantities of fallout radionuclides. Thus, Lowe (1978) reported that canopy leaves on Malaysian trees contained four times more ^{137}Cs when they were covered in epiphyllae. Similar evidence showing a major role of epiphyllae in increasing foliar levels of fallout caesium has been produced by Kline *et al.* (1973) for natural vegetation in Puerto Rico; this phenomenon was considered to explain the increase in ^{137}Cs levels observed at higher altitudes on the island. In the case of a nuclear accident, the presence of epiphyllae on the foliage of agricultural arboreal crops could have major radioecological significance.

In conclusion, there is an urgent requirement for both field and controlled experimental studies on the behaviour of radionuclides in tropical agricultural ecosystems, considering the various combinations of soils, crops and climatic conditions which pertain in these ecosystems.

4.3 SEMI-NATURAL ECOSYSTEMS

4.3.1 FORESTS

4.3.1.1 Introduction

The Chernobyl accident in April 1986 exemplified the importance of forest ecosystems as a link in the transfer of radioactivity to man (Desmet and Myttenaere, 1988). Before the accident, several studies had considered the transfer of radioisotopes (particularly ^{137}Cs) in forest ecosystems after introduction as tracers (Auerbach *et al.*, 1964) or after contamination of the soil (Auerbach, 1987). Nevertheless, the transfer atmosphere–canopy, the first important step after an atmospheric release (Ronneau, 1990) had not been studied.

Several mathematical models had been developed to predict the cycling of various radionuclides (Groom and Rasdale, 1980; Garten, 1980; Jordan *et al.*, 1973; Olson, 1965; Prokhorov and Ginzburg, 1971). In 1986, Auerbach (Auerbach, 1987) had summarized results on the behaviour of radionuclides in three forest ecosystems and emphasized the soil compartment as the major accumulator of caesium. Most of the radioactivity transported to the canopy leaves was shown to be leached, but in the case of radiocaesium the analysis of first fallen leaves and of the throughfall water revealed that part of the radioactivity was returned to the tree before leaf fall (Auerbach *et al.*, 1964).

These results underlined the need to study the biogeochemical cycle of deposited radionuclides within forest ecosystems. Also, since forests appeared to be very effective accumulators of atmospheric radioactivity (Sokolov *et al.*, 1990; Sombre *et al.*, 1990; Tikhomirov, 1990; Van Voris and Dahlman, 1976) and since they occupy slightly more than one third of the land surface, studies of the behaviour of the most important radionuclides released in the event of an accident are required. Moreover, understorey plants and wild animals which inhabit forests may be highly contaminated after the redistribution of radioactivity trapped by the canopy. The consequence of secondary transfers is a possible contamination of food and exposure of man and populations to ionizing radiation.

The use of forests for recreation may also expose man to direct gamma shine through contact with the ground, trees and other vegetation. The industrial uses of timber may also be a source of irradiation. The analysis of risk associated with deposition of radioactivity into forest ecosystems requires a forest dose assessment model to evaluate the importance of the forest ecosystem as a source or sink of radioactivity (Berg, 1990).

4.3.1.2 Radioactivity transfer and cycling in forest ecosystems—main pathways

In the case of an accident the canopy, at least during dry periods, constitutes an effective filter for atmospheric pollutants. In fact in areas contaminated after the Chernobyl accident the crown of trees trapped the deposited radioactivity.

Radionuclides deposited on foliar surfaces may be:

1. washed out by the rain and deposited as litter in the form of fallen leaves, bud scales and twigs;
2. absorbed and translocated in the plant leading to contamination of the wood;
3. resuspended by wind, fire and evapotranspiration;
4. absorbed by humic material (decomposed litter);
5. transported to deeper soil layers and to surface as groundwaters after deposition on soil or after decomposition of plant material;
6. reabsorbed by tree roots and understorey plants (e.g. herbs, berries, mushrooms);
7. transferred to wild animals and aquatic biota.

The above processes provide a basis for developing simulation models for forest ecosystems. The biogeochemical cycles of some radionuclide analogues are well established and may assist in designing scenarios for radionuclide transfer. A review of potassium and calcium biochemical cycles in forest ecosystems throws some light on the possible behaviour of ^{137}Cs and ^{90}Sr in these environments (Van der Stegen de Schrieck and Myttenaere, 1992).

4.3.1.3 Deposition and interception by the canopy

The majority of radionuclides are transported in the atmosphere as aerosols with various physico-chemical properties and this gives rise to different deposition mechanisms (Section 4.1.3).

The global deposition velocity strongly depends upon:

1. air turbulence which is the driving transport phenomenon from the atmosphere to the canopy;
2. the reactivity of gases and the deposition properties of particles;
3. the structure and nature of plant surfaces.

The final step in the deposition process occurs after particles or gases have been brought close to the canopy surface by mechanisms such as diffusion, gravitational settling and impaction.

The efficiency of capture depends upon particle diameter and wind speed in a somewhat complex manner, with a minimum deposition velocity of the order of 0.1 cm/s for particles of about 0.2–0.5 μm (Ronneau *et al.*, 1987). Moreover it has been shown that deposition velocity is higher for Scots pine than for grass. Recent results (Tveten, 1990a) show that the deposition velocities for trees are about one order of magnitude greater than onto smooth surfaces. For single trees it was also shown that the local deposition velocity is proportional to bulk mass in much the same manner as for grass.

Forest canopies may also be contaminated by radioactivity in rain, and part of the radioactivity trapped during the previous dry periods is leached mostly by throughfall. Rain acts as a powerful cleaning agent of both the atmosphere and canopies at the expense of soils which are the ultimate sink of air pollutants. Normally, during a full meteorological cycle, dry deposition accounts only for a small fraction (few per cent) of the wet deposition.

The wet deposition of radioactivity from Chernobyl was studied in a spruce forest in Belgium using gauges in clearings and under spruce canopies (Ronneau, 1990; Ronneau *et al.*, 1987). Table 4.11 gives accumulation factors (AF) observed after rain scavenged the atmosphere on 4 May 1986. The AFs were close to unity except for radiocaesium (0.2); such a low value was never observed in a several-year study on non-radioactive pollutants. This is attributed to the solubility of caesium which allows needles to absorb it very efficiently.

Table 4.11 Deposition of Chernobyl radionuclides in clearings (CL) and under the canopy (CAN) of a spruce forest (May 1986)

	Bulk deposition (Bq/l)		
Radionuclides	CL	CAN	AF(CAN/CL)[a]
^{103}Ru	226	180	0.80
^{131}I	584	620	1.06
^{132}I	763	965	1.26
^{132}Te	805	1075	1.34
^{134}Cs	97	19	0.19
^{136}Cs	23	5.4	0.23
^{137}Cs	178	38	0.21
^{140}La	29	21.6	0.75

[a] Accumulation factor (AF) $= \frac{\text{radioactivity collected under trees}}{\text{radioactivity collected in clearings}}$.

Deciduous trees were less at risk from the Chernobyl episode because, given the time of year, they were less able than the evergreen species to intercept water.

Observations two years after the accident gave different results (Table 4.12) because at that time leaching of the incorporated radiocaesium was responsible for the high values obtained under canopies. Moreover, determinations made on rain collected under spruces and oaks showed that there was a good correlation between the deposition fluxes of both radiocaesium and stable potassium, suggesting that potassium may be used as an indicator of the behaviour of ^{137}Cs (Olson, 1965).

Experiments conducted in controlled conditions with thermogenerated aerosols have shown that the interception was of the order of 80 per cent. Other experiments conducted recently have given higher figures (~99 per cent). At Chernobyl the crown of trees trapped 60 to 90 per cent of the radionuclides deposited (Sokolov *et al.*, 1990).

Table 4.12 Bulk deposition (Bq/l) of Chernobyl radionuclides in clearings (CL) and under the canopy (CAN) of a spruce forest (1988 and 1989)

		CAN	
Sampling dates	CL	Spruce	Oak
9 December 1988	0.02	0.49	0.096
2 January 1989	0.01	0.31	0.108
27 January 1989	0.01	0.42	0.103
14 March 1989	0.01	0.27	0.067

4.3.1.4 Retention and leaching of intercepted radioactivity

Retention properties are a function of the species (biotic layer) and of the inactive particles (natural or anthropogenic) previously deposited and trapped onto the surface of the leaves (abiotic layer). The surface of leaves is covered with particles of various sizes and the possibility of some interaction between radiocaesium and the soil-derived fraction of the abiotic layers has been suggested (Sombre *et al.*, 1990) to justify its preferential retention.

The weathering half-lives of intercepted radiocaesium have been calculated for oak and pine trees contaminated with a fallout simulant (^{134}Cs adsorbed on 88–175 μm quartz particles; Witherspoon and Taylor, 1969) and for deposited thermogenerated radiocaesium (Al Rayes *et al.*, 1988; Sombre *et al.*, 1990). The first experiment showed the influence of species (oak leaves lost caesium quicker). In both cases the retention curves showed that at least a two-compartment model has to be used. Spruces contaminated by thermogenerated aerosols and then exposed to natural rain showed two weathering rates (Tb_1 5 d ~ 50 per cent of the activity; Tb_2 50 d) (Sombre *et al.*, 1990). These results were confirmed by determination of caesium retained by leaves and by results obtained near Chernobyl four years after the accident (Sokolov *et al.*, 1990; Tikhomirov, 1990). The non-removable caesium may penetrate leaves and become available for transfer to other parts of the tree.

The translocation in woody plants has been examined by inoculating vegetation with radioisotopes. These experiments, as well as data obtained for stable elements (nutrients or not), have shown that potassium as well as caesium is very well transported in the plant and may be readily transferred to wood before winter. On the contrary, calcium as well as strontium is stored in older tissues and no recycling within the tree occurs (Van der Stegen de Schrieck and Myttenaere, 1992).

These observations may explain why, several years after deposition, radiocaesium levels in leaves do not decrease or might even increase (Ronneau, 1990). The last observation might also be due to an absorption by roots of activity deposited or redeposited on soil.

4.3.1.5 Behaviour of radionuclides in forest soils

In the long term, deposited radionuclides are removed to the forest floor where chemical adsorption, ion exchange and complexation occur on contact with the organic rich humic and moss layer (Witherspoon and Taylor, 1969).

The litter of temperate woodlands on acid soils tends to accumulate in thick layers in which three horizons may be distinguished:

1. the L horizon (litter) which consists of intact litter with few visible signs of decomposition;
2. the F horizon (fermentation) that is beneath and consists of fragmented litter;
3. the H layer (humus) between the F layer and the mineral soil containing little or no mineral matter.

In temperate woodlands the annual litter fall may take 3–5 years to decompose (Schell and Myttenaere, 1989). On the contrary, in tropical forests the leaves disappear completely within six months, whereas in coniferous forests the needles may take more than a decade in cool temperate or boreal sites (Packham and Harding, 1982). In extreme cases litter may be broken down and incorporated in the mineral soil within a year (mull) or it may accumulate for many years on top of the mineral substrate (mor) thereby returning nutrients (Witkamp and Van der Drift, 1961). The fate of radionuclides in forest soils may thus vary with climate conditions.

Post Chernobyl observations (Thiry *et al.*, 1990) have confirmed that forest soils accumulate and strongly retain radionuclides, particularly caesium, in the upper organic zones. The thickness and nature of the organic matter present at the surface of the floor seems to be the main factor which influences the vertical distribution of the deposits.

Such results may explain maximum concentrations of radiocaesium found in some compartments of the understory. Only a few data obtained *in situ* give information on bioavailability of radionuclides deposited on multilayered soils. Recent studies have underlined the necessity to review the methodology (Andolina and Guillitte, 1990) as well as the concepts and expression modes (Thiry and Myttenaere, 1991) for radionuclide behaviour in multilayered soils. Thiry and Myttenaere (1991) highlighted the potential for misinterpretation of activity measurements for forest soils if based on expression by weight (the different layers having a different density). On the other hand, the calculation of a K_d based on the soil solution activity and the volumetric soil activity gives a better indication of bioavailability and mobility of radionuclides in multilayered soils. This confirms the fact that estimates of soil-to-plant transfer require a better understanding of chemistry of soil solution (Nisbet and Lembrechts, 1990).

Radiocaesium half-lives in the top 5 cm of different forest soils were given by Belli *et al.* (1990). For some sampling sites the data showed a great spatial variability which may be accounted for by different physical and chemical characteristics of the soils.

4.3.1.6 Contamination of the understorey

Tables 4.13 and 4.14 give the ^{137}Cs activity of the components of a pine forest situated 15 km from the Chernobyl nuclear power plant (Thiry *et al.*, 1990). The understorey is much more contaminated than the overstorey. A calculation based on the activity of the different compartments of the same forest (soil included) showed that four years after the accident most of the ^{137}Cs radioactivity was found in the organic layer of the soil.

Radiocaesium in the organic layer is integrated very quickly into the caesium cycle. This is indicated by the high activities observed in mushrooms and in plants which take up nutrients from the upper organic layer (Guillitte *et al.*, 1987; Randa

Table 4.13 Distribution of ^{137}Cs activity (percentage) in a contaminated pine forest at Bourakovka, Chernobyl in 1990

	^{137}Cs (MBq/ha)	Percentage
Needles	47	
Twigs	189	
Bark	201	1.45
Wood	108	
Roots	?	—
Litter	1612	4.29
Soil		
0 (0–2 cm)	30 350	80.72
Ah (2–5 cm)	4048	10.76
B (5–10 cm)	1045	2.78

Table 4.14 Distribution of ^{137}Cs in a contaminated pine forest at Bourakovka, Chernobyl in 1990. Data expressed in Bq/g dry matter

Overstorey	
Needles/leaves	10.1
Twigs (wood)	7.6
Bark	25.1
Trunk (wood)	1.6
Understorey	
Mosses	270–515
Lichens	275–510
Heath	500
Fungi	380

et al., 1990). Incorporation in organisms prevents physical migration even if the K_d is low (Rommelt *et al.*, 1990). On the contrary ^{90}Sr migrates slowly due to high sorption of divalent cations to organic material.

Contamination levels of mushrooms vary with the soil layer explored by the mycelium. Moreover the migration of caesium in the soil profile with time may determine the value of any individual species as a bioindicator. Lichens, mosses and fungi which grow on forest soils and which explore different layers of the soil surface may be differently contaminated; this means that a species may be a good bioindicator at only a certain time after an accident.

Such restrictions require from the scientist a very good knowledge of the dynamic aspects of the movement of radionuclides in the soil and of the growing habits of the living forms considered.

4.3.1.7 Effects on forest plants and animals

After the Kyshtym accident a reduction was observed in the contaminated territory in the populations of invertebrates with a long life cycle and an extended phase of development in the forest litter (Spirin *et al.*, 1990). These effects were most marked for earthworms, myriapoda and mites and at contamination levels above 3.7 MBq/m. Less marked effects were observed for flying insects and for invertebrates with an exoskeleton even though they spend most of their lives on the surface of the forest litter. Birds and mammals received lethal doses in autumn and winter (1957–58) if they lived permanently in areas where contamination exceeded 37 MBq/m. For mammals, the most notable effects were observed in mouse-type rodents (increase in mortality and decrease in life expectancy for 37 MBq/m). After 15 years populations were comparable to those of other animals. No similar effects were observed in other animal populations.

A substantial proportion of the areas affected by Kyshtym and Chernobyl were covered by forest (respectively 20 per cent and 50–60 per cent). One of the main differences why the two accidents had a different impact on the environment was the time of occurrence (Kyshtym 29 September 1957, autumn—Chernobyl 26 April 1986, spring). Radioactive contamination occurred in Kyshtym when plants and many animal species were in their physiologically dormant state during which the process of radiation damage and regeneration were slowed down. Deciduous trees were less at risk after the Chernobyl episode because, at that time of year, they were less able to intercept radioactivity than were evergreen species (Arkhipov *et al.*, 1991).

4.3.1.8 Conclusions

The development of a general model of transfer for a defined radionuclide through a generic forest may be extremely difficult. A descriptive model of radionuclide pathways through a forest ecosystem has been developed recently (Van Voris *et al.*, 1990). However, many of the processes affecting radionuclide behaviour in forest ecosystems are not well understood and the models developed previously reflect this lack of knowledge.

The goal of future studies should be to determine and model the relevant processes involved in the transfer of radionuclides within and out of the forest. One of the main questions that such as a study has to answer is: How long will the forest maintain a level of radionuclides harmful to man?

4.3.2 ARID ENVIRONMENTS

The fate of man-made radionuclides in arid and semi-arid environments (defined collectively herein as arid sites), takes on a special significance when considering that over 30 per cent of the Earth's land surface, 22 per cent of the world's arable land, and 16 per cent of the world's population is tied to low precipitation areas. While there is an enormous amount of literature on radionuclide behaviour in the environment this discussion is based on research sponsored by the US Department of Energy (DOE) on plutonium in arid areas of the US (US DOE, 1980, 1985).

More than a third of the world's land surface is classified as arid (receives less than 250 mm of annual precipitation) or semi-arid (250 mm to 500 mm of annual precipitation). The continental interior deserts can be hot or hot and cold depending upon elevation and latitude. They are subject to both summer thunderstorms and winter frontal precipitation patterns where snow can be an important part of the annual water balance.

Of interest in radionuclide transport processes in arid areas are the relative amounts of wind and/or water erosion. Wind erosion is prevalent in areas where vegetative and rock cover is sparse or absent and where winds blow over dry surfaces. Under these conditions, much soil can be moved and, if contaminated, result in significant transport of radionuclides.

In many of the continental deserts, and especially in their associated semi-arid areas, high-intensity thunderstorms can result in flash-flooding and the resultant movement of large amounts of water and sediment in affected areas. This is a very efficient means of transporting sediment-associated contaminants. In areas subject to heavier winter rains and snow, runoff and erosion can occur over extensive areas. While the severity of these winter events may be less than their summer counterparts, their areal extent may be much larger and they may result in significant movement of sediment and contaminants.

Snow and snowmelt in the colder arid and semi-arid areas may result in more soil water recharge and relatively more movement of sediment-associated contaminants into the soil profile than in areas dominated by rainfall. The significance of these different processes in transporting radionuclides will depend upon specific site and radionuclide source conditions.

4.3.2.1 Radionuclide distribution in arid ecosystems

Spatial relationships

Most sources of radionuclides presently in arid ecosystems in the US were initially deposited on the ground surface decades ago as a result of single contamination events or as chronic events over long periods of time.

Results of field studies in arid and semi-arid areas in the western US show quite convincingly that soil will eventually be the primary repository of plutonium and many other radionuclides with long half-lives. For example, the percentage of the total plutonium inventory associated with the soil component in four different arid site ecosystems was, in all cases, greater than 99 per cent, despite the fact that the source of the plutonium at each site was different.

The relative amount of plutonium in the surface 2.5 cm of soil from four study areas in New Mexico, ranged from 4 to 41 per cent of the total measured in the sampled profiles (Table 4.15). Mortandad Canyon, at the time the data in Table 4.15 were collected, was actively receiving treated radioactive liquid effluent from Los Alamos research and development activities. Note that the plutonium inventory in sediment profiles was about equal in the various depth increments. In contrast, over two thirds of the inventory of plutonium in sediments from Acid Pueblo Canyon, which had not recieved liquid effluents for about 20 years when the data in Table 4.15 were collected, was in the 12.5–30 cm depth profile. Depletion of plutonium in the near surface profiles in Acid Pueblo Canyon reflects the combined influence of runoff, erosion and downward movement into the alluvial soil. At the Trinity Site over a 23-year period, the content in the 0.2–5 cm depth profile decreased by a factor of about 40 (Table 4.16).

The highest concentrations of many radionuclides in soil are usually associated with the smaller particle size fractions. Even though concentrations of radionuclides

Table 4.15 Mean percentage plutonium inventory in soil profiles from Los Alamos and Trinity Site study locations in New Mexico (adapted from Hakonson and Nyhan, 1980)

	Trinity Site[a]			Los Alamos[a]	
Depth (cm)	Area GZ	Area 21	Depth (cm)	Mortandad	Acid Pueblo
0–2.5	29 (0.78)[b]	41 (0.46)	0–2.5	20 (0.44)	4 (0.76)
2.5–5.0	18 (0.72)	19 (0.63)	2.5–7.5	36 (0.23)	10 (0.48)
5–10	21 (0.81)	6 (0.88)	7.5–12.5	22 (0.55)	20 (1.3)
10–15	15 (0.67)	8 (0.92)	12.5–30	22 (0.79)	66 (0.18)
15–20	10 (1.1)	9 (0.95)			
20–25	7 (1.3)	7 (1.0)			
25–33	ND[c]	10 (1.2)			

[a] n = 8 for Trinity Site data; *n* = 10 for Los Alamos data.
[b] Value in parentheses is the coefficient of variation, CV (standard deviation/mean).
[c] Not detectable.

Table 4.16 Comparison of plutonium concentrations (kBq/m^2) in surface (0–2.5 cm) soils from Chupadera Mesa as a function of time after the atomic bomb test at Trinity Site in 1945 (from Hakonson and Nyhan, 1980)

1950	1951	1973
27.6 (0.01)[a] $n = 6$	12.6 (0.03)[a] $n = 3$	0.67 (0.18)[a] $n = 8$

[a] Value in parentheses is coefficient of variation, CV (standard deviation/mean).

may be much higher in a particular soil size fraction, the total inventory of radioactivity in that fraction will depend on its mass relative to the whole soil. For example, despite the higher concentrations of plutonium in the silt-clay fraction of Los Alamos alluvial soils, less than 15 per cent of the total plutonium inventory in whole soil was in this fraction. In contrast, the silt-clay fraction at a Trinity Site location not only contained the highest plutonium concentrations, but it also contained over 70 per cent of the total soil plutonium inventory.

Soil will serve as the primary reservoir of many long-lived radionuclides in arid environments and complex spatial, chemical, and physical relationships will determine the mechanisms of transport and their relative importance. Wind and water sort soil particles during the detachment, transport, and deposition phases of erosion. Because radionuclide concentrations can be strongly related to soil particle size, there is a potential to enrich radionuclide concentrations in eroding soil. This potential for enrichment can affect the long-term redistribution of radionuclides and their transport to biological components of ecosystems.

4.3.2.2 Biotic processes

Transport to vegetation

Despite the host of chemical, biological and physical factors which can modify the physiological (chemical) availability of radionuclides and subsequent transport to internal plant tissues, field studies suggest that contamination of foliage surfaces with soil particles containing the radionuclides is a major transport mechanism under many arid site and radionuclide source conditions (Arthur and Alldredge, 1982). For example, comparative studies of plant uptake of plutonium under both field and laboratory conditions generally yield results of the type shown in Table 4.17. Laboratory studies focused on root uptake of plutonium from soils yield concentration ratios which are at least 1 order of magnitude (and often 2–3 orders of magnitude) lower than ratios observed under comparable conditions at field sites. The differences in concentration ratios between laboratory and field studies imply

Table 4.17 Comparison of plutonium concentration ratios for field and glasshouse conditions (Romney and Wallace, 1976)

Soil source	Field	Glasshouse
NTS[a] Area 11B	0.13 to 0.16	0.00015
NTS Area 11C	0.045 to 0.34	0.00018
NTS Area 13	0.078 to 0.44	0.00011

[a]NTS (Nevada Test Site).

that mechanisms exist in arid environments for delivering at least 10 times more plutonium to vegetation than does direct transport via roots.

Studies at Los Alamos, New Mexico, demonstrated that rainsplash of soil particles with subsequent deposition on foliage surfaces can easily contribute most of the plutonium measured in field-site vegetation (Dreicer *et al.*, 1984). More importantly, those studies, which employed a labelled-soil particle technique and the scanning electron microscope, have shown that relationships that govern translational movement of plutonium by soil erosion processes also govern transport of plutonium to foliage surfaces. For example, the energy of impacting raindrops caused an enrichment of the smaller soil particles on foliage surfaces. In general, only the highly transportable silt-clay particles, which often contain higher concentrations of plutonium at Los Alamos, were retained by plant surfaces. While absorption of plutonium through leaf surfaces has been demonstrated (Cataldo and Vaughan, 1980) it is considered to be of limited importance in the field, particularly for annual or deciduous vegetation.

Studies on the uptake of plutonium by vegetable crops grown in field sites at Los Alamos show that as much as 50 per cent of the plutonium in crop samples was surficial contamination that could be removed by standard food preparation procedures (White *et al.*, 1981). Plutonium that cannot be removed from vegetable crop surfaces in arid environments does not necessarily reflect plutonium incorporated into plant tissues. Cataldo and Vaughan (1980) showed that submicron particles on foliage surfaces are difficult to remove by either simulated wind or rain.

Transport to animals

The concentrations of plutonium and caesium in animals collected from field sites in arid environments indicate that gut availability of these elements in the environment is relatively low, as shown by the low concentrations in internal organs and tissues across a variety of sites and source conditions. In addition, highest concentrations of plutonium and caesium are often measured in tissues exposed to contamination with soil particles. For example, plutonium in the pelt, gastrointestinal tract and lungs accounts for nearly all of the animal's body burden (Hakonson and Nyhan, 1980; Moore *et al.*, 1977).

The high mobility of large herbivores coupled with natural elimination processes could provide a mechanism for radionuclide transport across the landscape. Although the amount of plutonium transported across the landscape by this mechanism is considered to be small in areas where the extent of contamination is large (i.e. fallout areas) relative to the home range of the animal, there are circumstances where this transport mechanism becomes important. For example, in a nuclear waste burial site at Hanford Washington, jack rabbits (*Lepus californicus*) gained access to buried waste, ingested radioactive strontium, and subsequently excreted it on the ground surface of the site and surrounding area (O'Farrell and Gilbert, 1975).

Studies on pocket gophers (*Thomomys bottae*) inhabiting a low-level waste site at Los Alamos, New Mexico (Hakonson *et al.*, 1982), showed that the burrowing activities of this animal can greatly perturb cover profiles placed over low-level radioactive waste disposal trenches. Over a one year period, gophers excavated about 11 metric tonnes of soil per hectare from within the trench cover and created about 3000 m of tunnel system in the cover profile. In addition to altering the hydrological properties of the soil, animal burrowing activities can also alter radionuclide distributions within the soil profile, as has been shown for pocket gophers in plutonium contaminated sites at Rocky Flats, Colorado (Winsor and Whicker, 1980) and for other small mammals at the Radioactive Waste Management Complex at Idaho National Engineering Laboratory (Arthur and Markham, 1983).

4.3.2.3 Summary and conclusions

Present data demonstrate that soils and sediments serve as the major repository of plutonium, caesium, and strontium in arid ecosystems and that processes which redistribute soils and sediments can also cause major changes in the environmental distribution of these elements. It is clear that there is a need to determine the relative importance of radionuclides associated with soil and sediment as a source of contamination to biota. Available data on long-lived radionuclides in arid ecosystems point to the potential importance of the physical movement of soil and sediment through food webs.

Any phenomenon which retains radionuclides in contact with the biosphere for extended times has the potential to increase risks to biota due to exposure to these elements. The interception properties of vegetation, mechanical disturbance by organisms living in the soil, and processes which resuspend soil all contribute to maintaining radionuclides in contact with arid site biota. The ultimate fate of long-lived radionuclides in arid ecosystems, including transport to biota, will ultimately depend on the changes that occur in their distribution in soils and sediments and, barring significant changes in the environmental 'solubility' of elements such as plutonium, the fate of the soil and sediments themselves.

4.3.3 UPLAND ECOSYSTEMS

4.3.3.1 Introduction

The term 'upland' can be used in a purely geographical sense to define land over a certain altitude or to describe a range of habitats and their associated ecosystems. With regard to radioactive contamination it is the latter definition which seems most often implied in the literature. 'Upland' ecosystems are determined by climate, there is a low temperature regime and often a high rainfall. Thus under suitable conditions 'uplands' may occur at sea-level; in many respects they are determined by latitude rather than by altitude. The vegetation is poor in agricultural terms and management is often minimal or non-existent.

The use of upland ecosystems varies from that where man harvests the native plants and animals to systems where grazing pressures are manipulated and some fertilizer is applied. The higher the latitude the more man is dependent upon products from the grazing animal until, as described in Section 4.3.4, there are populations totally dependent upon animals such as reindeer and caribou. Research on upland systems has been dominated by the effects of the Chernobyl fallout. The research emphasis varies from country to country depending on whether, as in the UK, agricultural grazings were affected (Beresford *et al.*, 1987; Coughtrey *et al.*, 1989), or, as in Scandinavia, where natural ecosystems contribute largely to the pathways to man (Hove *et al.*, 1990). Whereas work has concentrated on radiocaesium, other radionuclides such as plutonium isotopes from bomb fallout are present. There is still little information on the factors controlling the behaviour of other radionuclides in upland ecosystems.

4.3.3.2 Source term

The dominant source of radionuclides to upland areas is wash-out of aerosol material by rainfall. After the Chernobyl accident some dry deposition of ^{131}I took place but compared with wash-out this was small. Radiocaesium deposition in the UK has been shown to be closely correlated with rainfall (Clark and Smith, 1988) and direct deposition from clouds (occult precipitation) at altitude is another source which has not been given as much attention. However, the chemical form of the aerosols does not appear to have been well determined although it is thought that they range from soluble forms such as hydroxides and carbonates to refractory particles.

4.3.3.3 Soils

Geology affects soil type and vegetation mainly according to the hardness and chemical composition of the rock. In high rainfall areas on acidic rock types highly organic soils are formed, and as the rainfall increases this may occur even over limestone. These organic soils, in combination with waterlogging, lead to the formation of a range of wetland habitats ranging from nearly open water fens to *Sphagnum* dominated raised bog.

Table 4.18 Caesium isotope distributions in soil profiles. All data decay-corrected to 1 August 1989

Sample site	Depth (cm)	^{134}Cs (Bq/m^2)	^{137}Cs (Bq/m^2)
Italy			
Tarvisio Podzol	0–3	1610	9700
	3–8	135	1240
	8–13	<26	310
	13–18	<18	145
	18–23	<20	41
	23+	<24	26
Pozzis Brown Earth	0–3	4860	26 100
	3–8	1900	12 400
	8–13	160	1210
	13–18	89	570
	18+	41	200
Sella Nevea Podzol	0–7	10 300	56 800
	7–15	690	4140
	15–22	350	2500
	22+	200	1200
Stolvizza Calcareous Brown Earth	0–5	8020	47 000
	5–11	560	5880
	11+	110	990
Norway			
Geiteseter Ranker	0–5	7410	40 500
	5–10	395	2030
	10–15	170	830
	15–20	100	530
	20–25	46	210
	25–30	<28	120
Stuttgonglia Brown Earth	0–5	7350	39 900
	5–10	570	3040
	10–15	590	3300
	15–20	88	490
	20–25	35	310
	25–30	<30	180
Grinningsdalen Fen/Mire	0–5	420	2290
	5–10	560	3060
	10–15	<25	65
	15–20	21	130
Scotland			
Sherrabeg Brown Earth	0–5	1100	7630
	5–10	144	1540
	10–15	80	680
	15–20	35	350
Glen Shirra Podzol	0–5	1530	9320
	5–10	97	650
	10–15	90	1290
	15–20	<27	280

Table 4.18 *(continued)*

Sample site	Depth (cm)	^{134}Cs (Bq/m^2)	^{137}Cs (Bq/m^2)
	20–25	<22	86
	25–30	<30	<35
	30–35	<21	46
	35–40	<17	56
Creag Meagaidh	0–5	400	2330
Ranker	5–10	140	1520

Samples collected August 1989. Data from Livens *et al.* (1991).

Though radiocaesium is leached readily from organic soils (Munson, 1985; Coughtrey *et al.*, 1983–85) it is also apparent that the presence of organic matter enhances uptake by pasture plants according to clay mineral content and composition (D'Souza *et al.*, 1972, 1980). The dominance of organic soil types in Western Europe and the mobility of radiocaesium in these soils has led to high uptake by vegetation and a consequent contamination of animal tissues. A knowledge of the physico-chemical properties of soils allows the mobility of radiocaesium to be determined in a relative manner. The 'immobilization capacity' of a soil (Livens and Loveland, 1988) is a reflection of clay mineral content and type, organic content, pH, ammonium content and potassium status. Although this is not quantifiable it does permit ranking of soils, and has been used with some success in the field. Soils classified as most vulnerable in Cumbria, UK, correspond closely with areas where sheep movement is prohibited. A closer approach to quantification has been made by Cremers *et al.*, (1990). The potential of a soil to absorb radiocaesium is defined as the product of the number of absorption sites and the selectivity for caesium with respect to potassium. Knowing these values and the ammonium-potassium status of the soils good agreement with the field situation may be obtained.

The distribution of radionuclides in upland soils will obviously depend upon the soil type. This can vary from a pure waterlogged peat through podsols to a variety of brown earths. It has been found (Cawse and Horrill, 1986; Sandalls *et al.*, 1988) that radiocaesium originating from both weapons testing and Chernobyl fallout are moving relatively slowly down the soil profile. Chernobyl radiocaesium dominates in the upper parts of the profile (Table 4.18) and over the first few years after deposition resided nearly all within the rooting zone of the vegetation. Shortly after the Chernobyl accident radiocaesium from this source was more mobile than weapons testing fallout but a gradual decline in mobility has been recorded with time.

In contrast to the approach whereby the physico-chemical characteristics of the soils are considered the biological activity within a soil cannot be ignored. In many upland areas the soil microflora is important in determining the breakdown processes. In some types of highly organic upland soils Dighton *et al.* (1991) have suggested that the bulk of the radiocaesium could be held in the fungal mycelia. Clint *et al.* (1991) have shown that soil fungi can hold an appreciable fraction of the soil radiocaesium.

It would appear therefore that in upland ecosystems there is still considerable doubt as to the reasons why the Chernobyl radiocaesium is recycling around the system so readily, and it is possible that an integration of both physical and biological factors is needed to explain the problem.

4.3.3.4 Vegetation

The range of habitats found in 'upland' areas is wide and includes grasslands, heaths, rock communities, and wet and dry peaty habitats. Woodlands cannot be excluded as the tree line is determined by exposure and the potential limits can extend from sea-level to thousands of metres. As a consequence of this wide spread of habitats the range of species is wide. Most frequent, however, are those species of the wet acidic habitats including many lower plants represented mainly by bryophytes, lichens and fungi.

Radiocaesium uptake varies widely depending on plant species, habitat and season. However, there are some generalities which may be made. It was first recorded by Bunzl and Kracke (1986) that accumulation within the family Ericaceaea was marked. This finding has since been confirmed by other workers. Even within the Ericaceaea consistent differences have been found between species. Horrill *et al.* (1990) found similar rankings for ericoid species collected from a series of dry heathland sites (Table 4.19).

Bryophytes and lichens obtain the bulk of their nutrients by trapping aerosols or as they rarely have conducting tissues, the capillary movement of groundwater up the plant body. It is frequently found that these lower plants have higher concentrations of radiocaesium than other plants. This is illustrated by Livens *et al.* (1991) where data for species from 10 sites from Italy, Norway and Scotland show the bryophytes often have radiocaesium concentrations an order of magnitude greater than higher plants from the same habitat. Relatively high concentrations of radionuclides were reported in lichens during the period of maximum fallout from weapons testing. Cigna Rossi (1971), for example, reported data for ^{54}Mn, ^{90}Sr, ^{95}Zr, ^{106}Ru, ^{125}Sb, ^{134}Cs, ^{137}Cs and ^{144}Ce in samples collected in Italy and concluded that a selective absorption mechanism existed only for ^{90}Sr.

Radionuclide concentrations in different species from the same locality have been found to vary widely even when species are only separated by a few centimetres (Table 4.20). Many possible explanations for this have been put forward, including physiological and morphological differences, but the true reasons are still obscure. Kirton *et al.* (1990) demonstrated that the temporal pattern of radiocaesium concentrations varies according to the growth patterns of individual species, and it has been demonstrated that regrowth after grazing can result in enhanced concentrations of radionuclides.

Fungal fruiting bodies have been found to play an important part in determining the contamination levels of animals in upland regions. Selective uptake of radiocaesium can be shown by using the ^{90}Sr/^{137}Cs ratio which was about 0.01 in the fall-

Table 4.19 Comparative performance of ericoid species[a] on dry heathland sites (Bq/kg dry weight)

Site	Species	^{137}Cs	^{137}Cs in soil
Grasmoor	*Calluna vulgaris*	469	669
NGR NY167198	*Vaccinium myrtillus*	220	
360 m	*Erica cinerea*	93	
Hindscarth	*Calluna vulgaris*	366	730
NGR NY218177	*Empetrum nigrum*	107	
490 m	*Vaccinium myrtillus*	78	
	Erica cinerea	33	
Yewbarrow	*Calluna vulgaris*	227	76
NGR NY16709	*Erica cinerea*	40	
85 m			
Robinson	*Calluna vulgaris*	126	500
NGR NY193176	*Vaccinium myrtillus*	96	
335 m	*Erica cinerea*	65	
Gawthwaite	*Calluna vulgaris*	39	98
NGR SD263848	*Vaccinium myrtillus*	170	713
200 m	*Erica cinerea*	28	98

[a] All above-ground parts of the plant sampled. Collected in August/September 1988. Data from Horrill *et al.* (1990).

Table 4.20 Corney Fell—individual species[a] (Bq/kg dry weight)

	^{137}Cs	^{134}Cs
Higher plants		
Calluna vulgaris	6590	2075
Vaccinium myrtillus	3964	1194
Agrostis canina	1247	379
Juncus squarrosus	1132	303
Juncus effusus	936	295
Eriophorum angustifolium	595	121
Nardus stricta	514	153
Bryophytes		
Sphagnum spp.	7472	2813
Polytrichum commune	4026	1445
Polytrichum alpestre	3543	1187

[a] All above-ground plants sampled. Collected in spring 1987.

out and 2–3 orders of magnitude lower in mushroom species (Mascanzoni, 1990). Radionuclides can accumulate to high levels in certain species, up to 445 kBq/kg for ^{137}Cs (Hove *et al.*, 1990) and selective grazing by animals such as goats takes place. Goat milk in late August in Norway has been recorded as averaging 800 Bq ^{137}Cs per litre. In Scandinavia the unfortunate coincidence of the fungal fruiting and hunting seasons means that harvesting of wild animals takes place at the worst possible time with regard to contamination of meat.

4.3.3.5 Animal populations

Animal populations of the uplands of north-west Europe are those for which most data are available and the radionuclides best documented are ^{90}Sr and ^{137}Cs. Both have originated from atomic weapons testing, nuclear reprocessing and accidents such as those at Kyshtym and Chernobyl. The range of animals affected is wide and includes wild species and domestic stock. The type of land used by domestic animals varies widely from natural vegetation to enclosed managed pastures.

Wild species

Radionuclide uptake by wild species is highly dependent on many factors even within the same area. Lowe and Horrill (1991) found significant differences in concentration factors between food and muscle tissue depending on species, food, sex, breeding condition and age (Table 4.21). Here animals from the same general area show great differences in tissue concentration. Rabbits (*Oryctolagus cuniculus*) feeding on grasses on mineral soils, where radiocaesium mobility is low, always

Table 4.21 Analysis of samples of flesh and vegetation from Loyal Estate in Sutherland (Bq/kg fresh weight)

		Flesh		Vegetation	
		^{134}Cs	^{137}Cs	^{134}Cs	^{137}Cs
Red grouse	(cock)	325.0	962.5	147.1	408.3
	(hen)	602.1	1684.3	267.9	692.4[a]
Red deer	(hind)	111.9	310.7	78.3	202.7
	(calf)	186.3	535.4	70.6	189.7
Rabbit	(buck)	8.1	17.8	—	8.2
	(buck)	5.1	13.1	—	11.9
Fox	(vixen)	175.6	460.6		

[a] Heather shoots collected by hand, all other vegetation samples obtained from the gut of the animal. Samples collected in autumn 1988. Data from Lowe and Horrill (1991).

had the lowest tissue concentrations whereas the red grouse (*Logopus logopus*) feeding exclusively on heather had the highest.

The high concentrations found in many of the plant species of the uplands, particularly the lower plants, combined with seasonal scarcity of fodder, often means that dietary intake at certain times of the year can be high in radionuclides. The high accumulation by food species such as lichen means that the intake by reindeer can be 10 times the activity of radiostrontium and radiocaesium of grasses growing in the same area (Nevstrueva *et al.*, 1966).

The Chernobyl accident put a single pulse of radionuclides into the ecosystem. It became possible, by use of the ^{134}Cs isotope, to compare not only the movement of radiocaesium from two sources, weapons testing fallout and Chernobyl, but also to study the ecological half-life in some systems. Pre-Chernobyl ^{137}Cs concentrations reached over 600 Bq/kg fresh weight for red grouse and roe deer in the UK. The ecological half-life of ^{134}Cs in native roe deer has been calculated as 28 days (Lowe and Horrill, 1988).

Domestic stock

Due to their economic importance domestic stock have been extensively studied with regard to radiocaesium, as has the passage to man in meat and milk products. Control measures after Chernobyl have received much attention particularly from an economic point of view.

Radiocaesium uptake by animals in the uplands is rapid. Within a few days of the Chernobyl material being deposited, radiocaesium was found in milk and shortly afterwards in meat. Elimination of radiocaesium is equally rapid, the biological half-life being 11–15 days for sheep when removed from contaminated areas to lowland pasture (Howard *et al.*, 1987).

A wide range of radiocaesium concentrations have been recorded for domestic stock with values ranging from hundreds to thousands of Bq/kg fresh weight depending on location. Uptake of radionuclides is complex depending, as for wild animals, on species, food, age and breeding condition. For instance the transfer to goat's milk appears to be higher than other animals. The state of the radionuclide, whether incorporated in vegetation or bound to attached soil particles, is of great importance as the latter is much less available for uptake.

Ameliorative measures to reduce radionuclide contamination in upland systems have to be cheap and easy to use. They fall into two categories—management and diet additives. Management techniques rely on removing the animals from contaminated pastures before sale or use and allowing the short biological half-life of the caesium isotopes to operate. It is too expensive to consider treatment of large areas of land but diet additives can be economic. The use in Norway of salt licks or administering boli containing the binding agent ammonium-ferric-cyanoferrate (AFCC) has proved effective in the case of reindeer and sheep (Hove and Ekern, 1988).

4.3.4 ARCTIC ECOSYSTEMS

Considerable interest in artificial radionuclides in arctic and sub-arctic ecosystems, particularly those in circumpolar regions, began with simultaneous and independent reports of appreciable concentrations of worldwide fallout in successive links of the atmosphere–lichen–caribou/reindeer–man food chains of northern Alaska and Scandinavia during the 1960s (Liden, 1961; Palmer *et al.*, 1963). Several circumpolar research programmes were initiated to promptly evaluate the radiation exposures of Inuit (Eskimos), Indians, Sami (Lapps), and Soviet reindeer herders because of their dependence upon caribou and reindeer for food and livelihood.

This section is based on radioecological research on worldwide fallout from nuclear weapons tests in arctic ecosystems during 1959–1980, with ancillary consideration of radioactive debris from satellite re-entry episodes. Emphasis is placed upon ^{137}Cs because of its greater accuracy of measurement, consistent ratios to other fallout radionuclides, and radiological implications in arctic regions (Hanson, 1966).

4.3.4.1 Study areas

Radiation ecology studies were conducted over the general area of northern Alaska between 66° North latitude and the Arctic Ocean, constituting about 310 000 km^2. Within this area are five major physiogeographic provinces, the north-western terminus of the North American continental divide, four major caribou herds which constitute the largest aggregation of big game animals in North America, and several Inuiat Eskimo and Athabascan (Kutchin) Indian villages representing four major ethnic groups. A spectrum of ecosystems of varying complexity are utilized by the migratory caribou herds and by the native groups inhabiting the region, and were the subjects of the studies. The approach proceeded from the general description of radionuclides in the northern biosphere to specific investigation of the most promising components of critical food webs, for purposes of defining routes, rates and concentrations of worldwide fallout in arctic food chains. The Arctic Coastal Plain, a northern extension of the Great Plains of interior North America, consists of a 100 km wide smooth plain that rises imperceptibly from the Arctic Ocean to a maximum elevation of 180 m at its southern margin, where it abuts the Arctic Foothills. The Plain is poorly drained, marshy in summer, and covered by elongated thaw lakes. There are no glaciers, although the entire land area is underlain by more than 300 m of permafrost. A network of ice-wedge polygons covers the Plain; an active layer (depth of summer thaw) varies from 10–100 cm and allows limited percolation of moisture. Annual precipitation is 10–15 cm, including 50–90 cm of snow.

The Arctic Foothills consist of rolling plateaus and low linear mountains extending 150 km southward from the Coastal Plain. The northern section rises from 180 m where it joins the Plain to 350 m on the south, with broad east-trending ridges dominated by mesa-like mountains; the southern section is characterized by irregular buttes, knobs, and other landforms of about 350–450 m elevation, with local relief up to 750 m. In common with the Coastal Plain, there are no glaciers and the entire province is underlain by permafrost. The substrate contains ice wedges, stone stripes, polygonal ground, and other features of a frost climate. Annual precipitation ranges from 15 to 25 cm, with about 1–1.2 m snow.

The Arctic Mountains Province consists of a 125 km wide by 1000 km long area of features carved chiefly from folded and over-thrust Paleozoic and Mesozoic sedimentary rocks during Pleistocene glaciation. An irregular and indistinct northern boundary is shared with the Foothills Province, in contrast with the abrupt southern front. The central and eastern Brooks Range is an area of rugged glaciated east-trending ridges of 2100–2400 m elevation in the northern portion and 1200-1800 m in the southern portion. Small cirque glaciers are common in the higher parts of the Range, such as the 2700 m Schwatka and Romanzof Mountains in the northeast section.

In general, the Arctic Coastal Plain and Foothills Provinces are mostly covered by Intrazonal Tundra and Bog soils, usually strongly acid and saturated by water. These soils are poorly drained silt loam or silty clay loam and of least importance in the mountains, where they are confined to the valleys. Arctic Brown is the zonal soil in which the dominating influences of climate and vegetation are expressed; however, it is present on a small portion (1 per cent) of the area, being most common in dry, well-drained environments of the Foothills Province and on kame terraces on valley floors within the Arctic Mountains. With increasing ground moisture, Arctic Brown soil is successively replaced by Upland Tundra, Meadow Tundra, Half Bog and Bog soils. Upland Tundra soils appear most often on steep slopes which are better drained than the gentle slopes of the valley floors, where Meadow Tundra, Half Bog and Bog soils occur.

Vegetation of the Coastal Plain Wet Tundra is dominated by graminoid species, particularly *Carex aquatilis*; cottongrass tussocks (*Eriophorum vaginatum*) characterize the Moist Tundra environment of the Foothills; and low, mat-forming heath communities occupy extensive fellfields of Alpine Tundra in mountainous areas. Lichens and mosses are common components of all three tundra vegetation types, with lichen carpets most extensive on stream terraces in mountain valleys. Stream drainages are dominated by high brush, such as willows (*Salix* spp.), alder (*Alnus crispa*), with isolated stands of cottonwood (*Populus balsamifera*). Alpine Tundra and Moist Tundra share much of the southern exposures of the Brooks Range mountains, intergrading with closed spruce–hardwood forests composed of white and black spruce (*Picea glauca* and *P. mariana*), paper birch (*Betula papyrifera*), aspen (*Populus tremuloides*), and balsam poplar (*Populus balsamifera*) with a rich understorey of shrubs and extensive lichen carpets that are the major winter forage of caribou.

4.3.4.2 Soils

Inventories of ^{137}Cs in surface (top 5 cm) soils in northern Alaska during 1975–1979 declined at an effective half-life of 9 y, including minor (6 per cent) increases due to fallout deposition during snow-free summer months. Much of the radioisotope loss from surface soil is a combination of surface erosion during snowmelt and percolation to depths >5 cm. Sampling below lichen carpets showed that 5 per cent of the ^{137}Cs and 15 per cent of the ^{90}Sr inventories was in A_0 horizons (humus layer) and 3–4 per cent for both radioisotopes in A_2 horizons (organic mineral soil).

Plutonium in surface soils at northern Alaska and northern Greenland study sites decreased with a 0.4–0.5 y half-life and was measured with considerable difficulty, due to the large samples required to provide statistically valid measurements.

4.3.4.3 Vegetation

Lichens are particularly efficient accumulators of fallout radionuclides, of which ^{90}Sr and ^{137}Cs are the most important biologically. Concentrations of ^{137}Cs in lichens tend to be higher than in nearby unvegetated soils. This is due to a number of factors such as the ability of lichens to absorb mineral metabolites from rainwater and snowmelt, radionuclide loss from soil by surface erosion, and upward cycling of ^{137}Cs from underlying soil horizons. Thus, lichens provide an enriched, long-term, source of important radionuclides which tend to be concentrated in upper portions of the preferred winter forage of caribou and reindeer.

The possibility of geographical and ecological differences in ^{137}Cs and ^{90}Sr concentrations in lichen communities over the very large study area of northern Alaska was investigated during 1967 and 1972 by intensive sampling at 20 locations from north-western Canada to the Chukchi Sea, including three locations across the wintering range of the caribou that provided critical food to the native villages. Three-way analysis of variance of results for ^{137}Cs showed no significant difference between three ubiquitous species, three physiogeographic provinces from which they were collected, or sampling years (Hanson, 1973). Student's *t* test of combined means of all species tested between years indicated that there was usually a highly significant ($P < 0.01$) difference between ^{137}Cs concentrations of the nine species from the three provinces, with samples from the Arctic Coastal Plain and Arctic Foothills more similar than those of either province compared to the higher values in the Arctic Mountains (Brooks Range).

The upper 12 cm (active growing portion) of *Cladonia stellaris* lichen communities generally contain 70–90 per cent of the ^{90}Sr and > 90 per cent of the ^{137}Cs inventory in the lichen carpet, with some indication of fractionation between upper and lower 6 cm increments. ^{137}Cs is more readily translocated than ^{90}Sr, resulting in higher concentrations of ^{137}Cs in upper layers of lichen mats. Experiments with ^{90}Sr and ^{137}Cs in lichens indicate that translocation along lichen thalli is primarily diffusive in character but complicated by cation exchange, especially for strontium. Fixation of strontium to the thalli explains the greater mobility of caesium between compartments of the lichen carpet and underlying strata. Several circumpolar studies after weapons-testing fallout indicate effective half-lives of about 5 ± 2 y for ^{137}Cs and 1.0–1.6 y for ^{90}Sr in arctic and sub-arctic lichen communities.

Comparison of simulated and observed data in a deterministic model of ^{137}Cs concentrations in lichens collected at Anaktuvuk Pass during 1963–1973 (Thomas *et al.*, 1982) demonstrated the deficiencies in W.C. Hanson's model despite intensive sampling and analysis. A major need was accurate measurements of fallout deposition rates at the lichen sampling sites, which would have greatly improved model performance in the reality of variable inputs of fallout due to discontinuous nuclear weapons test series and weather systems.

Plutonium isotopes in lichens from northern Alaska during 1967–1979 occurred in pronounced peaks during 1968, 1972, 1974, and 1976 that correlate well with periods of high-yield nuclear tests. Both ^{238}Pu and $^{239,240}Pu$ isotopes showed the same pattern of concentrations, with ^{238}Pu consistently 0.1 times the $^{239,240}Pu$ values rather than the 0.022 ratio reported in fallout. This factor of 5 disparity was maintained throughout the series, indicating appreciable retention in the lichens. Results of studies in northern Alaska and central Sweden suggest that ^{238}Pu is more tightly bound in the upper 6 cm stratum of lichens than is $^{239,240}Pu$. Both plutonium isotopes demonstrate an effective half-life in the lichen carpet of about 6 y.

Plutonium-239 was reported to have a mean residence time of 4.3 ± 0.5 y in the top 3 cm of the *Cladonia alpestris* carpet in Sweden, compared to 6.1 ± 0.5 y in the entire 12 cm carpet (Holm and Persson, 1975), with the difference ascribed to ingrowth of lichen biomass and low solubility of the fallout plutonium oxide.

4.3.4.4 Herbivores

Concentrations of ^{137}Cs in caribou and reindeer flesh follow an annual cycle with low values during autumn months and maximum values in spring months (Figure 4.2). The low values occur in animals returning from their summer range on which sedges and other fresh forage provide a diet containing relatively low amounts of fallout ^{137}Cs and substantial amounts of potassium, which combines with increased summer body water turnover to produce the abrupt decline in net accumulation of ^{137}Cs in soft tissues. An effective half-life of 28 days was usually observed for ^{137}Cs in caribou muscle between late May and late August, when migration to winter ranges usually began. It is generally accepted that the rate of caesium loss is dependent upon the potassium intake rate; experiments with reindeer (Holleman *et al.*, 1971) showed that, for the slow ^{137}Cs component, the biological half-time was about 17 days with a dietary potassium concentration of 1 mg/g dry weight and about 6.7 days with a dietary potassium concentration of 5 mg/g dry weight. Extrapolation of a log–log plot gives a corresponding half-time of about 30 days for a potassium concentration of 0.37 mg/g, which was the mean value measured in 17 lichen samples collected from caribou winter range near Anaktuvuk Pass during 1964–1967.

During autumn, caribou and reindeer gradually shift to their winter diet composed mainly of lichens, and ^{137}Cs concentrations in soft tissues begin a steady increase through winter months; in Alaska the ^{137}Cs levels usually plateaued during

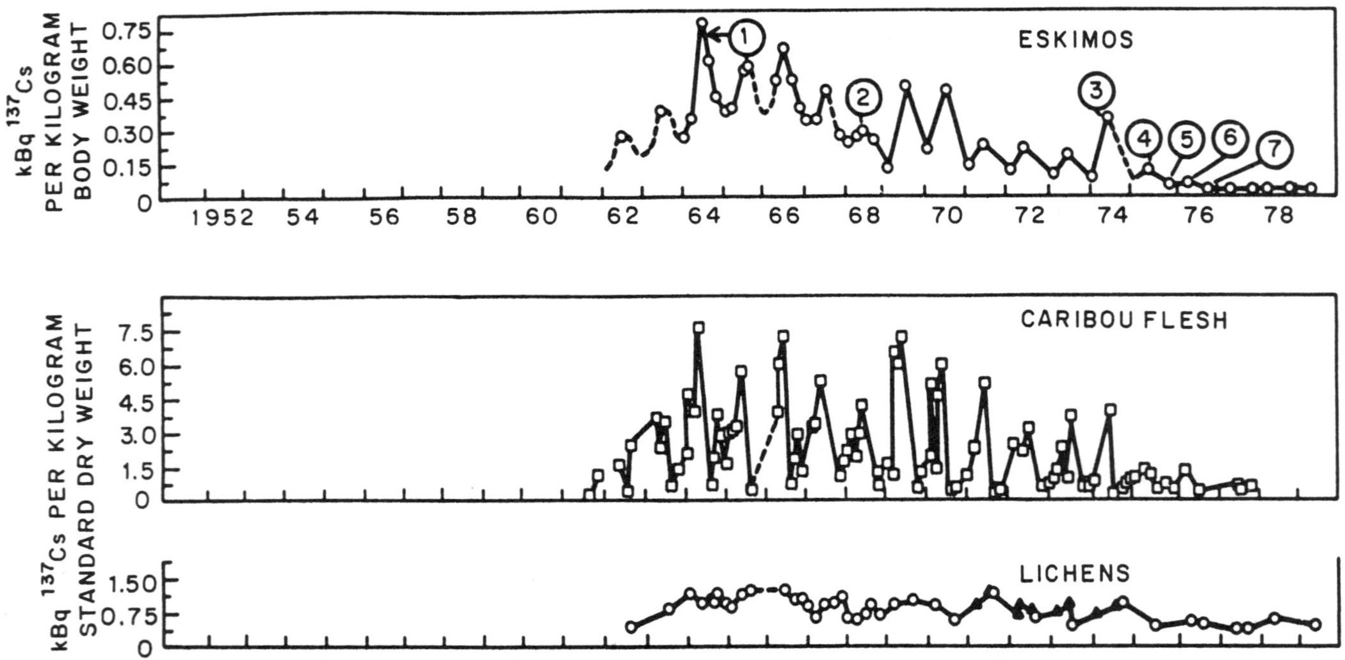

Figure 4.2 Fallout ^{137}Cs in biota of northern Alaska ecosystems during 1962–1979.

January–April of each year. An appropriate ^{137}Cs tracer kinetics model applied during periods of simultaneous sampling of caribou and lichens at the same location on winter range indicated that 4.5–5.0 kg dry weight lichens were ingested per day. ^{137}Cs concentrations in caribou flesh at the end of the winter were 4.0 ± 0.9 (SD, range 3.2–5.5) times the ^{137}Cs concentration of lichens.

Comparison of ^{137}Cs levels in other important herbivores such as moose (*Alces alces*) and Dall sheep (*Ovis dalli*) on nearby ranges to those utilized by caribou showed no significant seasonal pattern and were significantly lower than caribou values. Sheep muscle contained four times the ^{137}Cs concentration in moose muscle but only one-fourth to one-tenth of ^{137}Cs concentrations in caribou muscle. These differences were due to food habits; the sheep fed mostly on sedges, forbs, and modest amounts of lichens, while moose were observed to feed mostly on willows and other shrubs, with some minor amounts of grasses.

After prolonged exposure, ^{90}Sr concentrations in bone are about 1000 times those in muscle of caribou and reindeer. In northern Alaska, ^{90}Sr levels in caribou bone were relatively stable near 740 Bq/kg until 1962 and then began a sharp increase to 2600 Bq/kg in 1966–1969, after the major nuclear weapons test series of 1961–1962. The ratio of concentrations in caribou bone/lichen was about 7.6 during the 1966–1969 plateau period, when a condition of 99 per cent of equilibrium existed. During the period May 1964–November 1974 the values declined at a half-life of 56 months to levels near those measured in 1962. Strontium-90 in flesh followed a more gradual curve of increase, peaking in years of nuclear weapons testing, similar to the ^{137}Cs pattern (Hanson and Thomas, 1982).

Concentrations of plutonium in caribou bone samples during the 1971–1975 period of maximum values in lichens were barely detectable. Concentration ratios relative to lichens were usually in the range of 0.02 for ^{238}Pu and 0.001 for $^{239,240}Pu$ compared to values in lichens.

4.3.4.5 Carnivores

Concentrations of worldwide fallout ^{137}Cs in circumpolar food chains follow an annual cycle dependent upon the subsistence patterns of human populations and the food habits of carnivores, particularly those associated with caribou and reindeer.

Extensive sampling of wolves, foxes, and wolverines in northern Alaska showed a repeated annual ^{137}Cs pattern with a rapid increase through autumn and winter that paralleled the increase in ^{137}Cs concentrations in caribou. Wolves contained twice the concentrations observed in foxes and wolverines, both of which, in Alaska, are primarily scavengers on wolf-killed caribou and consumers of small mammals that contain much lower ^{137}Cs concentrations. The ratio of concentration for ^{137}Cs in wolf flesh/caribou flesh averaged 2.7 at a time at which 80 per cent equilibrium had been achieved.

A similar pattern occurred in ^{90}Sr concentrations in bone of the wild carnivores and the same relationship was observed; maxima occurred in late winter following utilization of caribou that were increasing their concentrations and wolves contained twice the ^{90}Sr concentrations of foxes and wolverines. Wolves consume more caribou bone than do foxes and wolverines and therefore ingest a richer source of ^{90}Sr than do scavengers. The concentration ratio for ^{90}Sr in wolf bone/caribou flesh at about 50 per cent of equilibrium at time of sampling was 0.4. An effective half-time of about 7 years for ^{90}Sr in bone was observed in all three species of carnivores.

Using the lichen forage ingestion model it was estimated that caribou on winter range ingested about 3 Bq ^{238}Pu and about 30 Bq $^{239,240}Pu$ per day during the mid-1960s to mid-1970s, yet bone samples contained barely detectable amounts. Assuming averages of 0.7 Bq/kg for ^{238}Pu and 7 Bq/kg for $^{239,240}Pu$ in lichens during the above period, concentration ratios of 0.02 and 0.001 are estimated. This indicates that ^{238}Pu was more readily transferred through the lichen–caribou food chain than was $^{239,240}Pu$; however, the tenuous nature of the values makes such a conclusion speculative.

4.3.5 TROPICAL ISLANDS

The island environments of the atolls in the Marshall Islands represent a unique ecosystem where radionuclides that were introduced between 1946 and 1958 have had nearly 40 y to equilibrate. Bikini and Enewetak Atolls were the sites of 66 atmospheric nuclear weapons tests. These atolls are composed of coral limestone, which has accumulated on old igneous seamounts that arise in the Pacific Basin. The depth of coralline deposits beneath these atolls is about 1.2 km at which point there is contact with primordial igneous rocks. An atoll usually comprises a discontinuous series of islands on a ring of coral reef that surrounds a relatively shallow lagoon. An actively growing reef of coral organisms is present on the seaward side of the atoll and coral beaches occur on the lagoon side.

The species diversity of the atolls is quite restricted because of the remoteness from continental masses that typically supply organisms for island colonization. Most of the vegetation on the atolls has arrived there by either water or animal transport. Most of the plant species that comprise the stable, climax vegetation of the atolls in the Marshall Islands are pantropical in distribution. The coconut palm, *Cocos nucifera*, and the pisona tree, *Pisona grandis*, are good examples of the pantropical floristic element, and these species occur throughout the Pacific Basin on both islands and atolls.

Areas that were disturbed in the 1950s have not returned to the stable tropical forest type and are present as open woodlands of *Messerchimidia argentea* and *Scaevola frutescens*, with vines, sedges and grasses occurring between the clumps of low trees and shrubs.

The amount and frequency of rainfall at the atolls are of paramount importance in the cycling of a mobile ion such as ^{137}Cs. Most of the annual rainfall occurs from May through November; the months from December through April tend to be very dry. About 50 per cent of the annual rainfall occurs from August through November.

Rainfall is variable and ranges between 114 and 140 cm/y in the northern Marshall Islands to as much as 380 cm/y in the southern Marshall Islands. The lack of rainfall during some years seriously affects the growth of vegetation, and in drought years some species die back. Most of the research discussed here has taken place in the northern Marshall Islands where native agriculture often does not succeed with rainfall as the only source of water.

4.3.5.1 Radionuclide distribution in the soil

The unique properties of the soil at the atolls dictate to a large extent the distribution and cycling of radionuclides in the ecosystem. The composition of coral soil at Bikini and Eneu Islands is listed in Table 4.22. The soil is composed primarily of calcium carbonate ($CaCO_3$) with some magnesium carbonate ($MgCO_3$) and no

Table 4.22 Composition of coral soil from Bikini and Eneu Islands

Island location and depth	Total[a]							Extractable
(cm)	pH[b]	Sr (%)	Ca (%)	Mg (%)	P[c] (%)	N (%)	Organic Matter[d] (%)	K[e] (ppm)
Bikini No. 1								
0–5	7.7	0.38	30.4	0.95	1.35	0.64	14.4	79
5–10	7.8	0.39	30.8	0.89	1.28	0.62	13.2	26
10–15	7.9	0.39	30.9	0.89	1.29	0.63	12.3	20
15–25	7.9	0.40	31.9	0.86	1.17	0.50	10.6	23
25–40	8.3	0.39	34.3	1.28	0.67	0.19	4.5	4
40–60	8.4	0.31	34.5	2.05	0.16	0.11	1.6	3
Bikini No. 2								
0–5	7.8	0.40	31.0	1.02	0.82	0.49	10.7	50
5–10	8.0	0.40	32.4	1.09	0.71	0.46	8.5	24
10–15	7.9	0.38	33.1	1.18	0.56	0.35	7.4	24
15–40	8.2	0.38	34.7	1.79	0.32	0.11	1.6	6
Eneu No. 1								
0–5	7.7	0.32	32.0	1.74	0.085	0.30	5.1	41
5–10	8.0	0.34	32.6	1.76	0.055	0.35	5.6	20
10–15	8.0	0.31	34.3	2.08	0.037	0.17	2.6	9
15–25	8.4	0.28	34.0	2.40	0.016	0.06	0.9	1
25–40	8.7	0.28	34.4	2.48	0.014	0.05	0.8	1
40–60	8.9	0.30	33.3	2.37	0.015	0.03	0.6	<1

[a] Stable caesium was below detection limit (1.3 ppm).
[b] pH in water.
[c] High phosphorus values indicate ancient guano deposition.
[d] Organic matter by wet oxidation.
[e] Extractable in N ammonium acetate.

silica-clay component. The pH is high, ranging from 7.7 to 9.0. The surface horizons are high in organic content (as much as 14 per cent), although the organic content of the soil drops markedly with depth in the soil column. The soil is low in exchangeable potassium and marginal in phosphorus and trace-mineral content. Some of the native plant species and most introduced species show definite signs of potassium deficiency. In fact, with most introduced food crops and ornamental plants, growth is very limited without addition of potassium as well as nitrogen, phosphorus and trace minerals.

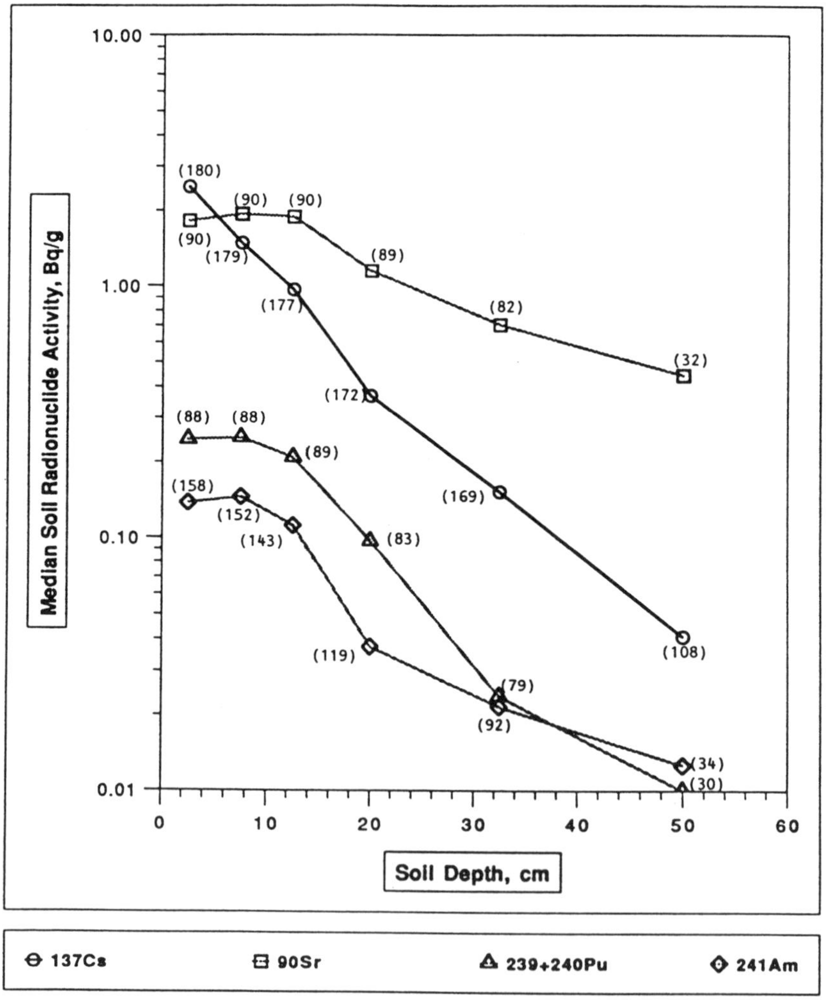

Figure 4.3 The median concentration of ^{137}Cs, ^{90}Sr, $^{239+240}$Pu and ^{241}Am as a function of depth in the soil column on Bikini Island.

The distribution of ^{90}Sr and ^{137}Cs with depth in the soil column is essentially exponential, with the highest concentrations in the top few centimetres of the soil column (Figure 4.3). The concentrations of ^{137}Cs and ^{90}Sr are generally very similar at the atolls. However, ^{137}Cs is the most significant radionuclide contributing to the dose to people inhabiting the atolls. The concentrations of $^{239+240}Pu$ and ^{241}Am are much less than those of ^{137}Cs and ^{90}Sr. Other radionuclides present a few years ago, such as ^{60}Co, have essentially disappeared due to radioactive decay. Most of the activity is within the top 25–40 cm of the soil column, which is where most of the organic material is located. Below the organic layer, the radioactivity concentration drops rapidly and is generally very low. The small amount of radionuclide activity present all the way to the groundwater represents the fraction of the soil inventory of ^{137}Cs, ^{90}Sr, $^{239+240}Pu$ and ^{241}Am that is solubilized and carried to the groundwater when rainfall is adequate to cause through-flow of rain water and recharge of the underground lens system.

4.3.5.2 Caesium-137 and strontium-90

The fundamental composition of the atoll soil produces dramatic differences in the uptake of ^{137}Cs and ^{90}Sr at the Marshall Islands compared to the uptake rates in the published literature, which are based primarily on silica-clay type soils. For example, the concentration ratio (CR), defined as the activity concentration of the radionuclide per gram of wet plant weight divided by the activity concentration per gram of dry soil, is about 0.1 for ^{137}Cs and about 1.0 for ^{90}Sr in silicate soils (Ng *et al.*, 1982). However, in the coral soils in the Marshall Islands the CR for ^{137}Cs (based on 0–40 cm) is about 5 while that for ^{90}Sr is about 0.0001 (Robison *et al.*, 1988; Koranda *et al.*, 1978). It is generally high for ^{137}Cs and low for ^{90}Sr in all the plant species that have been analysed (Table 4.23). Strontium-90 replaces Ca in the $CaCO_3$ matrix of the coral soil, and is relatively unavailable for uptake; ^{137}Cs is bound to the organic fraction of the coral soil and is relatively more available than in silica-clay soils where it is bound by mineral clays.

In many biological systems K is preferred over Cs and plants selectively take up K and discriminate against Cs (Handley and Overstreet, 1961; Middleton *et al.*, 1960; Nishita *et al.*, 1962; Wallace *et al.*, 1983).

Early studies and measurements were concerned with the distribution of radionuclides in the atoll ecosystem. Because of the importance of ^{137}Cs in the atoll environment, a landscape inventory was made at Enewetak Atoll on Engebi Island to determine the distribution of ^{137}Cs in the soil and above-ground biomass. The data indicated that approximately 3–5 per cent of the ^{137}Cs inventory in the soil was present in the vegetation at any time. This suggested that the soil-complexed pool of ^{137}Cs was tightly held and not available to the plant.

A laboratory study concerned with the availability of soil-bound ^{137}Cs was made to verify the low apparent availability in the ecological inventory. Five columns each with approximately 3 kg of soil were set up and leached with distilled water.

Table 4.23 Concentration ratio (CR) for ^{137}Cs, ^{90}Sr, $^{239,240}Pu$ and ^{241}Am at Bikini Island for several species of vegetation. Mean (median) CR (Bq per g in fruit, wet weight/Bq/g in 0–40 cm soil, dry weight)

	^{137}Cs	^{90}Sr	$^{239,240}Pu$	^{241}Am
Drinking coconut meat	5.5 (3.4)	6.1×10^{-3} (4.5×10^{-3})	2.8×10^{-5} (7.1×10^{-6})	5.3×10^{-5} (1.4×10^{-5})
Drinking coconut fluid	2.1 (1.3)	—	—	—
Copra meat	10 (4.2)	4.6×10^{-3} (3.1×10^{-3})	1.9×10^{-5} (6.0×10^{-6})	3.1×10^{-5} (8.2×10^{-6})
Breadfruit	0.8 (0.6)	0.057 (0.055)	1.3×10^{-5} (4.2×10^{-6})	2.4×10^{-5} (6.5×10^{-6})
Pandanus	15 (11)	0.12 (0.041)	2.3×10^{-5} (1.3×10^{-5})	6.3×10^{-5} (3.0×10^{-5})

The pore water was allowed to remain in the soil for varying lengths of time and then leached with 10 times the pore volume. The availability of the soil-bound ^{137}Cs was observed to be between 3 and 4 per cent in this laboratory experiment, which was in agreement with the ecological measurements made previously.

A field experiment involved a hectare of atoll surface in which the ground was irrigated with overhead sprinklers using seawater, and the presence of ^{137}Cs monitored in the groundwater lens beneath the soil. After delivering 20 m of seawater to the plot, it was apparent that a pulse of ^{137}Cs appeared in the groundwater compartment. The amount of ^{137}Cs appearing in the groundwater lens was on the order of 3–5 per cent of the soil inventory.

These experiments indicated that ^{137}Cs was tightly held in the organic stratum of coral soil on the atolls and, in general, did not respond to the leaching effect of distilled water or seawater. The small fraction of ^{137}Cs circulating in the biomass pool was found to be transferred to man via terrestrial foods.

4.3.5.3 Plutonium and americium

The atoll ecosystem is ideal for evaluating the root uptake of the transuranic radionuclides $^{239+240}Pu$ and ^{241}Am. Determining what fractions of the total plutonium and americium observed in plants is due to root uptake versus deposition on the plants via soil resuspension has been a difficult problem in most environments with the types of food crops generally used. However, in the atoll system most of the food crops (coconut, breadfruit, pandanus, etc.) grow to such heights that the fruit and canopy are well above resuspended soil aerosol and deposition is unlikely to be significant. In addition, the edible fruits have dense, thick layers that protect the edible portion of the fruit; these protective layers must be husked, peeled or removed in some manner in order to eat the fruit. Consequently, the edible portion is in a sense 'sealed' from any deposition type of contamination. The plutonium and americium observed in such foods can only be derived via root uptake.

The CRs for plutonium and americium (Table 4.23) generally agree with pot culture studies in glasshouses where experimental design eliminated any resuspension. The general magnitude of uptake of plutonium and americium seems to be about the same over a wide range of soil types, with the coral soils being at one extreme with high pH and nearly pure $CaCO_3$ plus organic material. This is in stark contrast to the very different CRs observed for ^{137}Cs and ^{90}Sr in different soil systems.

4.3.5.4 Remedial measures for ^{137}Cs uptake

The $CaCO_3$ nature of the soil and the low exchangeable potassium characteristic of coral islands provide the conditions for the observed high uptake of ^{137}Cs. On the other hand, these conditions would seem to provide a situation in which potassium added to the system might alter the uptake of ^{137}Cs. Consequently, experimental plantings of coconut palms, pandanus, breadfruit, and other food plants have been established at Enewetak and Bikini Atolls to study radionuclide uptake under semi-controlled conditions and to provide guidance on the resettlement of the atolls by the native population.

Several experiments using large areas of the coconut grove on Bikini Island have been implemented. The experiments include surface application of potassium ranging from 666 kg/ha to 5550 kg/ha, in the form of N,P,K fertilizer or KCl, and the number of trees per experiment ranging from 10 to 120. The total potassium has been applied over a 3-year period for some experiments and in one application for other experiments. Similar experiments have also been conducted on breadfruit and pandanus trees.

The results of all the experiments show a decrease in the concentration of ^{137}Cs by a factor of about 20 compared to the pre-treatment concentration in the fruits. The initial results from a recent experiment using 666 kg/ha indicated a reduced uptake of ^{137}Cs similar to that observed for the higher rates of added potassium. Examples of the results are shown for two of the experiments in Figures 4.4 and 4.5. In addition to blocking the uptake of ^{137}Cs, the treatment with KCl (or N,P,K) increases the growth rate and productivity of the plants as has been observed in other tropical locations (Vernon *et al.*, 1976).

The alternative remedial measure for rehabilitation of the islands is to excavate and dispose of surface soil to eliminate the radionuclides, particularly ^{137}Cs. Although this method is very effective in eliminating the radionuclides from the island, it has significant environmental consequences. All of the standing vegetation, which includes mature coconut groves that have taken 20 y to develop,

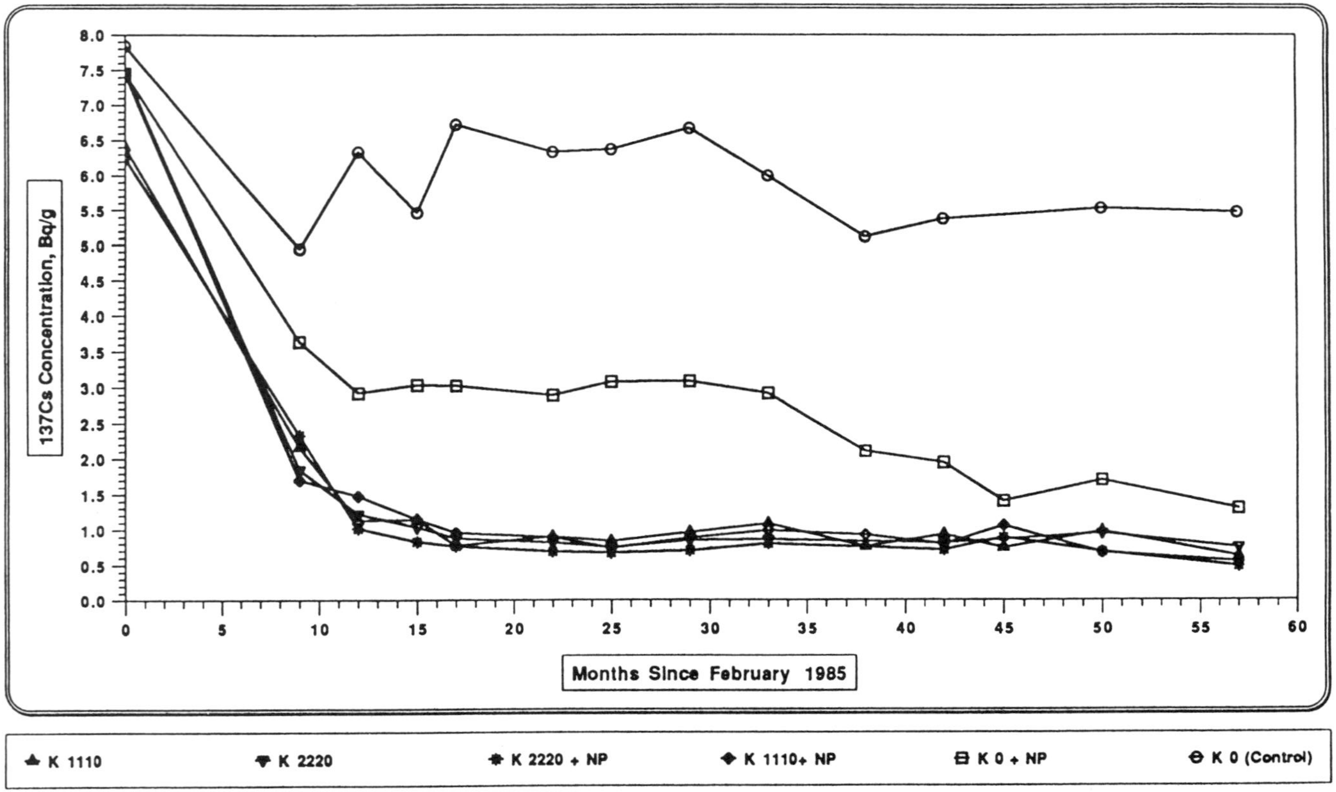

Figure 4.4 The effect of 1110 and 2220 kg K/ha, 500 kg NP/ha, 1110 kg K/ha + 500 kg NP/ha and 2220 kg K/ha on the uptake of ^{137}Cs in drinking coconuts at Bikini Island. (Reproduced from the journal *Health Physics* with permission from the Health Physics Society.)

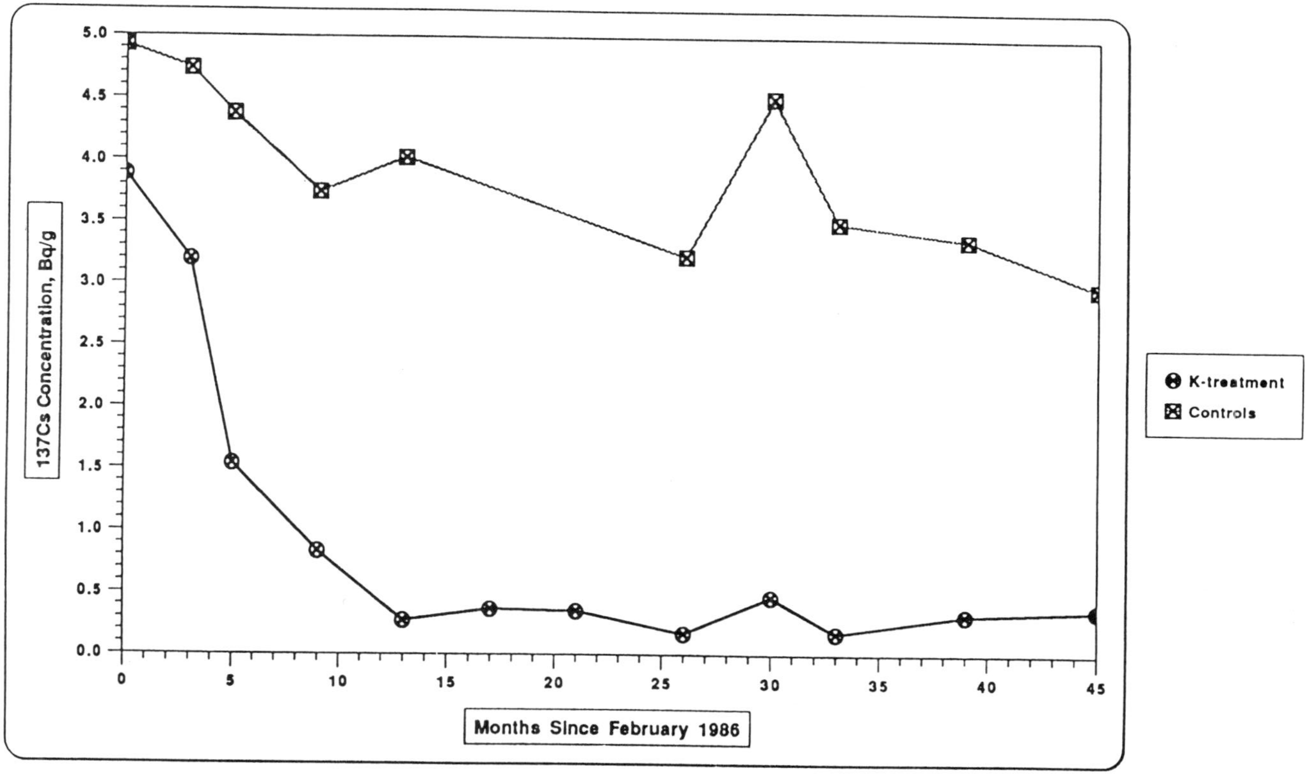

Figure 4.5 The effect of a single 5550 g/ha K application on the uptake of ^{137}Cs in drinking coconuts at Bikini Island

breadfruit, and pandanus trees, must be removed. Then the soil can be removed. This contains most of the radionuclides, but also contains most of the organic material that has taken centuries to develop on the coral islands; below the organic-material-containing layer is essentially beach sand. The soil-organic layer provides nearly all the nutrients needed for plant growth and contributes a significant fraction of the water retention capacity of the coral soil. The consequences for excavation of soil will have long-term ecological and agricultural effects; a very long-term commitment, probably the order of decades, would be required to rebuild the soil and to revegetate the islands.

4.3.6 COASTAL ECOSYSTEMS

Coastal ecosystems occur at the interface of land and sea. For the purposes of this section only systems which contain vascular plants and that are above the low tide mark are considered, thereby excluding the algal communities of rocky shores. Similar systems may also exist on the shores of large freshwater lakes. The wide variety of habitat types associated with coasts are described by Ratcliffe (1977) and include the following: coastal lagoons and swamps, salt marshes, beach systems, sand dunes, coastal grasslands and heath, and cliff vegetation.

Coastal ecosystems can be exposed to radioactive deposition resulting from man's activities (see Chapter 5). Resuspension from the water surface, either in the form of droplets or as suspended aerosols, is an important pathway for transport to coastal environments as is direct submersion by the tides. Mechanical movement, either by wind or living organisms, is also a potential contamination pathway. Coastal ecosystems far removed from the source may be affected by liquid discharges.

4.3.6.1 Coastal lagoons and swamps

Coastal lagoons and swamps often provide environments where, due to sheltered conditions, and trapping by vegetation, deposition of fine sediments is possible. In some instances such as the Great South Bay on Long Island, USA, small amounts of low-level radioactive waste can find their way into these habitats. Extensive use of these areas often occurs for fishing and recreation, and pathways to man include both the food chain and external exposure.

Mangrove swamps can constitute an important ecosystem in tropical areas but very few data exist on radionuclide behaviour in these systems.

4.3.6.2 Salt marshes

Salt marshes are typically found in river estuaries. However, even on exposed coasts salt marshes can occur, such as on the Norfolk coast of Great Britain. Plant species characteristic of salt marshes are specially adapted to tolerate inundation by salt-

Table 4.24 Mean, confidence limits and range of radionuclide concentrations in pectoral muscles and livers of a Greylag goose and wigeon from Ravenglass and Morecambe Bay (Bq/kg fresh weight) (From Lowe and Horrill, 1986)

Sample	Species	*n*	Years collected	Sample wt (g)	^{40}K	^{134}Cs	^{137}Cs
Ravenglass							
Pectoral muscles	Greylag goose	1	1981	342.9	184.6	3.3	57.7
	Wigeon	4	1981	151.3±33.2[a]	126.9±48.8	8.5±8.3	158.0±148.9
				123.8–172.6	85.8–153.5	3.7–15.5	80.7–289.3
Livers	Greylag goose	1	1981	52.9	124.0	—	27.8
	Wigeon	4	1981	18.8±3.3	28.1±60.0	—	99.5±52.7
				15.9–21.1	3.7–79.6	—	64.4–149.9
Morecambe Bay							
Pectoral muscles	Wigeon	2	1984–5	106.2	158.7	—	27.0
				69.7–142.7	146.5–170.6	—	18.9–35.2
Livers	Wigeon	10	1979–85	17.2±3.0	74.4±79.7	—	60.3±26.2
				12.7–25.7	3.7–348.9	—	12.2–141.0

[a]± = 95 per cent confidence limits about the mean.

water as well as freshwater at certain times of the year. Whilst some salt marshes still contain natural vegetation, many have been modified by grazing pressures. Salt marshes also serve as important feeding areas for large populations of wildfowl which may pick up radionuclides during winter feeding (Table 4.24).

In some cases, nuclear installations are situated on the coast in close proximity to areas of salt marsh. The accumulation of radionuclides in salt marshes is largely associated with sedimentation processes, naturally occurring in these habitats. Radionuclides are mainly sorbed onto sediments attached to the surface of the plants, rather than being incorporated in the tissues. In salt marshes of the Ravenglass Estuary in Cumbria, UK (near the Sellafield nuclear reprocessing facility) a wide range of gamma-emitting fission products can be detected in the silts, e.g. ^{60}Co, ^{95}Zr, ^{95}Nb, ^{103}Ru and ^{106}Ru, ^{141}Ce and ^{144}Ce, ^{154}Eu and ^{155}Eu, ^{134}Cs and ^{137}Cs. Also present are the actinide isotopes ^{241}Am, ^{238}Pu, $^{239/240}Pu$ and ^{241}Pu (Table 4.25).

All these radionuclides originate from the low-level liquid waste released from the Sellafield reprocessing plant. The deposition is related to both the time of tidal inundation and the structure of the vegetation (Horrill, 1983). Animals grazing these contaminated pastures ingest radionuclides attached to silt and quantifiable amounts have been found in their tissues (Howard and Lindley, 1985) (Figure 4.6). Studies of the transfer of radionuclides in sheep grazing the contaminated pastures show that lambs generally have higher concentrations of radiocaesium in their tissues compared with those of the dam. However, radionuclides with a long half-life, e.g. plutonium isotopes, are present in higher concentrations in the dam's liver than in the lamb. The availability of radiocaesium to animals grazing salt marsh is much reduced compared with that taken in the ionic form (Howard, 1989).

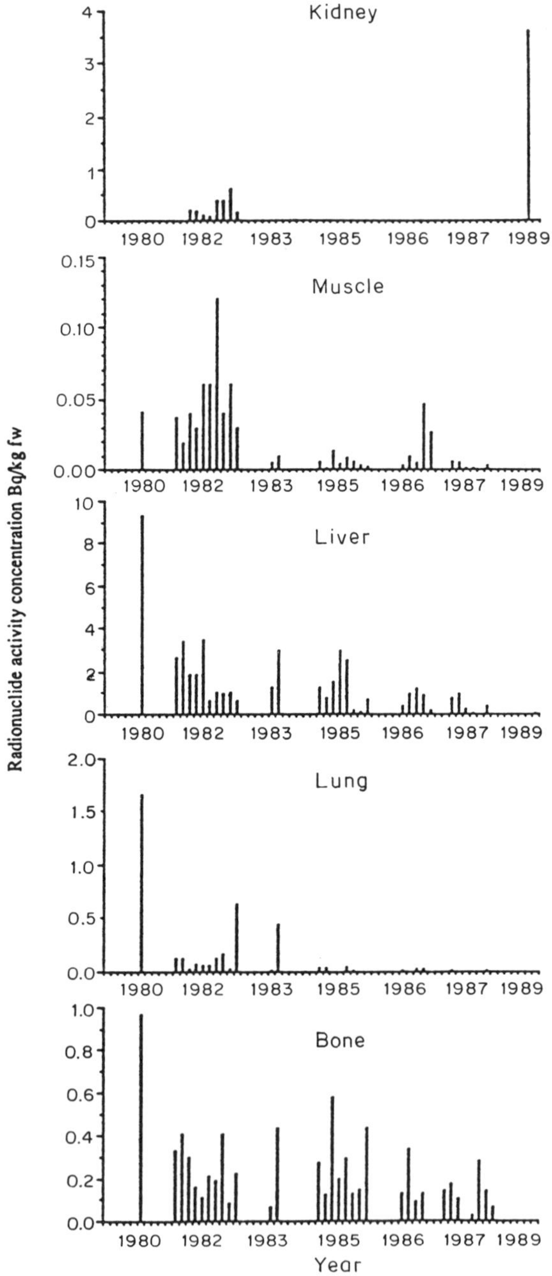

Figure 4.6 $^{239,240}Pu$ activity concentrations in tissues from sheep which had been grazing tide-washed pastures in estuaries bordering the Irish Sea. Note that samples have not been taken in every year since 1980 (from: Howard and Livens, 1991).

Table 4.25 Radionuclide activity concentrations found in sediments/soils of tide-washed pastures bordering the Irish Sea (from Howard and Livens, 1991)

Radionuclide	Activity concentration (Bq/kg dw)	Locality	Year
^{60}Co	97	R. Esk	1981[b]
	112	R. Irt	1982[c]
^{90}Sr	11 000	Eskmeals marsh	1974[a]
^{106}Ru	130 000	Eskholme marsh	1973[a]
	4350	R. Esk	1981[b]
	18 700	R. Irt	1982[c]
^{137}Cs	18 000	Eskholme marsh	1978[a]
	up to 30 000	R. Irt	1979
	10 260	R. Esk	1981[b]
	16 810	R. Irt	1982[c]
	10 000	R. Esk	1982
	500–3620	R. Duddon	1985
^{144}Ce	69 000	Eskholme marsh	1973[a]
	690	R. Esk	1981[b]
	4920	R. Irt	1982[c]
^{154}Eu	300	R. Esk	1981[b]
	441	R. Irt	1982[c]
^{155}Eu	260	R. Esk	1981[b]
	620	R. Irt	1982[c]
^{237}Np	4	R. Esk	1986
$^{239/240}Pu$	3560	R. Esk	1981[b]
	10 750	R. Irt	1982[c]
	8000	R. Esk	1982–1988
	230–1570	R. Duddon	1985
^{241}Pu	79 000	R. Esk	1989
^{241}Am	3650	R. Esk	1981[b]
	5650	R. Irt	1982[c]
	9000	R. Esk	1982–1988
	290–2400	R. Duddon	1985

[a] Highest value obtained.
[b] Mean value for an ungrazed salt marsh.
[c] Mean value for a grazed salt marsh.

4.3.6.3 Beach systems

Beach systems generally either consist of bare sand or shingle. As such they are almost totally devoid of vegetation and provide little scope for the biological accumulation of radionuclides. The larger systems become vegetated as a result of the accumulation of organic material along the strand line which serves as a foothold for vegetation. If sufficient shingle accumulation occurs then the effect of the sea is diminished and a vegetation type containing heathers can develop. Monitoring at the classic shingle beach at Dungeness, UK, has demonstrated that only small amounts of radionuclides, relative to discharge, are accumulated in this type of habitat.

4.3.6.4 Sand dunes

Sand dunes are sub-maritime coastal habitats. They are not inundated by the sea, nor are they strongly saline. The systems result from the prevailing onshore winds and trapping of windblown sand by the vegetation. The type of sand determines the flora of the dune system and can range from base-deficient silaceous to strongly calcareous material (Salisbury, 1952).

Dune systems are well drained with the exception of the hollows, or slacks, between the dunes. Studies of coastal sediments have shown that coarser sands, such as those associated with dunes, have fewer absorption sites than do clay minerals, silts and muds. However, sands with a high proportion of carbonaceous particles accumulate more radionuclides than do other coarse-grained sediments (Jones *et al.*, 1984).

The areas most likely to accumulate radionuclides in active dune systems would be the dune slacks where organic matter accumulates and binding sites for radionuclides will be more frequent. As the dune system ages it becomes more stable and the area behind often develops into a heathland. The soil obviously accumulates more organic matter and the retention of spray and aerosols becomes more effective. Dune systems, however, have not been found to be important habitats where large amounts of radionuclides accumulate (Nellis, 1990).

The source of building material for dunes, the intertidal sands, has been studied on the Cumbrian coast by Eakins *et al.*, (1990). These authors found that the intertidal sands contained about 1 per cent of the alpha-emitting activity released from Sellafield up to 1982 but that only in a few cases did the concentrations of man-made alpha-emitters exceed those of natural alpha-emitters.

4.3.6.5 Coastal grasslands and heath

As dune systems age and new dunes are thrown up nearer the sea the old dunes are stabilized by the accumulation of organic matter and development of vegetation. With the consequent grazing pressure, either by wild or domestic animals, grassland, and eventually by acidification, heathlands often develop. These systems will be subject to spray and aerosol deposition from the sea and the better-developed soil structure will enable greater retention of deposited radionuclides than the sands of the dunes. The greater accumulation of radiocaesium in heathland habitats as well as the ability of heath species to accumulate radionuclides has been well documented (Bunzl and Kracke, 1984; Horrill *et al.*, 1990).

Grasslands also develop behind the many areas of salt marsh on the coast or in river estuaries. These will be subject to aerial deposition and rare inundations by high tides. Whilst the short-lived radionuclides will soon decay, there could be an accumulation of long-lived radionuclides such as ^{137}Cs, $^{239,240}Pu$ and ^{241}Am as a result of periodic flooding.

4.3.6.6 Cliff vegetation

Vegetation associated with cliffs varies according to exposure to wind and driven spray, the bedrock type and angle. Vegetation is often severely affected by the presence of bird colonies, but on the other hand cliffs can be refuges for plants sensitive to competition or grazing pressure.

The main route for the input of radionuclides is by atmospheric deposition and by spume from breaking waves (Malloch, 1972). Deposition on cliff vegetation will be very variable and in the majority of cases investigated concentrations are only slightly above background levels (Toole *et al.*, 1990).

4.4 MODELS FOR RADIONUCLIDE TRANSPORT IN TERRESTRIAL ECOSYSTEMS

There are numerous models that can be used to simulate the transport of radionuclides in parts of or whole terrestrial ecosystems. Over the last 10 years much attention has been focused on the development of models that can be used to assess the radiological consequences of releases to atmosphere, land or water. This section summarizes some of the recent developments in modelling terrestrial ecosystems with particular attention to radiological assessment models and their underlying databases.

4.4.1 DEPOSITION TO GROUND AND INTERCEPTION BY VEGETATION

Though considerable use is still made of the concepts of deposition velocity, washout ratio (or rate) and interception fraction (Section 4.1), it became clear after the Chernobyl accident that such models have considerable uncertainties, not least due to the effects of varying physical and chemical form of the depositing radionuclides. There is clear evidence post-Chernobyl for variable concentration of radiocaesium and radioiodine in pasture herbage according to height within the sward (Coughtrey *et al.*, 1990a). Also, the multilayered nature of the canopy within forests leads to differences in interception and retention. The reasons for this relative partitioning are both physical and biological, and the implications for modelling radionuclide transport are very great. Various authors have previously proposed the use of multilayered interception models (e.g. Lassey, 1983); however it is only recently that such models have been tested to any extent (e.g. Pinder *et al.*, 1989) and there do not, as yet, appear to have been attempts to apply such models to post-Chernobyl data.

Much attention has been focused on the deposition and interception of particulates in dry conditions. Events after the Chernobyl accident identified a requirement for a more detailed examination of wet deposition and interception. This requirement also applies in the context of waste disposal assessment where contamination of herbage and other vegetation may occur via irrigation with ground waters (Watkins, 1990). Post-Chernobyl data have been analysed in a number of ways to determine empirical relationships between vegetation content and wet deposition as is apparent in the model summaries included in BIOMOVS documentation (e.g. BIOMOVS, 1991). Muller and Prohl (cited by Watkins, 1990) modelled interception for wet deposition (f_w) as follows:

$$f_w = k_1 L[1 - \exp(-k_2 R)] / R \tag{4.4}$$

where: L is the leaf area index;
R is the amount of precipitation (mm);
k_1 and k_2 are element-dependent constants.

Predicted values for a range of vegetation types were stated to be in good agreement with measured values (though predicted values for potatoes were consistently lower than observations).

4.4.2 ABSORPTION, TRANSLOCATION AND LOSS FROM VEGETATION FOLLOWING SURFACE CONTAMINATION

Methods for modelling absorption, translocation and loss from vegetation are highly dependent upon the model structures adopted. These range from a single compartment to represent above-ground vegetation to multi-compartment models that attempt to separate 'surface' contamination from 'internal' contamination (BIOMOVS, 1991). Nevertheless, there have been a few recent developments in modelling the processes summarized in Section 4.1. Particularly in conditions of wet deposition or when rainfall follows dry deposition, retention on vegetation (whether measured on a content or concentration basis) shows multi-exponential trends. In the case of data based on concentrations, one component of the observed decline curve will reflect dilution by new growth.

Retention half-lives recorded for radiocaesium in the first few months after the Chernobyl accident ranged from 1 to 30 d with some apparent distinction between values recorded for areas primarily affected by wet or dry deposition (Coughtrey *et al.*, 1990a). BIOMOVS (1991) separated data for four sites into two phases of decline with retention half-lives of 8.3 to 12 d for the first 30 d and of 24–173 d thereafter.

No process-oriented models appear to exist for retention, absorption and translocation. All existing models are based on empirical factors derived from experimental or field observations.

4.4.3 FIXATION AND TRANSPORT IN SOILS

The last decade has seen a rapid development of models describing radionuclide transport in soils. Dynamic, process-oriented models have replaced simple empirical or 'black box' models (Coughtrey, 1988). Prior to the Chernobyl accident, the rapid developments in modelling had not been accompanied by a similar expansion in data. Chernobyl provided an opportunity to test and further develop many of the models but the timescale of some soil processes implies that it could be several years before adequate data can be acquired for validation purposes. Nevertheless, Chernobyl highlighted two aspects of radionuclide behaviour in soil which require careful consideration. These are:

1. The rapid penetration of radionuclides to some depth in soil profiles at early times after deposition, particularly in conditions of wet deposition.
2. The problems of dealing with material that enters the soil as discrete particles often of low initial solubility.

Though no specific examples appear to exist in the literature, the former can probably be dealt with on the basis of the physical characteristics of the soil concerned (particularly infiltration capacity, see Section 4.1.2). The latter is more difficult to treat without detailed knowledge of the initial chemical characteristics of the deposited material and treatment of subsequent chemical behaviour within soil (Section 4.2.1). Nevertheless, considerable work has been undertaken, particularly in the former USSR. Konoplyov and colleagues (e.g. Konoplyov and Bulgakov, 1991) have used an approach of the type outlined in Figure 4.1 in which the interchanges between 'fuel particle', 'cationic', 'ion-exchangeable sorbed', 'soluble complexes' and 'irreversibly sorbed' fractions in soil are modelled as a function of time. Considerable data have been collected for both ^{90}Sr and ^{137}Cs to develop and test the model. In general, agreement between predictions and observations has been very good.

Models of the type described above are essential to provide long-term predictions for exposure of man via both ingestion and external pathways. For Chernobyl, simple exponential models or those based on diffusion alone have not provided an adequate description of the behaviour of radiocaesium in soils other than in the first few months following the accident. Dynamic models for soil which take account of different chemical fractions are also valuable in predicting the rate of migration from soils to groundwater, the availability for plant uptake, and the potential for resuspension or erosion.

4.4.4 RESUSPENSION

Data concerning resuspension of radiocaesium after the Chernobyl accident have been considered by Garland and Pattenden (1990) who demonstrated a negative correlation between the calculated resuspension factor for radiocaesium and the total level of deposit. In part, this relation could be interpreted as the effect of long-range transport of resuspended material. An alternative explanation is the effect of rainfall in moving the deposited radiocaesium to lower soil layers and thereby rendering it unavailable for resuspension. Observed declines in resuspension factor with time were compared with a range of resuspension models. The models showed a wide divergence over the time interval occupied by the data reflecting, to some extent, the different applications of the models.

Although resuspension factor or rate methods are commonly applied in radiological assessments, other methods have been applied, for example the mass-loading approach. In areas close to Chernobyl, opportunities exist for further measurements of resuspension and it is understood that these opportunities are being exploited by some groups from the former USSR to develop revised models.

4.4.5 SOIL-TO-PLANT TRANSFER

Though the traditional methods of assessing soil-to-plant transfer are extremely useful for long-term assessments assuming equilibrium between plants and soil (Coughtrey *et al.*, 1990b), they are not always useful when dealing with short times after input to soil or with conditions in which biological availability within soils is changing. Over the last few years there have been several developments to provide realistic alternatives to the traditional soil–plant transfer factor. In general, these approaches require coupled soil and plant models in which the soil model allows for processes leading to biological availability. This generally requires complex modelling of soil processes and the incorporation of simulation methods dealing with root uptake from solution. However, few data exist as yet for the radionuclide, soil and plant combinations of interest. A complicating factor in this type of approach is the determination of the extent to which root-absorbed radionuclides are subsequently transferred to those parts of the plant consumed by man or animals.

For radiocaesium and radiostrontium there has been further interest in the development and application of models based on potassium and calcium uptake respectively. In many plant species, potassium and caesium interrelationships are not well understood and whereas models can be developed for overall transfer to vegetation, these are not necessarily adequate for individual component species (Coughtrey *et al.*, 1990c).

4.4.6 PLANT-TO-ANIMAL TRANSFER

Considerable work has been undertaken in recent years on the models which deal with absorption and turnover of radionuclides in animals. Multi-compartment models are often applied in radiological assessments to provide time-varying concentrations in different products (Coughtrey, 1990). In several cases it is possible to extrapolate data for one species to another taking into account differences in body weight, metabolic rate and physiology. One of the critical factors in models of this type is the method for treating initial absorption from the gastrointestinal tract, particularly in cases where significant endogenous secretion occurs (Section 4.2.1.2). Current research programmes sponsored by CEC are directed towards categorizing the processes involved in gut absorption and subsequent turnover in animal tissues.

A major uncertainty in many of the existing radiological assessment models is the rate of ingestion of contaminated produce and the nature of its supply (e.g. comparisons between strip-grazing and free-ranging cattle). More attention is therefore being directed towards integrated soil-plant animal models that can allow for a variety of agricultural scenarios.

A further area of development concerns radionuclide turnover in growing animals such as pigs, calves, lambs and chickens. Models similar to those applied to adult animals have been developed in which transfers between compartments are modified according to animal age and weight and in which gastrointestinal absorption is allowed to vary according to age and/or weaning (Cuff, 1989). Such models can provide predictions which appear to be consistent with published data but much more work needs to be done to extend the database for model development and testing.

4.4.7 DOSE ASSESSMENT

The various developments summarized above relate to individual components of dose assessment models which deal with all pathways from source to man. In both waste disposal assessment and accident assessment, much more use has been made in recent years of probabilistic consequence models. For terrestrial pathways, some problems do exist in these approaches mainly because many of the parameters used in the models are correlated. This makes adequate parameter specification (range and distribution) a difficult process. Also, in the context of waste disposal assessment, major uncertainties still remain in the methods that can be used to couple biosphere and geosphere models and in the ways in which changes should be incorporated into biosphere models to simulate, for example, the effects of climate change.

Since models used for radiological assessment often incorporate a large number of potential transfer pathways, comparison between model predictions and measurements requires caution. Without a comparison of predictions and measurements at all intermediate steps in the pathways to man, the potential exists for the model to give an apparently correct estimate of the endpoint but for entirely the wrong reason. Self-compensating errors can apply either within an individual pathway to man or between several different pathways. Errors of this type are apparent in the results of the BIOMOVS intercomparison based on post-Chernobyl data (BIOMOVS, 1991). It is anticipated that exercises such as VAMP will help to elucidate the causes of self-compensating errors.

4.4.8 NATURAL AND SEMI-NATURAL ECOSYSTEMS

In general, natural and semi-natural ecosystems only provide a direct route for transfer of radionuclides to man via critical pathways. However, natural ecosystems also provide a long-term source of radionuclides to other systems. Particularly significant in this context are forests and alpine or upland grasslands, both of which can receive a higher-than-average deposition relative to surrounding land, and both of which have individual characterstics which tend to retain radionuclides within biological cycles. At the present time there are significant developments in the modelling of alpine or upland grasslands. Examples are work being undertaken in the UK (e.g. Mitchell and Davis, 1990; Crout *et al.*, 1990) in which the results of recent surveys of radiocaesium behaviour are being used to construct both retrospective and future scenarios for radiocaesium concentrations in vegetation and animals. Where forests are concerned, model structures have been proposed (Section 4.3.1) but the major impetus at present is in collation and interpretation of available data, much of which is being co-ordinated by VAMP.

4.5 CONCLUSIONS AND RECOMMENDATIONS

Radionuclides of varying half-life and ecological characteristics enter terrestrial ecosystems by many routes and become widely dispersed within the component parts of the ecosystem.

Terrestrial ecosystems are important to man since they provide the main source of his diet. The majority of the existing information on radionuclide behaviour in terrestrial ecosystems is for temperate agricultural ecosystems. In contrast, much less information exists for coastal systems (other than salt marshes), forests, upland and wetlands, and tropical systems (other than atolls). The effects of chemical form of input remain an overriding area of uncertainty in the behaviour of radionuclides in terrestrial ecosystems.

Dry deposition of particles and gases to vegetation or soil surfaces is well documented and relatively well understood. A possible exception is the role of the abiotic layer (mostly soil particles) on vegetation surfaces in capturing and retaining deposited materials. In comparison with pasture grasses, relatively few data exist for other types of vegetation, particularly agricultural crops and forests. In contrast to dry deposition, practically no information exists on the processes involved in wet deposition to either soils or vegetation, or on the subsequent redistribution of wet-deposited radionuclides between vegetation and soils. This is equally true for all radionuclides other than radiocaesium and is important when considering inputs via either rainfall or irrigation waters. Additionally, deposition during cloud cover (occult deposition) or via snowfall can be important and is not well documented.

Until recenty, relatively few data existed for resuspension other than in arid environments. Chernobyl is extending the database for temperate conditions but more information is required on the relative importance of processes such as soil erosion by wind or water, rainsplash, volatilization, and the effects of burning. The latter could be of particular importance for contamination of forests or natural grasslands where fire is a natural ecological process.

The behaviour of radionuclides in soils determines radiological impact via both terrestrial and aquatic pathways. However, some aspects of radionuclide behaviour in soils are still poorly understood, especially for sandy or organic soils. It is currently possible to simulate the redistribution of radionuclides in soil profiles but the degree of availabilty is largely unknown. Additionally, Chernobyl has highlighted the need for a better understanding of particle migration in soils.

Present studies on transfer between soil and plants take more account than previously of chemical and biological availability in soil, particularly for those radionuclides which show irreversible binding to soil particles and hence long-term trends in plant uptake.

The role of micro-organisms in both soil and foliar uptake has not been studied sufficiently in the context of radionuclide behaviour.

Very few studies have distinguished between the individual processes of interception, retention, absorption and translocation in terrestrial plants and few data exist on the dynamics of partitioning within plants following absorption. The extent to which soil contamination contributes to radionuclide concentrations of plants has been determined in only a few studies.

Seasonality is an important factor, both in terms of the time of year at which contamination occurs and in terms of plant and soil concentrations following the contamination event. Present data are only for accidents which occurred either in early spring or autumn—significant deviations can be expected for an event which occurs during summer or winter.

Present studies on plant-to-animal transfer have identified a requirement for a better understanding of processes involved in absorption and excretion within the gastrointestinal tract and the need to separate true absorption from apparent absorption. This issue can be resolved by further kinetic studies following input of radionuclides to blood and by the re-evaluation of existing data from less controlled studies. Though some recent studies have been undertaken with young animals (e.g. lambs, chicks), much more work is required before the behaviour of radionuclides in these animals can be predicted accurately. As with plants, few studies have considered the effects of intakes of radionuclides attached to soil. Also, detailed information is still required on the seasonal variation in dietary components of free-ranging animals to interpret seasonal changes in body tissues and burdens.

Considering specific radionuclides, most information exists for biologically important fission products (^{131}I, ^{90}Sr and ^{137}Cs). Less information exists for other fission products (e.g. isotopes of Ag, Ru, Ce, Zr) and, other than for ^{60}Co, for the majority of activation products (e.g. isotopes of Ni, Mn, Zn and Cd). In respect of releases to atmosphere, the behaviour of ^{14}C and ^{3}H is relatively well documented. However, especially for ^{14}C and for other radionuclides of potential significance in waste disposal (e.g. ^{129}I, ^{99}Tc, U-Th series nuclides) much more information is required, especially if these radionuclides enter soil via groundwater pathways.

In the past decade there have been substantial developments in the modelling of radionuclide transport in terrestrial ecosystems, and a hierarchy of models exists for different parts of the ecosystem and for a number of the processes of importance. Nevertheless many of the models rely on empirical evaluation of existing data without due account of the underlying mechanisms and processes. Such models can rarely be applied to circumstances other than those that applied to the original source data. Models can be developed rapidly relative to the timescale for acquisition of experimental data, hence there are few published examples of rigorous testing of those models that have been developed. In this respect, the model intercomparisons that have been undertaken contain little discussion of the effect of model structure on the end-result, despite the fact that agreement between model prediction and measurement can often be good but for entirely the wrong reasons. More emphasis should be placed on the development of agreed structures which incorporate processes known to be important and, in compartment models, in ensuring that the individual components of net fluxes between compartments have been specified correctly.

Important gaps in our knowledge of the behaviour of radionuclides in terrestrial environments can be summarized as follows.

1. Chemical speciation in relation to bioavailability to plants and animals.
2. The role of micro-organisms in the cycling of radionuclides.
3. The behaviour of radionuclides in forests and other natural and semi-natural ecosystems.
4. The behaviour of 'exotic' radionuclides (e.g. ^{110m}Ag).
5. The behaviour of radionuclides associated with particles.
6. The role of secondary sources of radionuclide contamination (e.g. resuspension).

Future research should be directed at filling these gaps in knowledge and should also include the following topics.

1. Validation of dynamic models for food-chain transfer with particular attention to the reliability of transfer parameters.
2. Development and testing of models for young animals.
3. The long-term behaviour of radionuclides in organic soils.
4. The influence of agricultural practices (e.g. irrigation) on the transfer of radionuclides.

5 Radionuclide Aquatic Pathways

Co-ordinators: **J. Hamilton-Taylor, M. Kelly, P. Kershaw and C. E. Lambert**

Contributors: **A. Aarkrog, D. P. Calmet, S. Charmasson, R. Carpenter, S. Fowler, M. Ivanovich, G. Kuznetsov, S. P. Luttrell, S. J. Malcolm, P. I. Mitchell, H. Nies, B. Patel, G. G. Polikarpov, I. Rjabov, D. Swift and U. Tveten**

5.1 INTRODUCTION

Aquatic environments occupy a major portion of the Earth's surface and, therefore, an understanding of the radionuclide and radiation exposure pathways within such systems is essential. The environments include rivers, lakes, estuaries, shelf seas, deep oceans, ice sheets and glaciers, groundwater and ground-ice. The aquatic environments all consist of an aqueous phase and a solid phase which is mainly sediment (particulates) in surface environments and the host bedrock in groundwaters. Living organisms, which make up the aquatic biota, can be involved in the geochemical cycle of radionuclides and play a role in their phase distribution. Radionuclides are present in the living and non-living components of each aquatic environment, both natural radionuclides of primordial and cosmogenic origin, and artificial radionuclides from nuclear and non-nuclear industrial wastes, accidental releases and nuclear weapons fallout (see sources in Chapter 1). Solid and liquid inputs of artificial radionuclides to the aquatic environments can be direct or indirect. Direct input mechanisms include fallout deposition onto the water surface, liquid discharges and releases from dumped solid wastes. Indirect inputs from secondary sources are also very important, as a result of the remobilization of contaminated material within an environmental compartment, e.g. erosion of soils contaminated by fallout radionuclides in a river catchment or of marine sediments contaminated by nuclear fuel reprocessing wastes. Indirect inputs also occur by transfer between environments of radionuclides in solution and on sediments.

Each radionuclide will be partitioned between the solid and solution phases. A variety of disparate processes may be involved in this partitioning, including sorption by an inorganic or organic environmental sediment particle, or the bedrock for groundwaters, precipitation–dissolution, colloid aggregation–disaggregation, microbial activity, and uptake into and release from the biota. The solid-solution

Radioecology after Chernobyl
Edited by F.E. Warner and R.M. Harrison
 Published by John Wiley & Sons Ltd

partitioning of a radionuclide is a very important parameter describing its behaviour and, except for biological uptake, it can be defined by the distribution coefficient (K_d) or the numerically equivalent ratio (R_d) introduced by NEA/OECD (1983):

$$K_d = \frac{\text{Activity concentration of the solid phase (Bq kg}^{-1}\text{)}}{\text{Activity concentration of the solution phase (Bq l}^{-1}\text{)}} \tag{5.1}$$

The widespread use of the K_d is due to the need for a simple parameter describing solid/solution distribution for the purpose of modelling the biogeochemical distribution of radionuclides. The concept of K_d implies true equilibrium and reversibility, and that the solid/solution ratio has no effect on K_d due to sufficiently low radionuclide concentrations. In practice, the first two requirements are frequently not demonstrated. Variations in K_d are likely to occur not only because of possible biological and colloidal effects but also due to changing solution and sediment chemistries and to the non-attainment of equilibrium.

K_d values quoted in the literature have been determined experimentally or from observation of the partitioning in the environment. Distribution coefficients vary by 9 orders of magnitude between different nuclides and 3 orders of magnitude for any particular one. This variation depends mainly on the solution composition and the nature of the solid substrate. It can be modulated by non-equilibrium conditions, resulting from slow kinetics, and by the effects of colloid or biological processes. Especially important in the aquatic environment are the differences in K_d due to the change in solution chemistry from seawater to freshwater, with K_d generally, but not always, higher in freshwater than seawater (Table 5.1). The fraction of the radionuclide present in each phase is often expressed in terms of the K_d and the solid (sediment) mass concentration (C_p), either as the ratio of the activity in the particulate fraction to that in the solution fraction (F_a):

$$F_a = K_d \times C_p \tag{5.2}$$

or by the percentage of radionuclide in the particulate fraction (P_p), per unit volume of suspension:

$$P_p = \frac{100}{[1 + 1/(K_d \times C_p)]} \tag{5.3}$$

Sediment concentrations vary widely in aquatic environments, both spatially and temporally, e.g. from <1 to >1000 g m^{-3}. Characteristically, they are highest in high-energy environments such as shelf seas, tidal estuaries and rivers, especially during events which have low frequency of occurrence, such as storms. This variation means that the relative importance of the solid and solution phases in terms of radionuclide behaviour also varies widely in space and time. At one extreme, with high particulate concentrations, a large proportion of the radionuclides will

Table 5.1 A comparison of K_d values[a] in freshwater and seawater

Element	Freshwater	Seawater
I	3×10^2	10^1
Na	10^2	10^1
Ru	10^2	5×10^3
Sr	10^3	10^2
Cs	10^4	2×10^3
Pu	10^5	5×10^4
Lanthanides	5×10^6	5×10^5

[a] Values are the best estimates from Coughtrey *et al.* (1985).

be particle associated, even for low K_d elements. At the other extreme the converse will be true.

The chemical characteristics of the water are important in determining the radionuclide behaviour and K_d. Radionuclide species in solution may depend on the ionic composition and ionic strength of the water, the presence of organic ligands, the redox state (Eh) and acidity (pH); an important contrast in nuclide behaviour is thus observed between aerobic and anaerobic waters. The major differences in ionic composition and ionic strength existing between saline water and freshwater, and within the latter between alkaline and acid waters, play an important role in determining radionuclide speciation and behaviour. Also, individual radioelements vary widely in the complexity of their solution chemistry, e.g. Pu may exist as three or four oxidation states and many ionic species, whereas Cs is a single monovalent ion.

The sediment grains (and bedrock grain surfaces) differ in their potential for adsorbing radionuclides, which results in a K_d variation with lithology, i.e. grain size and composition. In addition, the nature of the grain surface can be important, i.e. presence of organic and Fe/Mn oxide coatings. Minerogenic sediment grains, mainly silicate minerals or rock fragments, are derived from the crustal rocks and their derivative soils and sediments. These are eroded by rivers, in shallow seas and estuaries and are transferred from one aquatic environment to another. A proportion of this minerogenic matter is provided via atmospheric deposition. Another component is of biogenic origin, produced partly in the aquatic environment and partly introduced by terrestrial erosion. This includes skeletal minerals (mainly carbonates) and organic matter. Chemical precipitates form a hydrogenous component and can be locally important. Direct anthropogenic inputs of sedimentary material also occur, e.g. sewage solids. In addition, the solid phase may be primary, e.g. fuel debris. The size and density of these sediment grains must be taken into account for determining radionuclide behaviour. Typically, there is a wide range of grain sizes present in sediments. These include mud (silt + clay), which extends in size from an arbitrary boundary with the 'solution' phase (usually 0.2 or 0.4 μm) up to 0.06 mm, sand (0.06–2 mm), gravel, etc. (> 2 mm). In addition, the presence of colloids of sub-micrometre size can be important for radionuclide pathways. Because of the increase in surface area per unit mass as grain size decreases, the fine-grained sediments have a higher adsorptive capacity for radionuclides than coarse, i.e. muds have higher activity concentrations than sands. Grain size also determines the potential mobility of the radionuclide particulate fraction. In terms of their dynamic behaviour, sediments are grouped into granular sediments (sand and gravel) and cohesive sediments (mud). The grain size is broadly related to the energy of the environment. In many situations, the fine-grained sediment is present not as individual grains but as biological aggregates (faecal pellets) and chemical flocs, which substantially change its dynamic behaviour. Occasionally, to facilitate comparison of activity concentrations of sediments with varying grain size, the concentrations are normalized to the Al, Sc or ^{40}K content. This assumes that these elements are mainly contained in clay minerals and that their average concentration in these minerals remains roughly the same. These assumptions may not be valid in shallow marine and terrestrial aquatic environments, especially in glaciated terrains where Al- and K-bearing minerals may be common in coarse sediments (e.g. feldspars).

The characteristic features of radionuclide behaviour in the aquatic environment are the redistribution by transport of the solution and solid phases, the chemical interactions between the phases, and their biological cycling. Transport, in addition to leading to radionuclide redistribution, results in their dilution, fractionation and mixing, as well as vitally affecting the residence time in the aquatic environment. In addition, in the surface environments, accumulation of the particulate phase occurs by sediment deposition (sedimentation). This induces post-depositional chemical and physical changes, called diagenesis, which affect the radionuclide distribution.

The transport of the water results from a number of driving mechanisms, all ultimately a response to gravitational forces, modified by Coriolis and friction forces. Water velocities vary widely in space and time in aquatic environments, from as low as $\mu m\ s^{-1}$ in groundwaters to $m\ s^{-1}$ in rivers, tidal seas and estuaries and, also, in deep ocean turbidity currents. A basic contrast exists in surface waters between current regimes that, at least occasionally, are strong enough to erode sediment from the bed and those which are predominantly weak. The former include river flow, tidal currents and wave base oscillatory currents and the latter, ocean circulation, coastal and estuarine saline density currents and water surface elevation compensating currents, set up by wind shear and wave drift in coastal waters and lakes. Weak currents, however, can be important for transport of sediment introduced to the water column by another process. Transport of radionuclides occurs in these circulation/current/flow systems as a result of advection and dispersion. Advection is produced by the time averaged flow of water. Dispersion is due to a number of processes: molecular diffusion, turbulent eddy diffusion and dispersion due to velocity shear, i.e. the spreading that occurs in the direction of flow as a result of the vertical (and lateral) velocity gradients. The combined effects of the dispersion processes are described by dispersion coefficients applicable to the perpendicular axial directions: D_x, D_y, D_z. The magnitude of the dispersion coefficients varies with velocity, turbulence intensity and secondary characteristics of the different aquatic environments. Transport (flux) of a radionuclide by dispersion processes is related to the magnitude of the concentration gradient and the dispersion coefficient and takes place in the direction of the gradient.

Sediment and associated particulate phase radionuclides respond to the same advective and diffusive circulation processes as the solution phase, resulting in their transport and dilution. However, the sediment and particulate phase response is fundamentally different because a velocity-related threshold (bed shear stress) has to be exceeded before transport occurs, whereas this is not the case for the solution phase. Also, above this threshold, the concentration of mobile particles and their vertical distributions in the flow are dependent partly on velocity-related flow characteristics such as bed shear stress and turbulence intensity. These complex relationships to the flow depend on particle size and density. The state of aggregation of fine particles has a very important influence on this behaviour. As the capacity of the flow to transport sediment decreases, in space or time, above or below the erosion threshold, the excess sediment is deposited on the bed. However, because it takes a finite time for grains to settle through the water column there is a lag between the change in flow and deposition, which can have considerable importance for sediment behaviour. Settling rates vary from $\mu m\ s^{-1}$ to $cm\ s^{-1}$. Sediment can be transported either in suspension (suspended load), at velocities comparable to that of the water, or in contact with the bed (bedload) at a fraction of water velocity, as mobile bed forms such as ripples, dunes and bars. The sediment in suspension may be actively suspended by turbulence or may settle passively through the water after being introduced by another mechanism. The range of velocity regimes that can be found in aquatic environments means that either one mode of sediment behaviour may predominate, e.g. grain settling in oceans and lakes, or conditions may vary in space or time between various transport modes, e.g. in rivers. An important modifying factor is the availability of sediment: supply limitations may over-ride flow-imposed limits on sediment transport. Net deposition of sediment and associated radionuclides occurs where rates of sediment supply exceed those of transport. Such sediment deposits are extremely important as sinks or reservoirs for radionuclides. The radionuclide inventories preserved in aquatic sediment deposits vary widely, depending on radionuclide sources and on the deposit characteristics, such as lithology, sedimentation rates and residence times. The grain size dependence of sediment transport processes leads to the fractionation of sediment inputs and the production of deposits of different grain size and, hence, of different radionuclide concentrations. Certain environments are characterized by the lithological uniformity of their deposits, e.g. the mainly fine muds of oceans and lakes, whereas others, typically, have a more diverse range of sediment types depending on the local current regime, e.g. rivers and estuaries. Sediment accumulation (sedimentation) rates vary widely, from < mm y^{-1} to > m y^{-1} and usually correlate with grain size, with low rates in oceans and lakes and variable rates in other environments.

The sedimentation rates can be determined from radionuclide profiles. This can involve conventional radiometric methods based on the decay of an unsupported natural or artificial radionuclide such as ^{14}C or an unsupported fraction, e.g. excess ^{210}Pb. Alternatively, artificial radionuclide profiles can be matched with the record of release of the nuclide, e.g. ^{137}Cs from weapons test fallout and, increasingly of use in the future, from the Chernobyl accident. The application of these methods can be affected by processes such as diagenetic, chemical or physical redistribution of the radionuclide and, for artificial radionuclides, the effects of processes which modify the relationship between the initial release to the environment and the signal received in the sediment, such as mixing of sediments labelled at different times. The residence times for sediment deposits also vary widely, from >100 to < 1 y. In short-lived deposits, sediment remobilization leads to reintroduction of particulate radionuclides to the water column. If subaqueous, the deposits are saturated and the pore waters will contain solution phase radionuclides.

In contrast to the high energy environment of coastal zones, transport of radionuclides in the open ocean is generally mediated by biological cycling. The various materials that have adsorbed or incorporated these nuclides, such as biogenic debris and clay, are packaged by zooplankton and can be transferred to the sediment in a few months. Other biological processes, such as aggregation of phytoplankton material, also lead to a rapid transfer of radionuclides to great depths. In the water column, and sediment and bedrock porewaters, the biogeochemical interactions listed earlier can lead to repartitioning of radionuclides between the solid and solution phases, in response to changes in the chemical environment and/or to biological mediation. Particularly important in surface waters are sorption–desorption reactions involving the sediment-bound radionuclides, caused by changes in salinity or redox state, or to the introduction of unlabelled or partially labelled sediment. In addition, physical change in the particulate phase can lead to the same result, e.g. by colloid aggregation/disaggregation or by organic particle degradation. Gain by the solution phase will increase the mobility of a radionuclide.

Transport by diffusion can occur also across sediment/water column interfaces, with loss or gain of activity by the deposits. Other post-depositional (diagenetic) processes in sediments can lead to the physical disturbance or modification of the original sediment and include the effects of biological disturbance by organisms, called bioturbation, and physical processes such as slumping. The effects of bioturbation depend on the number, size and habits of the organisms; they include random mixing of sediment or grain size sorting. These processes also affect the chemical behaviour of radionuclides, mainly via oxygenation of interstitial waters in the burrows and changes in alkalinity due, for instance, to carbon consumption.

Advective movement of porewaters and solution phase radionuclides can occur in sediments (in addition to bioturbation) as a result of consolidation due to the lithostatic stress generated by the overlying sediment. This results in an increase in sediment bulk density, decrease in pore volume and upward migration of porewater. In sediments, and importantly in groundwaters, porewater flow is also a response to regional hydrostatic gradients.

Radionuclide uptake by biota occurs by a number of mechanisms from both solution and particulate phases. Uptake by the primary producers, e.g. phytoplankton, occurs from solution by surface adsorption and metabolic processes. Surface contamination by particulates can also occur with macro-algae. The primary uptake mechanism for invertebrate and vertebrate organisms is ingestion of food. However, for the many invertebrates which are detrital filter- and sediment-feeders, this directly involves particulate radionuclides in general. Respiration also involves intake of solution phase (and particulate) radionuclides. Radionuclide contamination of terrestrial organisms can also occur by feeding on aquatic organisms. Uptake will depend not only on the organism concerned but also on the element involved and its activity concentration. The ratio between the activity concentration in the organism and aquatic environment is defined as the concentration factor.

These different processes affecting the behaviour of radionuclides in aquatic media are illustrated in the following sections.

5.2 FRESHWATERS

5.2.1 TRANSPORT AND DYNAMICS IN RIVERS

5.2.1.1 Solution phase

Solute transport in rivers is generally described in terms of a one-dimensional, partial differential equation, incorporating advection and eddy diffusion terms, although alternative statistical or systems approaches are receiving increasing attention (Young, 1990). Because of the complex nature of river channel geometry, transport models require detailed calibration. Artificial radionuclides can be used as tracers for studying river flows and for model calibration. Tritium is ideal for this purpose but is rarely used in practice. The main reason for this is the generally dispersed nature of the source term (bomb fallout), with a well-defined point source being the exception. An example of the latter is where tritiated water from the Grafenrheinfeld nuclear power plant has been used to define the dispersion characteristics of a 320 km section of the Main River, Germany (Krause and Mundschenk, 1989). Flow times, flow velocities and longitudinal dispersion (eddy diffusion) coefficients (20–200 $m^2\ s^{-1}$) were determined as a function of river discharge from 0.4 to 4 times the mean runoff.

5.2.1.2 Sediments

While there is an extensive literature on the relations between the hydraulic characteristics of flow and sediment particle behaviour, there is no coherent, mechanistically based approach to describing sediment transport and dispersion in river systems. A major contributing reason is that sediment transport by rivers is subject to non-hydraulic as well as hydraulic controls. Important non-hydraulic factors include the geology and soils present in the catchment, catchment topography, hydrology, land use and vegetation cover. A further complication is that many of the hydrological factors are stochastic in nature. These include storm duration and spatial effects, rainfall intensity, and antecedent discharge, all of which influence storm-period particle transport. There are also significant stochastic processes occurring within the river channel, e.g. an important source of sediment supply is through bank collapse.

Studies of artificial radionuclides in rivers have frequently highlighted the complex nature of sediment transport and the individuality of rivers. The Great Miami River, Ohio, has been studied over a 25-fold range of river flows and a 6-fold range of particle concentrations (Sprugel and Bartelt, 1978). The activity concentrations of ^{239,240}Pu in solution (Bq m^{-3}) and in suspended particles (Bq kg^{-1}) did not correlate with either river flow or suspended particle concentration. The total Pu inventory was largely a function of suspended particle concentration. The mean concentration of ^{239,240}Pu in riverborne suspended sediment was 2 to 3 times that in the source material (arable soils containing fallout Pu). The enhanced activities were attributed to size fractionation in the catchment-river system, resulting in a greater fine-grained component in the river. Studies of periodic, pulsed inputs of plutonium to the Great Miami River from the Mound Laboratory in Miamisburg, Ohio, have also given important insights to the dispersion behaviour of Pu (Muller *et al.*, 1977). Mass balance calculations showed that under typical summer flow conditions, ~60 per cent of the effluent is lost through sedimentation within 10 km of the discharge point, and that resuspension of this material between pulses maintains a high 'background' ^{238}Pu flux in the river. The remaining ~40 per cent is transported downstream with each pulse.

Further understanding of the temporary nature of channel sediments comes from a study of the river Danube at Bezdan, near the Yugoslav–Hungarian border (Conkic *et al.*, 1990). The concentrations of ^{137}Cs, ^{134}Cs and ^{106}Ru in bottom sediments were studied following the Chernobyl accident. Decay-corrected activities showed exponential decreases with time that were the same for all three radionuclides, despite expected differences in chemical behaviour. This highlights the dynamic nature of the sediment deposits and suggests that the relative importance of physical processes was the same for all three radionuclides. Deployment of a sediment trap, higher up the Danube, at Vienna, provided evidence not only of a gradual reduction in Chernobyl activity with time in newly deposited sediments, related to the gradual downstream migration of the most contaminated sediments, but also of marked seasonal variations in activity (Maringer *et al.*, 1989). As expected, the total activity content was influenced by the grain size composition of the sediments, which in turn was related to river discharge. The activity concentration of the < 20 μm fraction was between two and four times that in the bulk sediment. High discharges, related to the passage of spring meltwaters, resulted

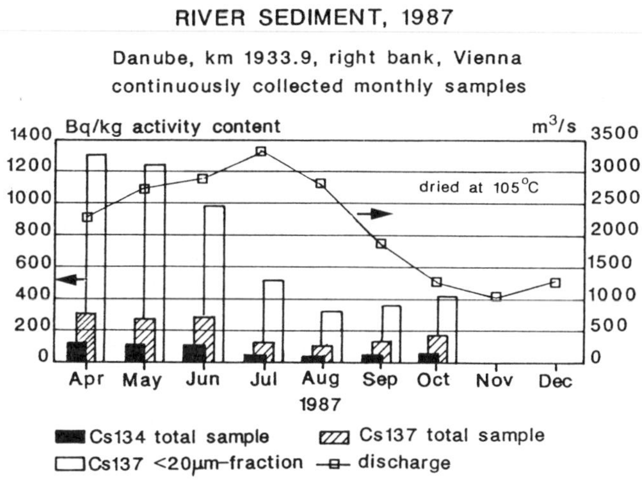

Figure 5.1 Relation between river discharge and the seasonal trend in radiocaesium activities in total sediment and the <20 μm fraction. From Maringer *et al.* (1989).

in a lowering of the ^{137}Cs activity in the newly deposited sediment, due in part to an associated coarsening in grain size but also due to a decrease with time in the radionuclide content of the fine-grained fraction (Figure 5.1). The subsequent fall in river flow produced a fining of the sediment in transit and a consequent increase in its radionuclide concentration.

Studies, using Chernobyl-derived radiocaesium as a tracer, have also begun to tackle some of the more intractable aspects of the dispersion of riverborne sediments, such as those related to bedload transport and the formation of floodplain deposits (e.g. Walling and Bradley, 1988).

5.2.2 TRANSPORT AND DYNAMICS IN LAKES AND RESERVOIRS

5.2.2.1 Solution phase

Mass transport of a solute through and within a lake may also be described in terms of advection and eddy diffusion, but the complexities of water motions are far greater than those in rivers. Dynamic processes can be classified according to origin: wind-induced currents including seiches, inflow-induced currents, and convective currents. For large lakes there are, in addition, significant currents produced by Coriolis force, gravitational forces, frictional forces, and the effects of meteorological variations. The need to take account of the various types of current likely to be discernible and the distinction between the advection and eddy diffusion terms are very much dependent on the time and length scales of interest. A particularly important feature of lakes is the degree and temporal dependency of vertical density stratification, which generally is related to temperature but can be related to differences in dissolved salt content (i.e. in meromictic lakes). Transport by molecular diffusion is rarely important in the water column, except possibly associated with the boundary layer (~1 mm) overlaying the bottom sediments and vertically in the thermocline region.

Artificial radionuclides have rarely been used in studies of water movements in lakes. A contributing factor is that the complex temporal variation of the input function of fallout ^{3}H, the radionuclide most frequently used in such studies, occurs on similar timescales to typical mixing times in lakes, thus greatly complicating interpretation. In a study of Lake Tahoe, no significant variation with depth in fallout concentrations was observed, despite it being a deep lake (max. 500 m) with a long hydraulic residence time (Imboden *et al.*, 1977). Far greater success has been achieved through employing experimental injections of tritiated water (Quay *et al.*, 1980).

A more widely applicable approach is to measure the parent–stable daughter tracer pair, ^{3}H–^{3}He (Torgersen *et al.*, 1977). The method is based on the fact that during periods of isolation, water bodies build up an excess of ^{3}He, relative to its equilibrium concentration with respect to the atmosphere, due to the *in situ* decay of ^{3}H. Consequently, an effective water mass age can be determined and this has been used to calculate (a) gas exchange rates between surface waters and the atmosphere or, more specifically, the piston velocity and hence the surface boundary layer thickness; (b) the extent of degassing or gas renewal at turnover; (c) vertical diffusivities in the well-mixed surface waters; (d) transport across the thermocline during stratification; and (e) renewal times for water masses at different depths. Vertical eddy diffusivities of 0.6–3.6 $cm^2\ s^{-1}$ have been estimated for the well-mixed surface waters of Lakes Erie, Huron and Ontario, and an apparent vertical diffusivity of 1 $cm^2\ s^{-1}$ for the thermocline region in Lake Constance. The data indicate also that, whereas degassing of the deep waters is virtually complete during turnover in the three Great Lakes, a degassing rate of only ~50 per cent per year occurs in the deep waters of Lake Constance.

Accidental releases of elevated ^{3}H from nuclear power plants have been used as tracers. For example, releases from Douglas Point and Bruce A power stations in Ontario have been used to study the counter-clockwise circulation of Lake Huron (Veska and Tracy, 1986).

5.2.2.2 Sediments

In contrast to rivers, lakes are generally regarded as being efficient and permanent sediment traps, related principally to their greater depths, smaller currents and longer hydraulic residence times. Thus lakes are short-lived on the geological timescale, disappearing from the landscape principally through the action of sedimentation. In all but very shallow lakes, bottom sediments away from the high-energy shoreline region are normally subjected to low current velocities and negligible wave action. The transport and fate of sediments within lakes, however, is far from simple because of the many competing processes regulating erosion, transport and deposition. In areas dominated by river action, grain size and the rate of sedimentation generally decrease logarithmically with distance from the river mouth. The distance over which river flow is important varies greatly, depending on factors such as basin topography, stratification, and river discharge. At a distance from the river mouth, wind and wave action become the dominating influence. The associated rate of net sedimentation increases with water depth from zero at some intermediate depth, due to sediment focusing, and involving the resuspension of shallow-water sediments and their transport to deeper waters.

The transport of sediment particles in lakes is studied predominantly through an examination of bottom sediments, rather than direct observation in the water column. Artificial radionuclides with high particle affinities have proved to be useful as tracers in such studies, especially as dating tools providing a measure of sedimentation rates. The first appearance of fallout ^{137}Cs, and to a lesser extent ^{239,240}Pu, in the early 1950s and peak fallout in 1963 have become particularly important stratigraphic markers in lake sediments. In Europe, ^{137}Cs resulting from the Chernobyl accident has become a useful additional time marker (1986). Edgington and Robbins (1975) employed the bomb fallout markers, together with ^{210}Pb dating, to determine sedimentation rates in Lake Michigan and found good agreement between the three methods. The sedimentation rate in the southern part of the lake varies by more than an order of magnitude from virtually zero in large areas of the western side to >50 mg cm^2 y^{-1} in areas close to the eastern shore, where several large rivers discharge (Figure 5.2a). Radionuclide fluxes at the sediment surface were normalized to fallout fluxes at the air–water interface, which represents the predominant source term. A map of the distribution of this function provides a measure of the relative sediment loss or gain at any point (Figure 5.2b). Particularly high sedimentation occurs offshore to the north and west of the major rivers, whereas bottom topography slopes continuously down to the central axis of the lake. No sedimentation occurs at water depths <50 m, even near the river mouths. This pattern of deposition, which is very different from the classic picture of focusing and delta formation described above, was attributed to the hydrodynamic characteristics of the lake and in particular the pattern of lake-wide currents.

More than 95 per cent of the ^{137}Cs and ^{239,240}Pu that has entered southern lake Michigan as fallout now resides in the sediments, indicating that the efficiency of sediment trapping is very high. This is not always the case, however, and studies with artificial radionuclides have provided data to assess the efficiency of trapping and its main controlling factors. Fallout ^{239,240}Pu has been studied in the sediments of seven lakes in eastern Ontario, Canada, with a large range of mean depths

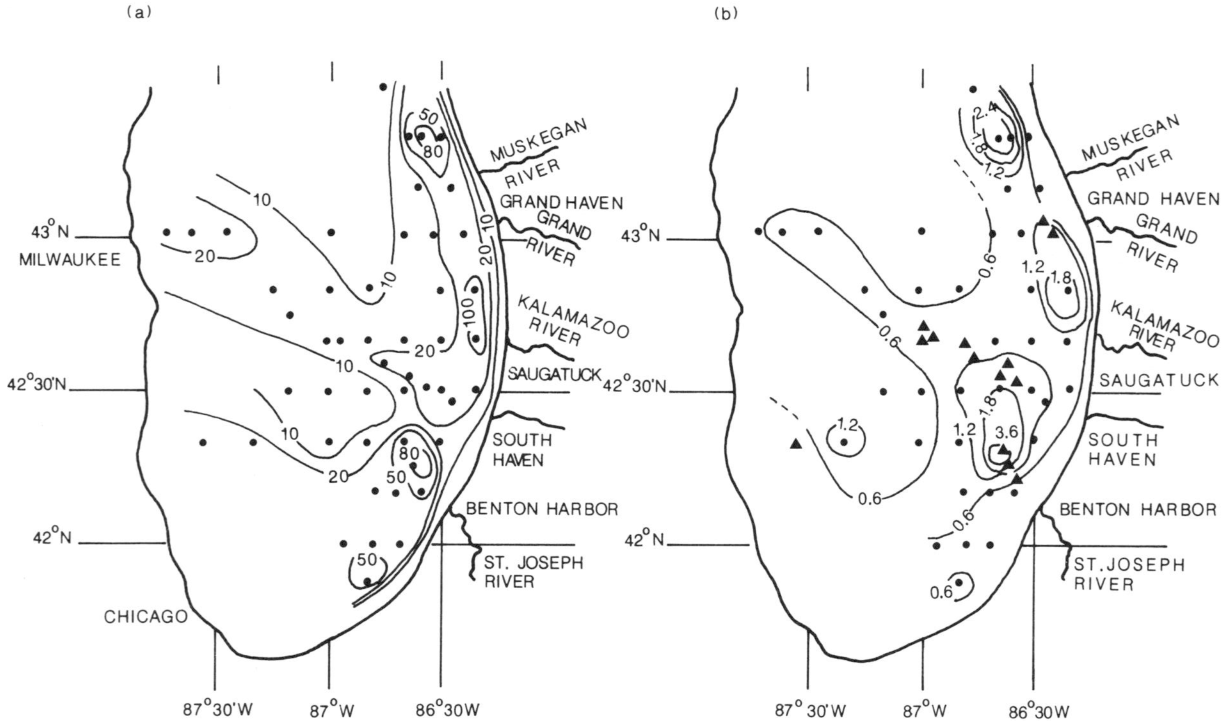

Figure 5.2 Sediment characteristics in southern Lake Michigan: (a) mass sedimentation rate (mg cm^{-2} y^{-1}); (b) flux normalization factor for ^{137}Cs (sediment flux/flux at lake surface). From Edgington and Robbins (1975); reproduced by permission of the IAEA.

(2 to 19 m) and hydraulic residence times (0.03 to 9 y) (Cornett and Chant, 1988). From 28 to 100 per cent of the ^{239,240}Pu input to the lakes was retained in the sediments, with the percentage retained being correlated with the hydraulic residence time. In other words, in the lakes with short hydraulic residence times, a significant fraction of the Pu and hence sediment inputs must have remained in the water column long enough or have been resuspended frequently enough to be lost via the outflow. The data also indicated the absence of any sediment focusing, which the authors linked to the shallow nature of the lakes.

Similar situations exist in Lake St Clair, located between Lakes Huron and Erie, and in a number of man-made reservoirs in the Susquehanna river valley. The latter study (Donoghue *et al.*, 1989) was based on a range of particle-associated tracers (^{60}Co, ^{134}Cs, ^{137}Cs and ^{65}Zn), derived from the nuclear power plants at Three Mile Island and Peach Bottom. It is apparent, therefore, that shallow lakes with short hydraulic residence times may be regarded as being intermediate in character between river channels and more typical lake basins in terms of their sediment dynamics.

Artificial radionuclides have also been used as sediment tracers to study other, diverse sedimentological problems. One example involves the sedimentological regime that exists at the eastern end of Lake Erie, which is a high-energy region with a substrate comprising bedrock and coarse-grained sediment. Fine-grained sediment is thought to be transported northwards to the Niagara river, and subsequently Lake Ontario, by the dominant advective flow in this part of the lake. An important radionuclide source, the Western New York Nuclear Service Center (WNYNSC), is located about 50 km south of Buffalo, NY. Water from the site eventually drains into Cattaraugus Creek and thence Lake Erie. Radionuclide inventories and ratios, involving ^{238}Pu, ^{239,240}Pu, ^{241}Am, ^{54}Mn, ^{60}Co, ^{65}Zn, ^{106}Ru, ^{134}Cs and ^{137}Cs have been used to demonstrate the presence of WNYNSC radionuclides in Lake Ontario sediments off the mouth of the Niagara River (e.g. Joshi, 1988). In this region, about 36 per cent and 80 per cent of the 1982 sediment inventories of ^{239,240}Pu and ^{241}Am respectively, are derived from WNYNSC, the remainder coming from atmospheric fallout. The studies also show that nearly all the Pu, ^{241}Am and ^{137}Cs activity, derived from WNYNSC, ends up in Lake Ontario sediments around the mouth of the Niagara River. These findings are an important illustration of the highly specific nature of artificial radionuclides as tracers. More particularly, they demonstrate that transport processes in surface waters can result in locally concentrated radionuclide activities at sites remote from source, even when separated by a lake basin, which would normally be associated with efficient removal of particle-associated radionuclides.

Another example involves the use of ^{137}Cs and ^{210}Pb profiles in sediments to understand the processes operating in a series of small arctic lakes with depths < 10 m (Hermanson, 1990). Interpretation is based mainly on the inventory in each core relative to decay-corrected fallout. Sediment redistribution processes are shown to vary widely over short distances and include ice rafting effects.

5.2.3 TRANSPORT AND DYNAMICS IN GROUNDWATER SYSTEMS

5.2.3.1 Introduction

The occurrence of groundwater may be divided into zones of aeration and saturation. The former zone consists of pore spaces occupied partially by water and partially by air. In the latter zone all pore spaces are filled with water under hydrostatic pressure. Over most of the land masses of the earth a single zone of aeration overlies a single zone of saturation and extends upward to the ground surface. In the absence of overlying impermeable strata, the upper surface of the zone of saturation is the water table. This is defined as the surface of atmospheric pressure and would be revealed by the level at which water stands in an open well. Water within the ground moves downwards through the unsaturated zone under gravity, whilst in the saturated zone it moves in a direction determined by the local hydraulic situation. Most natural groundwater discharge occurs as flow into surface water bodies but also occurs via the land surface, including transpiration from vegetation.

Due to advection, non-reactive solutes are carried at an average rate equal to the average linear velocity of the groundwater. However, there is a tendency of the solute to spread out from the path expected due to the advection hydraulics of the flow system. This is called hydrodynamic dispersion and causes dilution of the solute. It occurs because of mechanical mixing during fluid advection and because of molecular diffusion. The latter is significant only at low velocities. Dispersion caused entirely by the motion of the fluid is known as hydraulic dispersion. Dispersion is a mixing process and, qualitatively, has a similar effect to turbulence in surface-water regimes. For porous media, the concepts of average linear velocity and longitudinal dispersion are closely related. Longitudinal dispersion is the process whereby some of the water and solute molecules travel more rapidly than the average linear velocity and some travel more slowly. Thus, the solute spreads out in the direction of flow and declines in concentration.

Generally, tritium can be used to indicate the presence of young groundwaters (less than 30 years). At high continental latitudes, waters having more than about 4 TR fall in this category, while in low and middle marine latitudes the limit is nearer 1 TR. As a general guideline, it can be said that waters containing over 10 TR contain a thermonuclear test contribution, while 20 TR or more would suggest a component of groundwater recharged since 1961 (IAEA, 1983). On a local scale, ^{14}C can be used as a tracer of transfer processes from surface systems to either plants or groundwater. A large number of ^{14}C groundwater studies exist in the scientific literature (e.g. Fontes and Garnier, 1979). ^{36}Cl has virtually ideal properties as a tracer for solutes in groundwater and soil water. ^{36}Cl from the Gnome event near Carlsbad, New Mexico (the first nuclear detonation of the Plowshare series, detonated 10 December 1961) illustrates how ^{36}Cl can be used to study the redistribution of radionuclides in the soil profile (Phillips *et al.*, 1990).

Atmospherically derived ^{85}Kr activity is used in hydrology to identify admixtures of young groundwaters recharged during the last 35 years. The use of ^{85}Kr is particularly attractive in combination with ^{3}H, which has a similar half-life but a different history of release.

5.2.3.2 Tracing applications

There are a number of examples where tritium contamination from a localized source has been used to trace groundwater movement and to study surface water–groundwater interactions. Liquid-waste effluents are routinely discharged to the ground at the Hanford Site in Washington State (see site description in Chapter 2). Adsorption, chemical precipitation, and ion exchange attenuate or delay the movement of some radionuclides, such as ^{90}Sr, ^{137}Cs, and ^{239,240}Pu. Other radionuclides, such as tritium, ^{99}Tc, and ^{129}I, are not as readily retained by the soil. These radionuclides move through the soil column at varying rates and eventually enter groundwater. Radionuclide concentrations are reduced by dilution when they reach groundwater. The more mobile constituents move downgradient in the same direction and at a rate nearly equal to groundwater flow.

The maximum extent of radionuclide contamination in the groundwater beneath the Hanford Site can be defined using tritium. Figure 5.3 shows the distribution of tritium in the unconfined aquifer during 1989 (Evans *et al.*, 1990). The highest groundwater tritium concentration measured in 1989 was over 5×10^{6}pCi l^{-1} in a well in the 200-East Area. The large tritium plume from the 200-East Area moves generally east to the Columbia River, where it discharges. Separate tritium pulses, associated with two major episodes of the Plutonium–Uranium Extraction Plant (PUREX) operations, can be distinguished as lobes with concentrations exceeding 2×10^{5} pCi l^{-1}. The lobe near the Columbia River is a result of discharges to ground during the operation of PUREX from 1956 to 1972. The narrow lobe extending several kilometres south-east of the 200-East Area represents discharges from PUREX between 1983 and 1989.

The travel time of contaminated groundwater from source areas to the Columbia River and the mass of radioactive contamination that is reaching the river have not been determined with a high degree of accuracy. A recent estimate by the US Geological Survey (1987) stated the average travel time is in the range of 10 to 20 years. Contaminants also enter the river along the Hanford Reach as direct effluent discharges. In 1989, the average tritium concentration upstream of the Hanford Site was 63 pCi l^{-1}, while the average downstream concentration was 129 pCi l^{-1}, thus substantial dilution occurs when groundwater, containing tritium at concentrations greater than 2×10^{5} pCi l^{-1} discharges into the river.

Seasonal rises in the Columbia River stage result in bank storage, which affects groundwater levels and contaminant concentrations in wells near the river. In the eastern part of the site, where tritium exceeds 2×10^{5} pCi l^{-1}, the concentration of tritium in groundwater from wells near the river has been documented to fluctuate

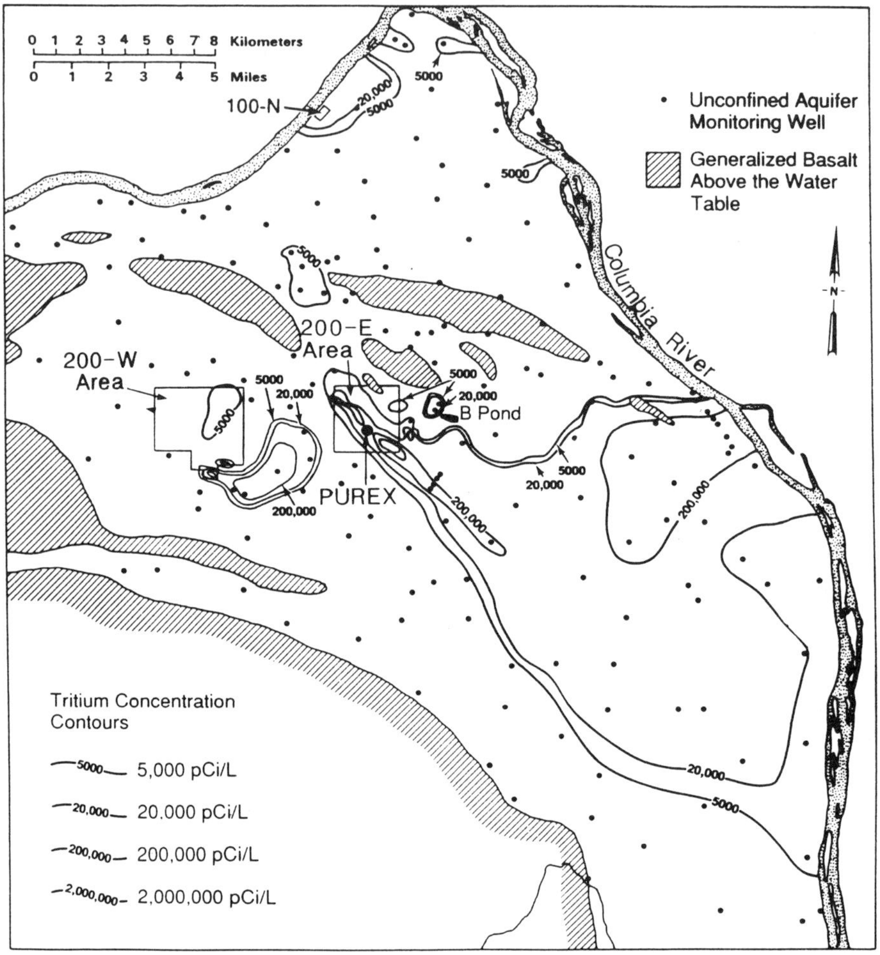

Figure 5.3. Distribution of tritium on the Hanford Site. From Evans *et al.*

as much as a factor of four over the course of a year. The highest concentrations correspond to periods of low river stage. The tritium concentrations are lower when the river stage is higher, during which times bank storage effects from the river dilute the tritium concentrations in the groundwater system.

Another radionuclide migration project was started in 1974 at the Nevada Test Site to determine the potential for movement of radioactivity away from underground nuclear explosions (Ogard *et al.*, 1988). The first field experiment in this project was a long-term single-well pumping test in which the activity of the nuclear explosion was treated as a slug-injection point and the explosion products as tracers. The radionuclides detected in the pumped water from the saturated zone of an alluvium aquifer were ^{3}H and ^{36}Cl. The elution curve of ^{36}Cl preceded that of ^{3}H, and the observed phenomenon was ascribed to an anion exclusion process (Thomas and Swoboda, 1970). Anions, being of the same charge as clays and zeolites in the soil, are repelled by these surfaces and are effectively prevented from entering into the intragranular porosity of the matrix.

A pulse of tritiated water, which was discharged accidentally from an isotope processing plant in the Glatt River Valley, northern Switzerland, was traced through a sewage treatment plant and various rivers and groundwater wells (Santschi *et al.*, 1987). Tritium concentrations were used to test predictions for the transport of conservative anthropogenic trace contaminants accidentally discharged into the sewage system. Mass balance calculations indicated that about 2–10 per cent of the tritium pulse infiltrated the groundwater and about 0.5 per cent of the total pulse reached eight major drinking-water wells in the area. In spite of the complex hydrology of the lower Glatt River valley, tritium breakthrough curves could be simulated effectively by a model developed from an experimental well field.

The accident at the nuclear power plant at Chernobyl also provided an opportunity to investigate the infiltration and migration behaviour of radionuclides in the Glatt River valley. The radionuclides ^{99m}Tc, ^{103}Ru, ^{131}I, ^{132}Te, ^{134}Cs and ^{137}Cs were measured several times during May 1986 in the river and an adjacent shallow aquifer (Waber *et al.*, 1987). The main radioactivity (> 75 per cent) of the river water was found to pass through a 0.05 μm filter. Iodine, Ru and Te were found in the groundwater as a result of river water infiltration, being subject to only slight sorption on particulates and other solid surfaces. This was explained to be the consequence of having formed anionic or neutral species. In contrast, Cs was retained completely by river sediments. Particulate (> 0.05 μm fraction) infiltration from the river into the groundwater system was found to be a negligible process.

5.2.3.3 Dating applications

Determination of groundwater residence time from radionuclide data is possible only if the initial concentration is known and the radionuclide contents do not change as a result of mixing with isotopically different waters. Increases in content and changes resulting from solution/deposition processes normally preclude estimation of residence times. The estimation of the initial content in the groundwater is difficult although, in principle, a plot of radionuclide content as a function of distance along flow lines in a piston-flow system may permit deduction of the initial concentration and thus, interpretation of changes in terms of decay of the radionuclide through time. The closed system requirement in a groundwater system, however, is the most difficult to satisfy.

The measurement of the parent–stable daughter pair, $^3H/^3He$ allows the calculation of groundwater age. Tritiogenic ^3He is added to the natural ^3He content of the groundwater, and if it is assumed that no tritiogenic ^3He is lost by diffusion across the groundwater table, the $^3H/^3He$ groundwater age is given by:

$$\tau = t_{1/2}/\ln 2 \times \ln(2 + [^3He]/[^3H]) \qquad (5.4)$$

where τ is the age in years and $[^3He]/[^3H]$ the concentration ratio. This determination is independent of the initial tritium concentration. This is an apparent age of a water parcel if ^3He sources other than ^3H decay can be excluded or corrected for and mixing of isotopically different waters is negligible. An example of the application of the method is given by Schlosser *et al.* (1988). For the bomb ^3H peak, the deviation of the $^3H/^3He$ age from the age determined by identifying the groundwater layer recharged between 1962 and 1965 was about 3 years (15 per cent). The deviation was explained by diffusive ^3He loss across the water table and by flow dispersion.

In practice ^{14}C dating can be applied in similar fashion to that of ^3H dating but in a longer time range because of its longer half-life. Applications have included identifying ancient recharge, calculating velocity of groundwater movement by dating the water at different horizons in the aquifer, and calculating contributions of different components to groundwater blends. The age obtained is again an apparent age only and is subject to substantial corrections due to exchange of carbon in solution with carbon of rock minerals and due to solution and precipitation of carbonate compounds during the transit of water through the aquifer. Nevertheless, a groundwater containing no ^3H and 30 PMC (per cent modern C) (apparent age 1×10^4 y) can be designated as ancient recharge with some confidence. Furthermore, differences in apparent age at different points in an aquifer system can be used to calculate minimum groundwater flow rate even though the absolute ages may not be known precisely.

If there are no internal sinks or sources for dissolved chloride and ^{36}Cl in an aquifer system, apart from loss of ^{36}Cl to decay, the groundwater may be dated using the equation:

$$t = -1/\lambda \ln((R-R_{se})/(R_o-R_{se})) \qquad (5.5)$$

where R is the measured $^{36}Cl/Cl$ ratio, R_o the initial $^{36}Cl/Cl$ ratio, and R_{se} secular equilibrium $^{36}Cl/Cl$ ratio due to hypogene production. The method was comprehensively tested on old groundwaters from the Great Artesian Basin, Australia, against a hydraulic model (Bentley *et al.*, 1986). The ^{36}Cl groundwater ages ranging from less than 1×10^5 years to over 1 million years were obtained in excellent agreement with a hydrodynamic model, but only in areas of the aquifer in which no groundwater mixing occurred.

5.2.4 BIOGEOCHEMICAL TRANSFORMATIONS AND CHEMICAL SPECIES IN FRESHWATERS

5.2.4.1 Solid-solution partitioning

One of the principal reasons for the common usage of the distribution coefficient (K_d) is the need, in radiological and transport/dispersion models, for a simple term describing the biogeochemical behaviour of radionuclides. The K_d provides a reasonable first approximation of this behaviour, but its associated limitations (see Section 5.1) must be recognized in any subsequent modelling exercise. More sophisticated solid-solution models (e.g. surface complexation modelling) do exist and have recently been applied to the behaviour of artificial radionuclides under environmental conditions (Turner *et al.*, 1991).

The importance of suspended particle concentration, when assessing solid-solution partitioning, should not be overlooked. The partitioning of three representative radionuclides is used to illustrate this point, based on typical particle concentrations (Table 5.2). Suspended particle concentrations in lakes are generally lower than those in rivers, so that a concentration of 1 mg l^{-1} is commonplace. At this concentration, even a K_d as high as 10^5 would result in less than 10 per cent of a radionuclide being associated with the solid phase.

Table 5.2 The fraction of three radionuclides associated with the particulate phase, based on K_d values of 3×10^2 (^{131}I), 10^4 (^{137}Cs) and 10^5 ($^{239,240}Pu$) and representative particle concentrations

Environment	Particle concentration (mg l^{-1})	Particulate fraction (%)		
		3×10^2	10^4	10^5
Fine-grained bottom sediment	2.5×10^5	99[a]	>99.9[a]	>99.9[a]
Rivers				
extreme range	<1–>10^5	<1–97	<1–>99.9	<9–>99.9
common range	10^1–10^3	0.3–23	10–90	50–99

[a] Based on a sediment porosity of 90 per cent and a sediment dry density of 2.5 g cm^{-3}.

Thus, what seem to be contradictory statements in the literature can be explained in terms of sediment concentration. A high particle affinity is widely attributed to ^{137}Cs and, by way of illustration, its extensive use as a sediment tracer has already been discussed in Section 5.2.2.2. In a study of ^{134}Cs released to the Susquehanna River during operation of the nuclear power plants at Three Mile Island and Peach Bottom, the nuclide budget was converted to a sediment budget by assuming all the ^{134}Cs was in a sorbed state on sediment particles, which was justified by the high affinity of caesium for particles in freshwater (Donoghue *et al.*, 1989). In contrast, following the Chernobyl accident, more than 90 per cent of the atmospherically derived ^{134}Cs in various European lake waters was found to be in solution

(Santschi *et al.*, 1987), with the relatively low sediment concentrations in these lakes undoubtedly being an important factor.

The solid-solution partitioning of radionuclides is also a function of the environmental history of particles, since this is likely to influence the form of the radionuclide in the solid phase and, consequently, the reversibility of the solid-solution reaction. Again this is well illustrated by reference to Cs. Sorption by micaceous clays (illite), when present in significant amounts, dominates the environmental behaviour of Cs. This association is attributed to the presence of frayed edge sites, corresponding to partially weathered 1 nm interlayers, which are highly selective for Cs^+ because of its size and charge. With time, a significant fraction of the Cs becomes 'irreversibly' bound by the illite, probably due to fixation at interlayer sites through collapse of the frayed edges and migration along the interlayers. For instance, less than 25 per cent of the ^{137}Cs is leached by 1 M HNO_3 from stream channel sediments, contaminated by effluents from the Savannah River Plant (Brisbin *et al.*, 1974).

A number of approaches have been adopted to provide insights into the related issues of the nature of solid-solution reactions, the chemical forms of sediment bound radionuclides, and their remobilization characteristics under a range of environmental conditions. Some of the unique advantages of artificial radionuclides, as chemical tracers in the environment, are well demonstrated by the following examples. One approach has been the application of sequential extraction procedures. Förstner and Schoer (1984) employed the technique in a study of various river sediments, subjected to discharges from nuclear installations in Europe and the USA. They compared the leachability of the associated radionuclides with their naturally occurring stable counterparts and with artificial radionuclides added to the sediments in the laboratory. The results indicated that the extent and ease of leachability for each element decreased in the order: laboratory spiked radionuclides > environmental radionuclides > stable isotopes. This suggests an ageing process, perhaps involving incorporation into lattice positions and recrystallization, and emphasizes the need to consider the environmental history of sediments when predicting solid-solution behaviour. A sequential extraction procedure has also been applied to bottom sediments and sediment trap material from Lake Michigan (Alberts *et al.*, 1989). ^{238}Pu, $^{239,240}Pu$ and ^{241}Am were associated mainly with a citrate-dithionate extract, corresponding to the nominal hydrous oxide fraction and probably occurring as mineral coatings. In contrast, ^{137}Cs was almost totally associated with the most inert fraction of the sediment, attributed to fixation by clay minerals.

The specific role of Fe–Mn coatings has been examined by a different approach, involving a series of novel field experiments (Cerling and Turner, 1982). The study involved placing contaminated gravels in uncontaminated streams, and vice-versa, within the White Oak Creek watershed, Tennessee. Fe–Mn coatings were known to form on stream gravels throughout much of the watershed, part of which is contaminated by several nuclear plants and a low-level radioactive waste disposal site. ^{60}Co was rapidly (e.g. days) sorbed from solution, principally by the Fe–Mn coatings.

Under oxidizing conditions, the sorption process was not reversible over a period of months, although some ^{60}Co was lost by abrasion of the Fe–Mn coatings. Under mildly reducing conditions, ^{60}Co dissolution accompanied reductive dissolution of the Mn oxide fraction of the coatings, while the Fe oxide component continued to form. Thus the behaviour of ^{60}Co was dominated by its well-known geochemical association with Mn oxide, probably through co-precipitation. ^{90}Sr was held primarily as an exchangeable cation, although some was associated in a non-exchangeable form with newly precipitated Mn oxide. Adsorption, together with desorption of the exchangeable fraction, was rapid (days), whereas only about half the remaining fraction had been lost after 8 months. ^{137}Cs was adsorbed irreversibly by illite, the dominant mineral, as discussed previously.

The examples described are only illustrative, so that the patterns of behaviour that emerge for ^{137}Cs and ^{90}Sr should not be regarded as being universal. Under the variety of conditions that exist in freshwater systems, different solid-solution behaviours must be expected. There is evidence, for instance, that the solid-solution behaviour and sedimentation of ^{137}Cs in marl lakes is dominated by scavenging, associated with calcite precipitation (Lindner *et al.*, 1989). Any process leading to a change in the solid phase composition is likely to affect K_d, and surface water systems are particularly prone to such changes.

The integration of ideas on solid-solution partitioning and other important processes into whole-lake studies is rare. This deficiency has most successfully been addressed by Santschi and co-workers through the use of experiments in which limnocorals and whole lakes have been spiked with artificial radionuclides (e.g. Santschi *et al.*, 1986). Such studies have highlighted that radionuclides are removed from lakes not only by hydraulic flushing and scavenging by sedimenting particles but also through direct adsorption to surface sediments. The latter pathway is dominant for less particle-reactive elements, and it is interesting to note that Cs can fall into this category at the low suspended particle concentrations normally present in lakes (see also above discussion). Whole-lake studies, employing Chernobyl ^{137}Cs as a tracer, have confirmed the potential importance of direct adsorption as a removal pathway (Santschi *et al.*, 1990).

5.2.4.2 Chemical speciation

The chemical speciation of an element in aqueous solution is important in determining its solid-solution partitioning and general biogeochemical behaviour. As in the case of Cs, which occurs predominantly as the hydrated Cs^+ ion, speciation may vary only slightly in natural waters and consequently not require special consideration. On the other hand, knowledge of speciation, and more importantly how it varies, might be critical to understanding and predicting the behaviour of an element. Pu is a good example and is therefore discussed at some length.

Pu can exist in aqueous solution in any of four oxidation states—III, IV, V, and VI. A number of novel methods have been developed specifically to determine the oxidation state, the most widely used being that based on the separation of 'reduced' Pu(III+IV) from 'oxidized' Pu(V+VI) by co-precipitating the former with neodymium fluoride. Dissolved Pu in freshwaters is invariably a mixture of the two forms. Consideration of the available thermodynamic data led to the suggestion that the oxidized form occurred mainly as Pu(VI). More recently, there has been a growing consensus, based in large part on studies involving the selective adsorption of Pu(VI) by silica gel and on the selective co-precipitation of Pu(V) with calcium carbonate (Orlandini *et al.*, 1986), that Pu(V) is the stable 'oxidized' form in natural waters, principally as PuO_2^+. A comparable situation exists in the case of reduced Pu. Pu(IV) is generally regarded as being of greater environmental significance, although thermodynamic data suggest that Pu(III) may be stable under anoxic conditions, particularly at low pH.

We saw in Section 5.2.4.1 that Pu has a high sediment affinity. There are a number of features of the sorption process that are linked to Pu speciation. Firstly, adsorption to natural particles is almost exclusively via the reduced state, and secondly, natural particles appear to take an active role in redox reactions. A series of experiments with Lake Michigan waters (Nelson *et al.*, 1987) showed that when natural particles were present redox equilibrium was attained within two days and that the same oxidation state distribution was achieved whether Pu was added initially in a reduced or oxidized state. In the absence of the natural particles, equilibrium was not achieved after four weeks. With particles present, not only did the oxidation state distributions converge, irrespective of initial oxidation state, but the final equilibrium distribution was the same as that *in situ* in the lake, associated with fallout Pu. The role of natural particles in the reduction of Pu(V) to Pu(III,IV) has been studied in more detail by Penrose *et al.* (1987). Rapid reduction occurred only in the presence of natural sediment (from Lake Michigan), with the rate appearing to be nearly first order with respect to Pu(V) and being proportional to sediment concentration. A small number of sites with homogeneous properties seemed to be responsible for the reaction. Direct microbiological mediation is unlikely, since reduction was unaffected by preheating the sediments to 105°C. Ashing the sediments at 500°C, however, decreased the rate of reduction by 98 per cent, suggesting that organic matter may be involved in the reduction process.

The other dominant factor in terms of Pu speciation is dissolved, or more correctly colloidal, organic matter (COC), comprising mainly humic substances derived from decaying organic matter of terrestrial and aquatic origin. Field measurements and laboratory experiments indicate clearly that Pu(III,IV) is complexed by COC. Figure 5.4 shows the K_d of reduced Pu as a function of COC concentration, determined in laboratory experiments with spiked Pu, a natural sediment and a variety of freshwaters. The critical COC concentration in each curve, above which K_d starts to decrease significantly, ranges from about 0.1 mg l^{-1} to about 3 mg l^{-1}. The other, less well-defined effect of COC is that it appears to influence the redox behaviour of Pu, probably by acting directly as a reducing agent. Field and laboratory studies

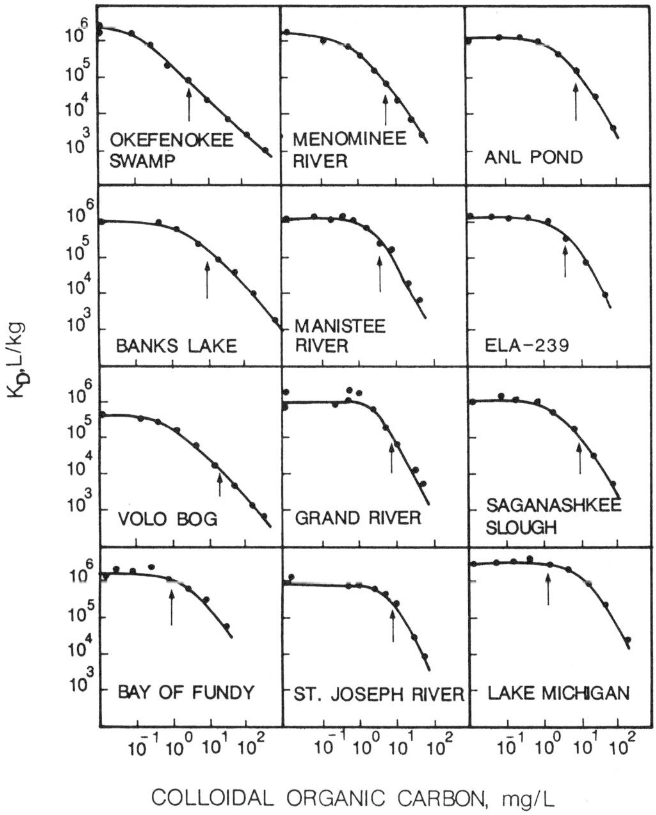

Figure 5.4 Variation of the K_d of Pu(III/IV) as a function of colloidal organic carbon concentration in a series of laboratory experiments, using natural freshwaters, as indicated, and sediment. The natural COC concentrations are indicated by arrows. From Nelson *et al.*, 1987.

indicate that the concentration of oxidized Pu is high only when colloidal organic matter is low. The effects of Pu(III,IV)–COC complexation, Pu(V,VI) reduction and the contrasting adsorption affinities of the two oxidation forms work in combination to produce a fairly coherent pattern of behaviour (Figure 5.5). Up to about 10 mg l^{-1} COC, dissolved concentrations and hence the K_d of total Pu stay more or less constant, since the effects of Pu(III,IV) complexation and Pu(V,VI) reduction seem to cancel each other out. Above 10 mg l^{-1} COC, Pu(V,VI) has all but disappeared from solution so that Pu(III,IV) complexation dominates and the total Pu K_d begins to decrease markedly (by up to two orders of magnitude, as has been observed in freshwater systems).

Less attention has been paid to other artificial radionuclides with regard to the direct measurement of speciation. Nelson and Orlandini (1986) have shown experimentally that ^{241}Am, which invariably occurs as Am(III) in natural waters, is com-

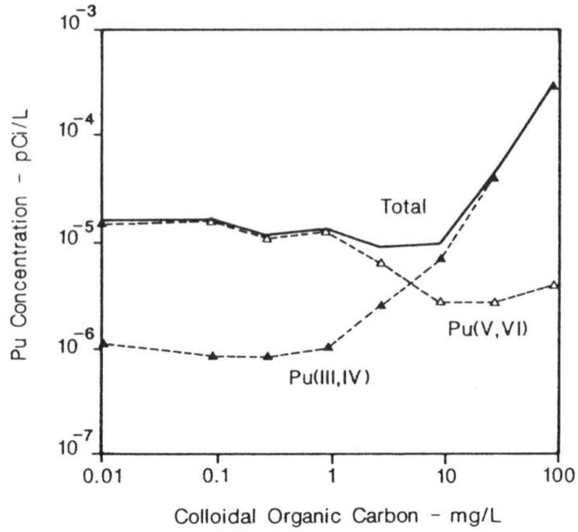

Figure 5.5 Concentrations of dissolved forms of plutonium as a function of colloidal organic carbon concentration in a laboratory experiment, using data from ANL Pond experiment shown in Figure 5.4. Pu concentrations are normalized to a constant plutonium concentration in the solid phase of 1 pCi g^{-1}. From Nelson *et al.*, 1987.

plexed by colloidal organic matter in the same manner as Pu(III, IV). Orlandini *et al.* (1990) studied Lake Trawsfynydd in North Wales, which receives waste discharges from a Magnox reactor plant, and found that all the dissolved actinides measured (^{239,240}Pu(IV), ^{241}Am, ^{244}Cm and ^{232}Th) were associated mainly with the colloidal fraction. Pu(V) made up less than 25 per cent of the dissolved Pu and, in contrast, appeared to be in true solution. A common approach to the question of speciation in environmental settings is to rely on thermodynamic predictions, e.g. with ^{131}I and ^{103,106}Ru. Iodine may occur in the –I, 0 and V oxidation states and Ru as Ru(III) or Ru(VI). Both elements can occur as oxyanions (IO_3^- and RuO_4^{2-}) and I as iodide. Their frequently observed mobility in surface and groundwater systems is generally attributed to the low adsorbance of anionic species, since natural particles themselves generally have a negative surface charge, especially in the presence of organic coatings.

5.2.4.3 Redox boundaries as critical features in the biogeochemical behaviour of artificial radionuclides

As we have seen, redox reactions in part determine the speciation and hence the mobility and fate of many artificial radionuclides in freshwater systems. The most important, but by no means only, driving force for such reactions is the microbiologically mediated, oxidative breakdown of organic matter by various oxidizing agents. The radionuclides may themselves undergo redox transformations or may be affected indirectly (e.g. by being associated with redox sensitive phases, such as Fe and Mn oxyhydroxides). The redox conditions in freshwater systems depend largely on the rates of reaction and the mixing processes affecting the availability of reactants. Reducing conditions are more likely to occur in soil waters and sediment porewaters, than in surface waters, due to high net respiration rates and restricted oxygen supplies. In river waters, reducing conditions are generally restricted to badly polluted stretches of river, in which the net rate of respiration exceeds reaeration. In lakes, reducing conditions exist mainly in bottom waters isolated by stratification, and in groundwaters they may be linked to high rates of respiration or the long hydraulic residence times. It is not uncommon for the redox potential in groundwaters to decrease with distance along the flow path, in other words with the length of time since the water was last in contact with the atmosphere.

The best-studied situation is the redox boundary that generally occurs in the vicinity of the sediment–water interface in lakes and reservoirs, principally because it is relatively stable and easy to observe with minimal disturbance to the ambient redox conditions. The results of such studies, involving artificial radionuclides, not only can be applied to other environmental settings but can be used as models for understanding better the behaviour of other elements. One of the best examples is ^{137}Cs. Although Cs itself does not undergo any redox transformations, elevated concentrations of ^{137}Cs have been observed in the isolated bottom waters of Par Pond, a seasonally anoxic reservoir on the site of the Savannah River Plant in South Carolina. Initially, it was tempting to link the mobilization behaviour of ^{137}Cs to the well-known reductive dissolution of Fe oxyhydroxides. However, more detailed studies indicated that mobilization was due to ion-exchange displacement of Cs^+ from sediments (principally clay minerals) by a range of cations, produced as a result of anaerobic diagenesis (Evans *et al.*, 1983). Combined field and laboratory experiments identified three types of ^{137}Cs binding sites: (1) surface sites on clays and other minerals, where binding is relatively non-selective being related to the charge density of the ions; (2) sites at the edges of clay interlayers of 1 nm spacing (frayed edges of partially weathered micas), where Cs^+ is displaced only by cations of similar charge and size; (3) true interlayer sites from which Cs^+ is not readily exchanged. NH_4^+ appears to be particularly efficient at displacing Cs^+ from the frayed edge sites, and this ties in with the fact that its concentration varies seasonally and with depth in the sediment pore waters and bottom waters of productive lakes. The fraction of ^{137}Cs present in each of the three forms identified by Evans and co-workers will vary, depending on the mineral composition of the sediments and to a lesser extent the time since release of ^{137}Cs, due to the slow fixation in interlayer sites. The former explanation, or more specifically the higher ratio of kaolinite to weathered micaceous clays, has been used to account for the greater leachability of ^{137}Cs from Par Pond sediments than from lake and river sediments from Oak Ridge, Tennessee. That is not to say, however, that ^{137}Cs significant remobilization, associated with seasonal anoxia, occurs only in situations where micaceous clays are unimportant. One year after the Chernobyl accident, the same process represented the major source of ^{137}Cs in the bottom waters of Esthwaite

Water, a lake in Cumbria (UK), the sediments of which have a clay mineral composition dominated by illite and chlorite (Davison *et al.*, 1992).

The effect of the same redox boundary on the biogeochemical behaviour of Pu is more controversial, there being disagreement as to whether significant release from bottom sediments actually occurs under reducing conditions (e.g. Sholkovitz *et al.*, 1982; Alberts *et al.*, 1986). This situation highlights the fact that determining the remobilization behaviour of elements under such conditions is far from being straightforward, and that the problem is best tackled by several complementary approaches. More specifically, it is inadequate to examine water column profiles alone, since this approach is open to misinterpretation.

5.2.5 LAKE SEDIMENTS AND POST-DEPOSITIONAL CHANGE

The study of lake sediments as a means of providing information on the sedimentary processes operating throughout a lake basin is considered in Section 5.2.2.2. Of more general interest is the way in which fine-grained sediments in the deeper waters of a lake may represent a more or less continuous historical record of the processes operating in a lake basin, its catchment and beyond. Thus, lake sediments have become the principal natural medium for providing historical records of environmental change. The ideal requirements are that: (a) the environmental change is reflected in some way in the nature or composition of the sedimentary particles; and (b) the record is not altered by the various diagenetic processes typically operating in a sediment, such as remobilization or degradation due to biogeochemical transformations, compaction, physical disturbance due to resuspension and reworking by bottom currents or benthic organisms. Within limits, the historical records can be corrected for the effects of diagenesis by mathematically modelling the constituent processes. Such diagenetic models have become an end in themselves and are capable of providing quantitative information on the various processes operating in a sediment (Berner, 1980).

Artificial radionuclides may act either as the 'pollutant', for which the discharge or emission record needs to be reconstructed or, more commonly, as the 'tracer' by which the diagenetic model is calibrated. The latter approach applies particularly to studies of the vertical distribution of ^{137}Cs in sediments, with the source term, due to fallout, being well defined. One of the earliest and most frequently quoted studies of this type is that of Robbins and Edgington (1975), who looked at sedimentation rates in Lake Michigan by means of ^{210}Pb and ^{137}Cs measurements. Figures 5.6a–c show the data and modelling results for the core with greatest surficial mixing. Firstly, a best fit to the ^{210}Pb profile was obtained for a model incorporating sedimentation and compaction only (Figure 5.6a). This provided a timescale for converting the known historical record of ^{137}Cs atmospheric fallout in Lake Michigan to a depth scale with respect to the sediment surface (Figure 5.6b). The ^{210}Pb results for all eight sites examined indicate that the sedimentation rates have remained unchanged at each of the sites for at least 100 years and perhaps

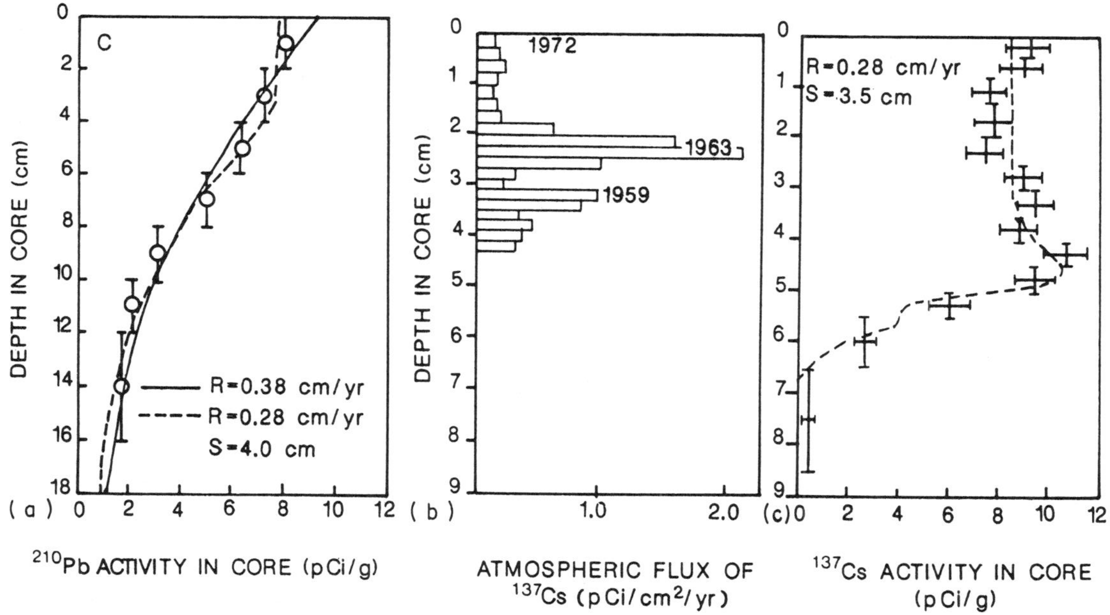

Figure 5.6 Vertical profiles in a sediment core from southern Lake Michigan. (a) Measured values and calculated profiles of ^{210}Pb. (b) The annual flux of ^{137}Cs to the lake surface from bomb fallout, plotted using a timescale established from the ^{210}Pb measurements. (c) Measured ^{137}Cs activities and the calculated profile, assuming rapid surface sediment mixing. (Reprinted with permission from Robbins and Edgington 1975; copyright (1975) Pergamon Press.)

for as long as 7×10^3 years. Assuming this to be true, there should be a straightforward correspondence between the profile in Figure 5.6b and that of ^{137}Cs activity concentration in the sediment (Figure 5.6c), which is clearly not the case. The first appearance and the 1963-64 maximum of ^{137}Cs fallout are deeper and the activity concentrations in recent years are greater than predicted. This is accounted for by additionally incorporating a sediment mixing term into the model. Best-fit values for the sedimentation rate at the sediment surface (R_0 in cm y^{-1}) and the mixing depth (S in cm), together with the calculated profile, are shown in Figure 5.6c. The model, with the mixing term, was then applied to the ^{210}Pb profile, resulting in best-fit values of $R_0 = 0.28$ cm y^{-1} and $S = 4.0$ cm. Thus fitting the model to the two independent profiles yielded the same values for sedimentation rate and mixing depth, providing supporting evidence for the validity of the model. Edgington and Robbins (1975) have also successfully applied this approach to $^{239,240}Pu$ profiles in Lake Michigan sediments.

The above approach yields the mixing depth and indicates implicitly that mixing is rapid in comparison with the sedimentation rate. When mixing is insufficient to produce a homogeneous mixed layer, the distribution of activity becomes sensitive to the details of the mixing process. Thus concepts, such as 'conveyor belt' mixing or 'biotransport', have been developed in addition to the more common approach of random mixing (see Berner, 1980).

Biogeochemical reactions may also be an important influence on radionuclide profiles, especially where concentration gradients are created in the sediment porewaters. Under such conditions, diffusional transport occurs even in the absence of mixing, i.e. by molecular diffusion. Lerman and Lietzke (1975) were amongst the first to consider the role of porewater diffusion on sediment profiles of artificial radionuclides. They developed a model to describe the vertical profiles of ^{137}Cs and ^{90}Sr in the sediments of Lakes Erie and Ontario. The cores were collected in 1969–70, so that the 1963–64 maxima of ^{137}Cs and ^{90}Sr were preserved at or close to the sediment surface. The model, which incorporated adsorption, diffusion in the porewaters and sedimentation, produced a good fit to the data, suggesting indirectly that sediment mixing was not important at the coring sites. The need for a porewater diffusion term was manifest most clearly in the much greater penetration of ^{90}Sr in the sediment cores, compared to ^{137}Cs, attributable principally to the two order of magnitude lower K_d for ^{90}Sr.

Later studies have extended the above concepts and have included the use of other radionuclides e.g. ^{106}Ru (Sickle *et al.*, 1983). In addition, an increasing number of studies have focused attention on major inconsistencies between sediment profiles of ^{137}Cs and local fallout histories for a large number of disparate and geographically dispersed lakes. This has raised questions as to the validity of ^{137}Cs dating, at least under certain conditions, and emphasizes the need to adopt more than one dating technique when a reliable chronology is required. Such behaviour has been observed in lakes in North America, Europe and Australia. Explanations for the inconsistencies include mixing effects, changes in sedimentation rate, delayed but significant inputs from the catchment, recycling in the water column, sediment focusing, and redistribution due to porewater diffusion. The latter appears to be important at low pH conditions and where there is a lack of micaceous clay minerals. Under such conditions, ^{137}Cs may be associated mainly with organic matter and consequently be more susceptible to biogeochemical recycling.

5.2.6 A CASE STUDY DEALING WITH EXPOSURE PATHWAYS IN SCANDINAVIA

5.2.6.1 Runoff characteristics and surface water and sediment activity

The case study focuses on the behaviour of ^{137}Cs and ^{90}Sr and a feature of the two radionuclides is the greater geochemical mobility of ^{90}Sr, as described in Sections 5.2.4 and 5.2.5. Pre-Chernobyl measurements in the large rivers of Finland show an average removal by runoff of 41 per cent for ^{90}Sr and 7 per cent for ^{137}Cs of the total amount deposited in the catchment areas (Salo, 1983). As the water surface is about 9 per cent of the Finnish area, these results indicate that ^{90}Sr is significantly removed from land areas by runoff, while ^{137}Cs is not. Measurements of meltwater runoff during spring 1989 in a Norwegian mountain area (Haugen *et al.*, 1991), combined with estimates of the total amount of snow, indicate that runoff of mainly Chernobyl-derived radiocaesium from the drainage area was 0.01–0.1 per cent.

Large variations have been observed in the relative activity concentrations of ^{137}Cs in solution and associated with suspended sediment (Carlsson, 1976; Haugen *et al.*, 1991). These probably reflect real differences in conditions through the seasons. In particular, sediment bound activity is relatively greater when suspended sediment concentrations are high, due to spring meltwaters (see Section 5.2.4.1). Finnish measurements showed that lake sedimentation resulted in a sharp decline in the surface waters concentrations of Chernobyl-derived caesium by the autumn of 1986. There was a further decline in 1987, and by the end of the year concentrations in the two drainage areas, where Chernobyl fallout was highest, were only 3–15 per cent of the maximum values in May 1986. A slightly more modest decrease (down to 10–25 per cent) was measured in the larger rivers. In Sweden, Hammar *et al.* (1989) observed a similar rapid decrease in concentrations in surface waters, together with a more than two-fold increase in bottom sediment concentrations during the period October 1986 to September 1988. The amount retained in lake sediments varied, e.g. with clay mineral content, but was invariably greater for ^{137}Cs than ^{90}Sr. In addition to the effects associated with spring meltwaters, there appear to be many other complicating factors, related to winter conditions at high latitudes, that are important in terms of the atmospheric deposition–runoff characteristics of radioactivity. Although poorly understood, some insight into these factors was gained from a study using weapons fallout as a tracer (Lund *et al.*, 1962).

From a population exposure point of view, radioactivity in surface waters and bottom sediments is unlikely to be important. Direct exposure of bottom sediment may occur, especially during dry spells. The critical areas will be the shores of lakes and rivers. Use of these areas can lead to radiation exposure directly from the radioactive materials on the ground and via inhalation of resuspended sediment. The activity content of the sediment may vary strongly, even quite locally. Variation is particularly marked in rivers, where high concentrations are often found on the inner side of river bends. Such locations may be used for swimming, sunbathing and camping. However, both the size of the potentially exposed population and the exposure time over a year will be quite limited, at least in a North European climate. Boating and fishing activities are also unlikely to pose any significant risk. Pre- and post-Chernobyl studies of ^{137}Cs have shown that irrigation is an insignificant exposure pathway unless it takes place in an area otherwise unaffected by contamination.

5.2.6.2 Drinking water

Post-Chernobyl measurements in drinking water were performed in Finland at a large number of waterworks from 2 May 1986. The dominating radionuclide was ^{131}I until about the beginning of June. Later on the main contaminants were ^{134}Cs and ^{137}Cs, but smaller amounts of several other radionuclides were also found. The concentrations of ^{131}I were up to around 30 Bq kg^{-1}, but rapidly decreased to about 1 Bq kg^{-1} at the end of May. ^{134}Cs and ^{137}Cs were rarely above 1 Bq kg^{-1} and decreased toward autumn. The 1987 values were lower than the 1986 values, and the decrease continued through the summer. The effect of water purification plants upon strontium and caesium concentrations in both pre- and post-Chernobyl studies was found to be of moderate importance. Iodine and strontium concentrations are hardly affected, and the reduction of caesium concentration is at most 50 per cent.

It is easy to show by simplified, pessimistic calculations that the concentrations in water resulting from deposition upon a water surface will not even in the worst circumstances lead to doses via drinking water which could rival those via other exposure pathways, unless the drinking water is used in areas otherwise not affected by contamination.

5.2.6.3 Freshwater fish

The total amount of weapons fallout ^{137}Cs contained in fish in a typical lake is reported to be as low as 1 per cent of the total amount present in the lake water, although the concentration in fish is much higher (Carlsson, 1976). Typical concentration factors for ^{137}Cs in fish relative to water are 10^3–10^4. The usefulness of concentration factors of this type, however, has been questioned after Chernobyl, since uptake does not take place directly from water but indirectly via numerous nutrition pathways. A direct correlation between concentration in fish and the average fallout in the drainage area (in Bq m^{-2}) seems to be just as relevant and useful. For trout in the mountain lakes of Norway this 'transfer factor' [(Bq kg^{-1} fish)/(Bq m^{-2} deposited ^{137}Cs)])] ranges from 0.01 to 0.2 (Tveten, 1991).

Post-Chernobyl measurements both in Norway (Ugedal and Blakar, 1991) and Sweden (Andersson *et al.*, 1990) indicate that the radiocaesium concentration in fish is closely linked to the content in sediment of the same lake. Predictive models perform quite well, in which concentration in fish is expressed as a function of the amount of radiocaesium in sediment (Bq m^{-2}) and various sensitivity parameters are used to characterize the lake. The sensitivity parameters may be height above sea level, area of the lake, residence time of water in the lake, and the potassium concentration in lake water. A certain division into type classes of lakes is necessary before it is possible to determine valid predictive models, however. In particular, mountain lakes seem to be in a class by themselves.

Both from earlier measurements on weapons fallout and from post-Chernobyl studies, it was found that strontium in an exposure context is less important than caesium, as strontium concentrates in bone. Most of the freshwater fish consumed are of species where most of the bones are easily removed, and it can be assumed that less than 10 per cent of the fish bone will be consumed, making the contribution from strontium negligible in most cases. The concentration of ^{137}Cs in fish depends, among other factors, upon the feeding habits of the fish, which are different for different species, but which also change with age and size of fish of the same species. Saxen and Aaltonen (1987) found post-Chernobyl that when fish were divided into three classes (benthic, predators and intermediate), the activity rose first in benthic and intermediate classes. In July to September 1986, activity in fish rose sharply, but was in autumn still lowest in predators. By 1987 the activity was highest in predators and was still rising, while for non-predatory fish it had started to decrease. Similar results were found in Sweden.

It was found in Finland in 1986 that for the same species (perch in particular), the smaller fish had the higher concentrations (Saxen and Rantavaara, 1987), and this was explained by differences in diet for different size and age of fish. The youngest perch eat mainly plankton. Later on they turn to insects, worms etc., and even later they start eating small fish. This inverse correlation between concentration and size did, however, not continue to be valid in 1987, when it was found both for perch and pike that the concentration was higher in larger fish. The difference between the years is explained by the fact that in 1986 the caesium had not had time to propagate through the food chain to the larger fish. Swedish investigations (Andersson *et al.*, 1990) in 1987 also found that the concentration was larger in larger fish.

Another correlation, observed in Finland (Saxen, 1990) and in Norway (Ugedal and Blakar, 1991) is that in the same drainage area the concentration in fish is higher in smaller lakes. The concentration is also higher in lakes where the residence time of the water is longer, when the lakes are otherwise similar (Andersson *et al.*, 1990). These authors found this factor to be the most important next to fallout levels. They also found that concentrations in fish are lower in hard water or water with high concentrations of phosphorus or potassium. Experiments on dumping large quantities of potassium in lakes to decrease caesium uptake to fish have been performed in Norway (a slight decrease could perhaps be seen) and Sweden (no decrease found).

Experiments to determine the 'biological caesium half-life' have been performed in Norway. It was found that excretion of caesium is very dependent upon water temperature, and one basin-type experiment (Tveten, 1990b; Christensen, 1989), performed over a whole winter season, showed no decrease at all. Long-term measurement of caesium half-life has been performed in one mountain lake in Norway continuously since Chernobyl (Brittain, 1991), showing that the half-life decreased from about 8 to about 2 years from 1986/87 to 1988/89. For the season 1989/90, however, it had increased once more to more than 9 years. The half-life measured in this experiment is a 'half-life' of the whole environment, as reflected in fish concentrations, since new activity from the drainage area will enter the lake, particularly during the spring melt. In the basin-type experiment, the fish had been transferred from its original contaminated environment to an environment free of radioactive contamination.

Typical freshwater fish consumption in Finland is reported to be about 4 kg y^{-1}. It is concluded (Saxen and Rantavaara, 1987) that the average intake of radiocaesium in 1986 via fish was of the same order of magnitude as via beef. In 1987, the relative contribution from fish was larger (about 65 per cent of the total average intake via foodstuffs) than in 1986; mainly because the activity levels in agricultural products decreased more rapidly than in freshwater fish. The radiocaesium concentration in cultivated rainbow trout in Finland was quite low (average of about 40 Bq kg^{-1} in the most affected areas). Norwegian cultivated trout and salmon had no increase in radiocaesium content after Chernobyl, as they are cultivated in ocean waters.

The Scandinavian case study provides a representative picture of the state of knowledge concerning radionuclide behaviour in the freshwater environment. The main radionuclide and exposure pathways are reasonably well defined, especially for common nuclides such as ^{137}Cs and ^{90}Sr, but a detailed knowledge is frequently lacking. Therefore the effects of short- and long-term environmental perturbations on the fate of radionuclides in the future are often difficult to predict.

5.3 ESTUARIES AND INTERTIDAL ENVIRONMENTS

5.3.1 INTRODUCTION

Radionuclides enter estuaries as both solution and particulate solid phases, together with the water and sediment, from any or all of four source environments: sea, river, atmosphere and the estuary itself. The magnitude and radionuclide composition of these inputs varies widely between estuaries (Table 5.3), as well as varying with time in any one. For many estuaries the dominant sources of artificial radionuclides

Table 5.3 Examples of mean activity concentrations for radionuclides in solution and particulate phases in estuarine waters

	Solution phase (Bq/m^3)			Particulate phase (Bq/kg)				
Estuary	^{137}Cs	$^{239,240}Pu$	Others	^{137}Cs	$^{239,240}Pu$	Others	Origin	Dominant source
Connecticut, 1983 (n = 7–10)		0.002			0.006		Weapons test fallout	Fluvial + marine
Savannah, 1986) (n = 5)		0.003		41.8	0.439		Weapons test fallout + reprocessing	Fluvial
Seine, 1979 (n = 11–13)	12.6	0.003		39.2	2.4	^{106}Ru 134	Weapons test fallout + reprocessing	Fluvial/ marine
Dnieper, 1988 (n = 3)	116		^{90}Sr 410				Chernobyl fallout	Fluvial
Esk, 1981 (n = 17–171)	12660	18.5	^{241}Am 53 ^{106}Ru 1050	16260	12790	^{241}Am22150 ^{106}Ru 790909	Reprocessing	Marine

Refs: Connecticut (Sholkovitz and Mann, 1987); Savannah (Olsen *et al.*, 1989);Seine (Jeandel *et al.*, 1980); Dneiper (Polikarpov *et al.*, 1991); Esk (Assinder *et al.*, 1985)

is fallout, introduced by rivers from the land and by currents from the sea, as well as by direct deposition on the estuary. Where nuclear industry discharges occur, they can be the dominant source, as river, estuarine or marine inputs, depending on the location of the discharge point.

Within an estuary, the physical and chemical pathways followed by the radionuclides in solution and particulate phases are determined by three main groups of processes. Firstly, transport in the water column of both phases by the estuarine circulation results in dispersion, dilution, mixing and, for the particulate phase, fractionation of the radionuclides, as well as enabling chemical interactions between the two phases. Secondly, deposition of the particulate phase in the estuarine sediment deposits results in the long-term accumulation of radionuclides in the estuary. In these deposits further chemical interactions between phases are possible. Thirdly, uptake by biota of radionuclides can occur. The last category is not considered in this section.

5.3.2 THE ESTUARINE ENVIRONMENT

In essence, an estuary is a narrow coastal inlet in which the sea is in contact with a river, resulting in a variation in salinity between seawater and freshwater. The seaward boundary is nominally set by the opening out of the coastline, although the geochemical effects associated with the plume of estuarine water extend the influence of the estuary into the shelf seas. The landward boundary can be put at either the limit of penetration of salt or at the limit of tidal changes in water level. The distinction is important, since there can be a considerable difference in the relative position of the two limits, e.g. in the Amazon, tidal rise and fall of water levels can be detected 850 km upstream whereas salt penetration is limited to the area of the plume outside the mouth. An estuary, therefore, can include three zones which have been given a variety of names: freshwater, also limnetic or riverine; brackish or mixohaline; and seawater or euhaline zones.

Estuaries can be subdivided on the basis of their morphology into a large number of types which reflect their origin. Important types amongst these include:

1. coastal plain estuaries, which are shallow sinuous estuaries formed by the drowning of river valleys by the post-Ice Age (Holocene) rise in sea level, which was mostly accomplished by 6000 years ago;
2. fjords, which are deep linear estuaries created by the drowning of glacially overdeepened troughs;
3. delta channels, which are extensions of the river system built out from coast by deposition of river sediments.

In addition to morphology, the other major determinant of the estuarine environment is the water circulation system. This has two basic components, a tidal and a non-tidal or residual circulation. The tidal circulation consists of reversing, landward and seaward tidal currents generated by the tidal change in sea level beyond the estuary mouth. These reversing currents are accompanied by changes in water level in the estuary itself (high and low tide). This results in a cyclical change in the volume of water stored in the estuary, the tidal prism, and a cyclical inundation of the marginal intertidal parts of the estuary.

The non-tidal circulation, in turn, includes two components: an outflowing current due to the river discharge and a saline density current generated by the density contrast between freshwater and saltwater. Together, these contribute to a circulation in a vertical plane, with a seaward current at the surface and a headward current at the bed. A horizontal non-tidal circulation can exist also, especially in wide estuaries, generated by the Coriolis force due to the Earth's rotation, i.e. anticlockwise in the northern hemisphere. These non-tidal circulations can be affected and even reversed by wind stress. For further discussion of estuarine circulation and hydrodynamics see, for example, Bowden (1980).

The relative importance of the three circulation components varies between estuaries, principally due to the variation of tidal amplitude around the globe, the size of the river discharge and the estuary morphology. Several different types of estuarine circulation can occur, defined by the longitudinal distribution of salinity in the estuary averaged over a tidal cycle, as shown by the isohalines for the three types in Figure 5.7.

1. Salt wedge estuaries are found typically where there are high river flows and small tidal inputs, e.g. Mississippi. The estuary is vertically highly stratified and saltwater penetrates as an arrested density current, from which there is only a small loss of water to the overlying freshwater by internal wave breaking and

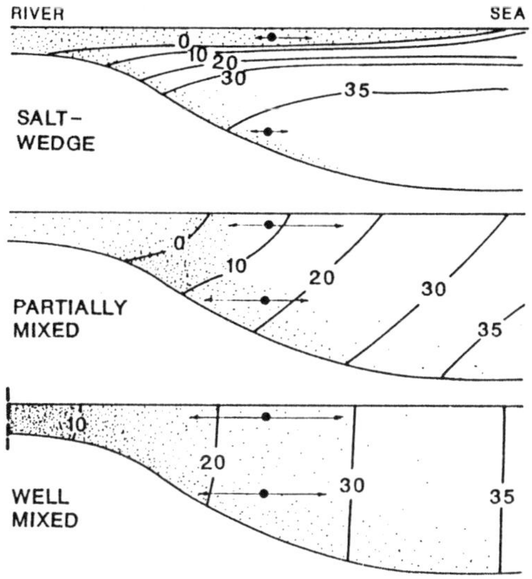

Figure 5.7 Estuarine circulation types based on salinity distribution (contours in ppt) and showing mean flow vectors over a tidal cycle (arrows) and suspended sediment distribution (dots). From Postma, 1980; reproduced by permission of John Wiley.

molecular diffusion. The circulation is, therefore, a strong surface seaward flow of freshwater and slow headward flow of saltwater below. However, the position of the saline wedge can migrate over considerable distances due to variations in the river flow, e.g. 150 km in the Mississippi.

Fjords are special cases of salt wedge estuaries where, due to their morphology, there is an intermittent landward flow of saline water over the shallow sill at the mouth into the deep basin of the fjord. Geochemically, fjords have similarities to offshore basins in terms of radionuclide behaviour and they are not discussed in detail in this section.

2. Partially mixed estuaries occur where significant tidal currents cause mixing between freshwater and saltwater, resulting in the establishment of gradual, tidally averaged, longitudinal and vertical salinity gradients. Consequently, the circulation consists of tidal reversing currents at all depths, superimposed on a relatively strong non-tidal circulation, with seaward flow at the surface and landward at the bed.
3. In well-mixed estuaries an intense degree of turbulence associated with high tides and shallow estuaries results in small vertical salinity gradients and a horizontal gradient limited to the head of the estuary. With a uniform vertical salinity profile, the non-tidal flow would be seawards at all depths but, in practice, a gradient of a few parts per thousand is often observed. Hence, the circu-

lation consists of tidal reversing currents at all depths, superimposed on a weak non-tidal circulation which may be seawards over most of the depth.

It is important to remember that the division of estuaries between these circulation types is not rigid, and a single estuary may exhibit variations in its circulation regime, both spatially and temporally, e.g. a well-mixed estuary may be stratified in its upper reaches and this zone may become more extensive on the early ebb tide.

5.3.3 TRANSPORT OF RADIONUCLIDE IN SOLUTION

The tidal and non-tidal components of the water circulation system transport the radionuclides within the estuary by advection and diffusion. This results in changes in the distribution in space and time of radionuclide activity concentrations in the water column of both the solution and sediment phases.

The dynamic behaviour can be considered over various timescales, i.e. on timescales appropriate to either the tidal cycle or to the non-tidal circulation (tidally averaged), or longer.

Advective transport is produced by the depth mean flow of water and the fluxes (or discharges) of radionuclide solution phases are proportional to the water discharge (mean velocity times cross-sectional area). In partially and well-mixed estuaries, advection by the tidal circulation produces a landward solution phase transport on the flood tide and seaward on the ebb and, consequently, it is an important mechanism for the input of radionuclides which have a marine source. Velocities and depth and, hence, water discharges and radionuclide fluxes show cyclical variations in magnitude, as well as a reversal in direction, falling to zero around high and low water. Peak velocities are related to the tidal amplitude and estuary morphology and, in macro-tidal areas, are often up to 3 m s^{-1} and occasionally reach 5 m s^{-1}. Figure 5.8 a shows the variations in $^{239,240}Pu$ solution phase flux and associated variables over a tidal cycle at a mid-channel station in the shallow, macro-tidal Esk Estuary, UK (also called Ravenglass Estuary in the scientific literature). In this case, although the estuary is small, the fluxes are high because of the strong tidal currents and its proximity to the Sellafield reprocessing plant which discharges directly to the sea.

Averaging the velocities with time over a tidal cycle removes the tidal component and reveals the non-tidal advective element of the circulation. This is responsible for the non-tidal discharge of water to the sea which is the combined result of the river flow and the compensation for the density-driven non-tidal discharge from the sea. Partially mixed estuaries have the most vigorous two-layer circulations, with a landward flow towards the bed and seawards towards the surface with velocities of 0.01–0.1 m s^{-1}. Beyond the landward limit of this circulation, at the head of the estuary, i.e. the null point, the non-tidal flow is seawards at all depths. Salt penetrates beyond this point, however, by the tidal dispersion processes described below. In contrast, in salt wedge estuaries, the two-layer flow consists of a relatively fast seaward surface flow and a virtually stationary lower saline layer. In well-mixed estuaries, whilst the non-tidal flow is theoretically seawards with low velocities at all depths, there is usually a weak landward flow at the bed.

The non-tidal circulation, in general, is important in influencing the net output of radionuclides to the sea from the estuary, as well as in providing another transport mechanism for the input of marine radionuclides.

Dispersion results in both transport and dilution of the solution and particulate phases in an estuary. In addition to dispersion due to turbulent eddy diffusion and to velocity shear, dispersion in estuaries is produced by the lateral trapping effect, by which there is a delayed release of saline water draining from marginal intertidal areas back to the main channel during the ebb tide. The magnitude of the dispersion coefficients varies between estuaries and, with time and position, within individual estuaries. It normally has to be determined empirically, from the salinity distribution or from tracer studies. Typically, the longitudinal dispersion coefficient D_x varies from 10 to 100 $m^2 s^{-1}$, increasing seawards and with river discharge.

The dispersion processes mix the water masses and their radionuclide solution phases present in the estuary, producing spatial gradients in salinity and radionuclide concentration, modified by biogeochemical processes for non-conservative radionuclides. This results in the dilution of the radionuclide inputs and their diffusive transport in the direction of the gradient. For marine or river inputs, a longitudinal gradient in radionuclide concentration in the brackish zone is produced, with activity concentrations decreasing away from the end of the estuary which is the dominant source (Figure 5.8c). The geometry of this distribution depends on the intensity of the dispersion. For radioactive discharges direct to an estuary, instantaneous and tidally averaged concentration gradients are set up in the three axial directions away from the discharge point.

Over a tidal cycle, the combined effect of advection and dispersion in a tidal estuary results in a salinity and solution phase distribution which migrates up and down the estuary during the cycle. In addition, non-tidal excursions can occur over longer timescales, because of river discharge variations and the cycles in tidal amplitude. The latter includes the semi-monthly spring–neap cycle, which gives bigger tidal ranges and more intense currents on spring than neap tides, and the semi-annual spring tide cycle. These effects can also be defined by considering the salinity and radionuclide distributions at the same point in time of a tidal cycle, e.g. high tide. Changes in river discharge can produce major changes in distribution which are often seasonally related. In climates with marked seasonal changes in rainfall large non-tidal excursions nearly equal to the length of the estuary can occur regularly, as in monsoon climates.

A net flux of solution phase radionuclides to the sea may occur over tidally averaged periods (or longer), depending on the dispersive and non-tidal advective fluxes. Where the sea is the dominant radionuclide source, i.e. in the simplest case of a one-dimensional steady-state distribution in a well-mixed estuary with concentrations decreasing towards the head, the seaward advective flux is balanced

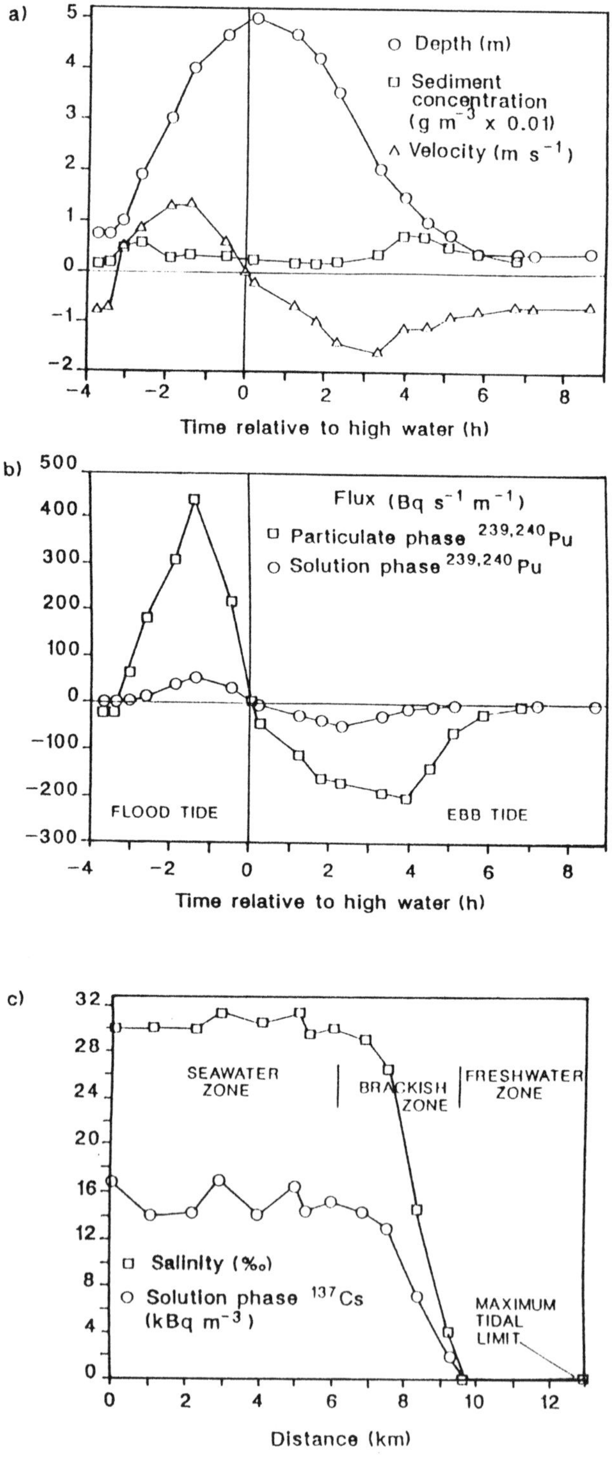

a)
Depth (m)
Sediment concentration (g m^{-3} x 0.01)
Velocity (m s^{-1})
5
4
3
2
1
0
−1
−2
−4
−2
0
2
4
6
8
Time relative to high water (h)
b)
Flux (Bq s^{-1} m^{-1})
Particulate phase ^{239,240}Pu
Solution phase ^{239,240}Pu
500
400
300
200
100
0
−100
−200
−300
FLOOD TIDE
EBB TIDE
−4
−2
0
2
4
6
8
Time relative to high water (h)
c)
32
28
24
20
16
12
8
4
0
SEAWATER ZONE
BRACKISH ZONE
FRESHWATER ZONE
Salinity (‰)
Solution phase ^{137}Cs (kBq m^{-3})
MAXIMUM TIDAL LIMIT
0
2
4
6
8
10
12
Distance (km)

by the headward diffusive flux and the net flux is zero. For dominant fluvial or estuarine radionuclide sources, i.e. with seaward decreasing concentrations, a net seaward radionuclide flux occurs which is the sum of the advective and diffusive fluxes. For example, the relatively large Severn Estuary, UK, receives inputs of ^{137}Cs predominantly from reactor discharges to the estuary, with subsidiary inputs from reprocessing waste discharges to the sea and from fallout on the river catchment. This results in a stable averaged longitudinal concentration gradient which decreases seawards. The large tidal amplitude results in a significant diffusive transport of ^{137}Cs out of the estuary, which exceeds the advective non-tidal transport by factors of 3–6 (Uncles, 1979).

The flushing time of an estuary is an important property affecting the fate of radionuclide inputs which vary with time. Several definitions exist for the flushing time, e.g. the ratio of the volume of the freshwater in the estuary (calculated from the salinity distribution) to the freshwater discharge into the estuary. This is the same as the residence time for conservative radionuclides in solution. Flushing times for partially and well-mixed estuaries vary with their topography, river and tidal water inputs and are generally in the range of days to a year. For highly stratified salt wedge estuaries and fjords, separate flushing times need to be given for the upper and lower layers, with the latter likely to have extremely long times.

In conclusion, the circulation of partially and well-mixed estuaries causes inputs of radionuclides in solution from fluvial, marine or estuarine sources to be dispersed throughout the estuary, up to the limit of the penetration of salt (for the last two sources). For estuarine or fluvial inputs, there will be a net flux to the sea over periods longer than the tidal cycle. For a dominant marine input, there is no net flux when the input is constant, a net flux to the estuary with increasing input and a net flux to the sea with decreasing inputs. The radionuclide inputs will be diluted in the estuary by the mixing processes associated with dispersion.

5.3.4 TRANSPORT AND DEPOSITION OF RADIONUCLIDE PARTICULATE PHASE

The absolute and relative magnitudes of the different sediment inputs vary widely between estuaries and, with time, in any one estuary. At one extreme are delta channel estuaries in which the fluvial source is overwhelmingly dominant and large, e.g. Mississippi, 3.3 Gt y^{-1}. At the other extreme, with a dominant marine source, will be estuaries with high tidal ranges and low river discharges. Partially mixed estuaries typically have important inputs from both fluvial and marine sources, as well as estuarine inputs, e.g. Chesapeake Bay (Schubel and Carter, 1977): river 1.1 Mt y^{-1}, estuarine 0.6 Mt y^{-1}, marine: 0.2 Mt y^{-1}; and the Seine Estuary, France

Figure 5.8 Radionuclide distributions in the water column, Esk Estuary, UK. (a) and (b) Variation with time over a tidal cycle of $^{239,240}Pu$ fluxes and associated variables (Pu fluxes are per metre width in mid-channel). Based on Kelly *et al.*, 1991. (c) Longitudinal variation at high tide of salinity and ^{137}Cs and estuary zonation. From Assinder *et al.*, 1985; reproduced by permission of Elsevier Science Publishers Ltd.

(Avoine, 1986): river 1.5 $Mm^3\ y^{-1}$ (mud), marine 6 $Mm\ y^{-1}$ (half mud and half sand). Fluvial inputs to a particular estuary will vary with river discharge, which may have a pronounced seasonal control. River sediment discharge may be dominated by flood events with a low frequency of occurrence, e.g. Chesapeake Bay rivers received 50 years of average sediment discharge in one storm, leading to extensive sediment deposition (Schubel, 1974). However, whilst such discharges might be mainly contained in a large estuary they would be flushed through a small one. Marine inputs will vary with the tidal regime. Again, low-frequency storm events may increase marine inputs significantly, both through increased wave activity and sediment suspension and from increases in tidal amplitude due to wind stress and low pressures.

The advective transport of the sediment and radionuclide particulate phase is complicated because, as in other aquatic environments, the mobility of the sediment is affected by a number of factors, both flow and non-flow-related (see Section 5.1). Since the concentration of the mobile sediment, mainly suspended load, and its grain size distribution are both a function of flow and because a flow related threshold for transport exists (erosion threshold), the particulate radionuclide fluxes and their characteristics vary much more widely with flow than they do for the solution phase. In tidal estuaries, the cyclical change in depth mean velocity over a single tidal cycle results in the advective transport of sediment, landward on the flood and seaward on the ebb, accompanied by deposition during decreasing velocity periods and erosion with increasing velocity. However, because of lags between the sediment response and the velocity changes (settling and scour (erosion) lag), concentration does not fall to zero in the low-velocity periods (Figure 5.8b). The sediment behaviour will define the behaviour of the particulate phase radionuclides. In general, outside of the turbidity maximum (see below), depth mean suspended sediment concentrations in estuaries range from <1 to 10^3 g m^3, depending on the velocity regime and the availability of sediment. The low sediment concentrations in the Esk (Figure 5.8b) are due to supply limitations, as the measuring station was in a gravel bed reach. A detailed discussion of estuarine sediment transport is given by Dyer (1986).

Just as for radionuclides in solution, an important aspect of the behaviour of the particulate phase radionuclides in estuaries is the mixing by dispersion processes of the different inputs, leading to their dilution and diffusive transport. However, for the particulate phase, significant mixing is not limited to brackish water but can occur throughout the estuary as a result of dispersion due to the effects of settling and scour lag, i.e. sediment deposited from the flood tide will tend to be eroded by a different body of water on the ebb, joining another suspended sediment population. Also, mixing will occur with older contaminated or uncontaminated sediment which is being eroded at the estuary margin or bed. Older contaminated sediments can be an important secondary source of radionuclides in shallow macrotidal estuaries.

Estuarine waste discharge or atmospheric particulate inputs will also be mixed with the other inputs by the dispersion processes. Depending on their relative magnitudes, inputs from any of these estuarine sources will either enhance or dilute the activity of the suspended sediment population.

For tidally averaged conditions, sediment will be transported by the non-tidal advection and dispersion transport systems. Net dispersion transport will be determined by the gradients in sediment and/or radionuclide particulate phase concentrations. This is likely to be seawards in the case of rivers with high suspended sediment discharges and landwards with small rivers and/or shelf systems with strong tidal and wave currents and relatively high suspended sediment concentrations. There will also be an advective sediment flux to the sea, due to the net outflow of water with the non-tidal circulation. However, this may be more than offset by net inputs of sediment by the density component of the non-tidal circulation, which occurs because of the higher suspended sediment concentration towards the bed, and/or by tidal asymmetry in shallow estuaries. In the latter case, a shortening of the period of the flood tide relative to the ebb can result in higher velocities on the flood and higher sediment discharges. In the Esk Estuary, UK, this resulted in a net sediment input of 18 t/tide (Kelly *et al.*, 1991). In any one estuary, the relationship between sediment inputs and outputs will change with time depending on the variations in sediment discharge of the marine currents and river.

In partially and well-mixed estuaries, the two mechanisms that can cause a net landward transport of sediment/particulates, i.e. the non-tidal circulation and the tidal asymmetry, lead also to the development of a zone of enhanced suspended sediment concentration in the upper part of the estuary, i.e. the turbidity maximum. There, suspended sediment concentrations can be as high as 10^3–10^4 g m^{-3} in macrotidal estuaries, but are generally lower in mesotidal estuaries (10^2 g m^{-3}). This zone extends beyond the landward limit (null point) of the non-tidal circulation, because of dispersive mixing. The sediments of this zone are muds, due to the decrease in velocities towards the head of the estuary. Consequently, this should be a zone of high radionuclide activity concentrations for the suspended particulate fraction. Such turbidity maxima have been described, from a sedimentological point of view, from many estuaries in Europe and North America, e.g. Gironde, France (Elbaz-Poulichet *et al.*, 1982). The position of the turbidity maximum migrates with the tidal and non-tidal excursions. The passage of a turbid zone downstream on the ebb tide, unrelated to high velocities, can be seen in Figure 5.8b.

Sediments in estuaries are essentially non-conservative in behaviour and a fraction of the sediment and particulate radionuclide inputs will be deposited in the estuarine sediment. Net sediment deposition (over tidally averaged periods) is due to the existence of longitudinal and lateral gradients in mean velocity in the estuary, decreasing towards its head and sides.

Combined with the settling lag effect, the gradients lead to sediment of a particular grain size being transported landwards beyond the mean velocity required to actively resuspend it. In addition, the gradients in carrying capacity, reinforced in tidal estuaries by lower velocities on the ebb than flood, mean that less sediment of a particular size can be transported.

The net effects of advective and dispersive processes on recent particulate phase radionuclide inputs are best seen in the spatial distribution of activity concentration in the surface layer of the bottom sediments. For example, the longitudinal distribution of ^{239,240}Pu in surface sediments of the Western Scheldt Estuary, Netherlands (Figure 5.9a) (Duursma *et al.*, 1985) is the result of the effects of the circulation system on three different inputs: a dominant marine input of Pu from reprocessing discharges in France and UK, a local estuarine source of Pu from nuclear wastes and a fluvial input of fallout Pu. The activity concentrations have been normalized in order to minimize the effects of grain size variation, by standardizing them against the Al content. In another example, from the Ulhas River Estuary (Trombay), India (Figure 5.9b), Patel *et al.* (1975) show the net effects of the circulation system on the distribution of ^{137}Cs discharged from a single source in the estuary. The shape is distorted as a result of a horizontal tidal circulation system.

The radionuclide flux to the sediment surface and, hence, the radionuclide inventory in the deposits depends on both the activity concentration, which is a function of grain size, and the mean sedimentation rate. In addition, the frequency of deposition events will also affect inventories for short-lived radionuclides.

The size dependence of the estuarine sediment transport processes leads to the size fractionation of the original sediment inputs and the concentration of these fractions in deposits which are separated in either space or time within the estuary. The fractionation process is not precise and particle populations, both in transport and deposited, have a wide range of grain sizes. The width of this distribution, or sorting, varies with the nature of the transport process, being narrower in granular than cohesive sediments. Because grain density will also affect dynamic behaviour, relatively dense grains will be concentrated together with larger normal (light) density grains, e.g. as seen with natural radionuclide-rich 'heavy minerals', such as thorium-bearing monazite. This should also be the case with the uranium-based fuel debris 'hot particles' in the Esk Estuary, UK, derived from nuclear fuel reprocessing wastes (Hamilton, 1985).

The location of the sediment deposits of different grain size in an estuary will be determined by the current regime or, in general terms, by the energy of the sedimentary environment, i.e. sands are found in higher energy environments and muds in lower. Thus, muds accumulate in relatively deep channels and basins, and sands in shallower channels. Muds are also deposited in the shallow marginal intertidal areas which are covered only during the low-velocity period at the end of the rising tide. The upper parts of these intertidal banks are covered infrequently enough for salt-tolerant vegetation to grow, forming salt marshes. This vegetation has an additional role in reducing the velocity of the water over the marsh, encouraging sedimentation. The distribution of sediment grain sizes over an estuary will be reflected by the distribution of radionuclide activity concentration, with lower activity sands and higher activity muds.

Relatively shallow estuaries should be in a state of near equilibrium, with sedimentation only occurring in response to the migration of the channel systems after

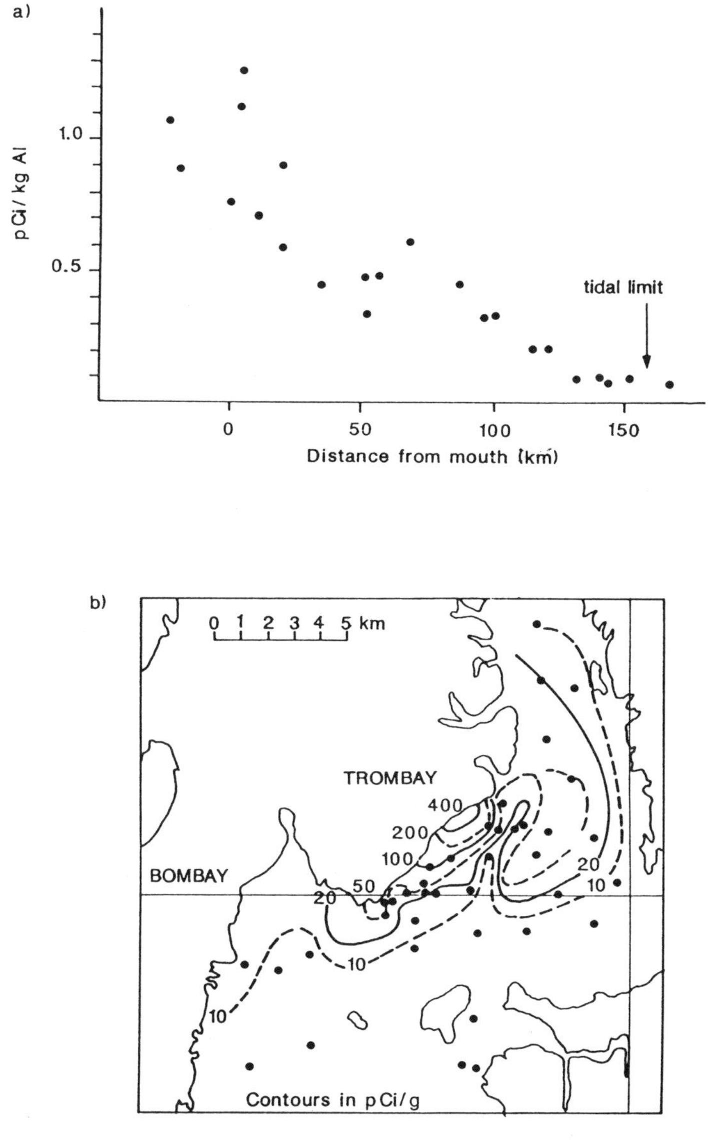

Figure 5.9 (a) Longitudinal distribution of $^{238,240}Pu$ (normalized to Al) in surface sediments in the Western Scheldt Estuary, Netherlands (1979–84; pCi/kg = 37 mBq/kg). Duursma *et al.*, 1985; reproduced by permission of the Commission of the European Communities. (b) Distribution of ^{137}Cs in surface sediments, Ulhas River Estuary, India (1970–71). (pCi/g = 37 Bq/kg). From Patel *et al.*, 1975; reproduced by permission of Academic Press Inc.

occasional erosional episodes, due to river floods and storm tides, or because of changes in the regional environment, e.g. land subsidence, sea-level rise and coastal evolution. Frequently today, sedimentation is the result of human disturbance of this equilibrium, e.g. sedimentation in marginal areas following construction of training walls and groynes or dredging. In general, sedimentation rates may be locally high on short timescales but are mainly low and interrupted by periods of reworking of deposits. For example, in the Hudson estuary, US, long-term average sedimentation rates of 1–3 mm y^{-1}, determined from radionuclide profiles, appear to be keeping pace with regional subsidence rates (Olsen *et al.*, 1984–85). However, in areas which are recovering from dredging, rates are as high as 700 mm y^{-1}.

Direct measurements of sedimentation on the intertidal banks in the Esk estuary, UK, showed how radionuclide deposition rates can vary widely over short (monthly) timescales at a single site, e.g. 50–750 Bq m^{-2} d^{-1} of ^{241}Am, due to the effects of changes in tidal amplitude and storm occurrence. These factors affect grain size, and therefore activity concentration, as well as the sedimentation rate (Kelly and Emptage, 1991). There was also a variation between sites in the same intertidal bank environment. Over a two-year period, the sedimentation rates ranged from 4 to 100 mm y^{-1} (or 5–82 kg m^{-2}) and ^{241}Am deposition rates (at a smaller number of sites) from 21 to 91 kBq m^{-2} y^{-1}. In deep estuaries and fjords, sedimentation will be more continuous, but often sediment supply rates are lower and the overall sedimentation rates are also low.

A variety of radiometric methods have been applied to determining estuarine sedimentation rates from profiles of artificial and natural radionuclides (Figure 5.10), including decay of unsupported natural radionuclides, such as excess ^{210}Pb (e.g. Sharma *et al.*, 1987: Clifton and Hamilton, 1982) and the matching of the observed profile with the history of release to the environment. The latter method may be a simple matching of peaks such as with weapons test fallout (Delaune *et al.*, 1978), or may involve matching against the complex history of a nuclear industry operational discharge by statistical fitting techniques (Hamilton and Clarke, 1984). Analytical models have also been used to generate a signal for comparison with the profile from the source term and incorporating factors to allow for the effects of processes such as bioturbation and transport lag terms (Stanners and Aston, 1981). Such radiometric methods of dating sediments and determining sedimentation rates have their limitations, due to uncertainties in the source terms and the possibility of pre- and post-depositional mixing of sediments labelled at different times. In the case of estuaries, a particular problem is likely to be the mixing of freshly contaminated sediment with eroded older contaminated sediment in the marine, terrestrial or estuarine environments.

The residence time for the radionuclides in these sediment stores is obviously an important factor in terms of their radiological consequences. Residence time will be determined largely by the radioactive half-lives of the radionuclides and the physical stability of the deposits.

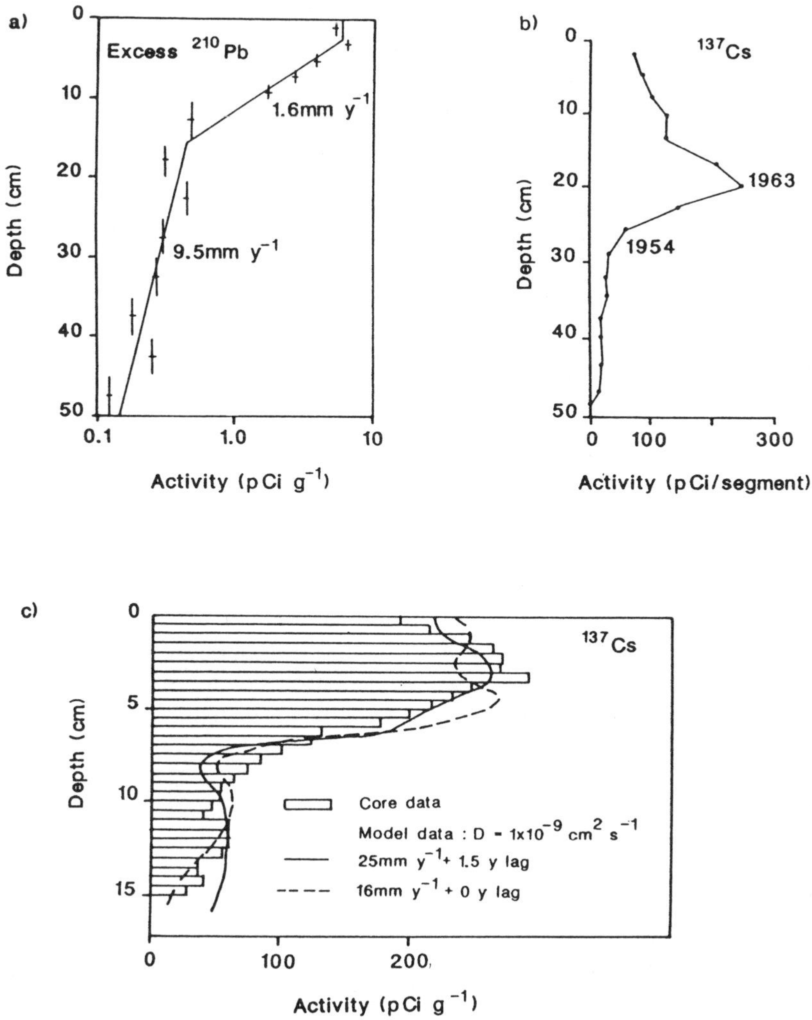

Figure 5.10 Radiometric dating of radionuclide profiles in intertidal sediments. (a) Excess ^{210}Pb S. Carolina saltmarsh, US (Sharma *et al.*, 1987). (b) Weapons test fallout ^{137}Cs, Louisiana salt marsh, US. Delaune *et al.*, 1978; reproduced by permission from *Nature*, vol. 275, pp.532–533. Copyright © 1978 Macmillan Magazines Ltd. (c) Fuel reprocessing discharge ^{137}Cs, sediment profile (blocks) and modelled profiles (lines), Esk Estuary, UK. From Stanners and Aston, 1981; reproduced by permission of Academic Press Inc. (pCi/g = 37 Bq/kg).

In a steady-state estuary, channel and lateral deposits will be reworked by the migration of the channel systems on time scales of 1–10^2 y and longer for large delta systems. Thus, sediment inputs will be balanced by sediment outputs to the sea and the fluvial sediments are effectively bypassed to the sea. This is the condition for the Mississippi and other delta channel estuaries. However, for most estuaries, the steady-state condition does not exist, due to changes in relative sea level caused by the post-glacial worldwide rise in sea level and local tectonic and isostatic crustal movements, both up and down. Consequently, deep estuaries, e.g. Chesapeake Bay and fjords formed by the deglaciation and Holocene rise of sea level, are still accumulating sediment in their basins. In shallow estuaries and parts of estuaries, continued sedimentation may be keeping pace with subsidence, as has been suggested for eastern US intertidal marshes (Sharma *et al.*, 1987).

In contrast, on short timescales, individual estuaries can vary from a condition of net gain to one of net loss, as the magnitudes of the sediment inputs and outputs change. In strongly tidal estuaries this can give a spring–neap cycling in the net balance, whereas in estuaries where there is a seasonal imbalance in river water discharges the cycle will be seasonal, e.g. net gain when river discharges are low and net loss when discharges are high. Such a relationship is important: it can result in the under-representation of the river sediment in the estuarine deposits, i.e. the fluvial sediments are bypassed direct to the shelf sea during high river flow periods.

5.3.5 PARTICLE–SOLUTION REACTIONS

In estuaries, many elements and, consequently, radionuclides, exhibit non-conservative behaviour, i.e. they undergo a change in the distribution between the particulate and solution phases, due to sorption or flocculation/precipitation reactions, or to biological uptake. Non-conservative behaviour needs to be differentiated from the effects of mixing of waters or sediments without particle–solution reactions. Such a mixing process will result in concentrations in each phase lying along a straight line between those of the two end members for that phase. Non-conservatism is most clearly demonstrated by comparing the ratio of solution phase activity concentration to the concentration of a conservative component such as salinity. The resultant distribution (Figure 5.11) shows departures from the mixing or dilution line if non-conservatism occurs, indicating solution loss or gain relative to the particulate phase. It is essential, therefore, that the end members for any mixing processes going on are well defined, in order to rule out mixing as an explanation. Theoretically, it should be possible to demonstrate non-conservatism from particulate phase data (suspended or bottom sediments) but it is often difficult to isolate this from mixing effects. This problem, mixing versus non-conservatism as an explanation of particulate phase pollutant distributions, has been debated for some time with respect to stable element behaviour in estuaries. It is now considered that

Figure 5.11 Particle–solution sorption reactions. (a) Desorption from fluvial sediment of ^{51}Cr and ^{65}Zn, Columbia River Estuary, US. From Evans and Cutshall, 1973; reproduced by permission of the IAEA. (b) Desorption from marine sediment of $^{239,240}Pu$ Esk Estuary, UK. From Assinder *et al.*, 1984; reproduced by permission of Selper, Ltd. (c) Absorption scavenging by (fluvial?) sediment of $^{239,240}Pu$, Chesapeake Bay, US. From Sholkovitz and Mann, 1987; reproduced by permission of Academic Press, Inc. (pCi/l = 37 Bq/m^3; dpm/100 kg = 0.17 Bq/m^3).

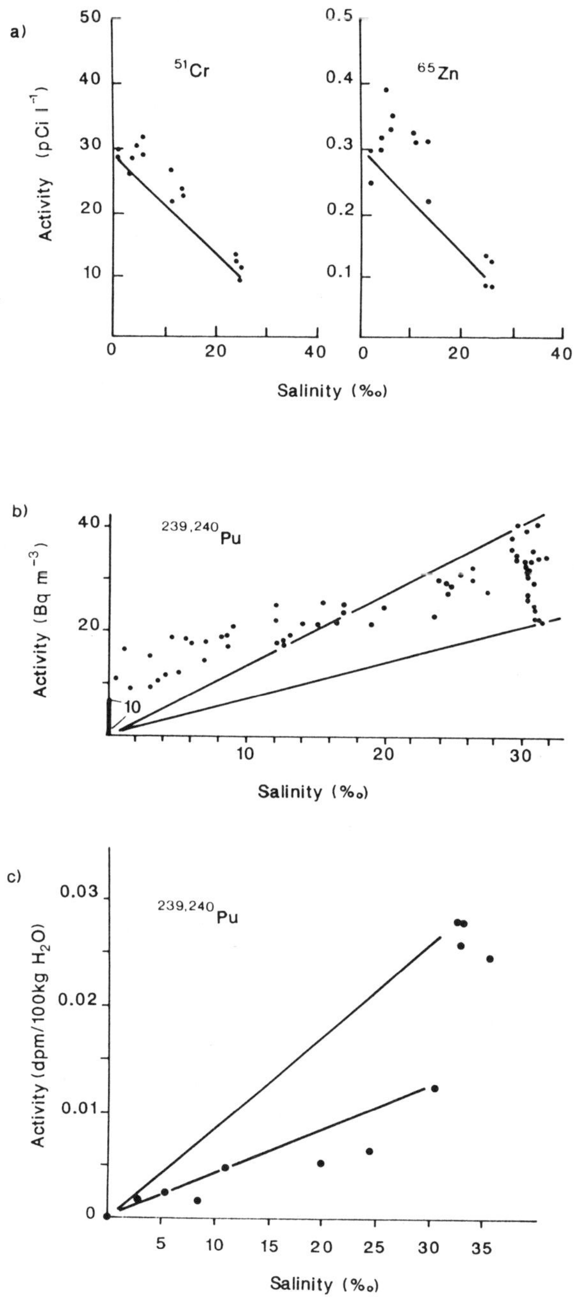
a)
^{51}Cr
^{65}Zn
Activity (pCi l^{-1})
50
40
30
20
10
0.5
0.4
0.3
0.2
0.1
0
20
40
Salinity (‰)
b)
^{239,240}Pu
Activity (Bq m^{-3})
40
20
10
10
20
30
Salinity (‰)
c)
^{239,240}Pu
Activity (dpm/100kg H$_2$O)
0.03
0.02
0.01
0
5
10
15
20
25
30
35
Salinity (‰)

the best evidence for non-conservatism comes from field data on solution phase distributions or laboratory data on partitioning of elements between particulate and solution phases using environmentally representative polluted sediments and waters (e.g. Duinker, 1980). Evidence from suspended particulate phase and K_d distributions is particularly difficult to interpret because of the existence of grain size gradients in the same direction as the salinity gradient, with finer sizes towards the head of the estuary giving higher particulate activity concentrations.

Four processes can be recognized in estuaries which lead to non-conservatism, i.e. mixing of fresh and saline water, scavenging by sediment, degradation of organic matter, and biological uptake. Sorption reactions generally characterize the first two situations. In water mass mixing, these reactions are a response to the change in the geochemical environment which is associated with a corresponding change in K_d. The reactions are able to take place because particles equilibrated with a solution of one salinity come into contact with a solution of another, either by tidal mixing of suspended sediments or as a result of the tidal and non-tidal excursions of the boundary between saltwater and freshwater in the estuary. A particular radionuclide may undergo either desorption or adsorption, depending on the relative levels of particulate and solution phase radionuclides in the inputs of different salinities. Sediment adsorption of radionuclides (scavenging) can occur in response to a decrease in the K_d or with constant K_d when uncontaminated or under-contaminated sediment is added to the suspended sediment population. Organic degradation can result in a change of radionuclide species via redox reactions or organic complexation which affects their partitioning between the solution and particulate phases.

Non-conservative behaviour can vary with radionuclide and between different estuaries, as well as with space and time in the same estuary. Important secondary factors will be those which determine the availability of the sediment bound radionuclide on the appropriate timescale, e.g. the suspended sediment concentration and the fraction of the sediment bound activity which has reaction kinetics of the appropriate rate, i.e. rapid on the timescale of the residence time of sediment in the water column (hours, days) and the physical and biological processes which lead to sediment resuspension. The relative importance of these has been studied with radiotracers in artificial enclosures in a shallow marine environment (e.g. Li *et al.*, 1989).

Desorption of radionuclides from a fluvial sediment input has been demonstrated for the Columbia River Estuary, US, which received large amounts of particulate and solution phase radionuclides from the Hanford plutonium production plant between 1944 and 1972 (Evans and Cutshall, 1973). Field data showed that, for the longer-lived neutron activation products, the behaviour of ^{54}Mn and ^{65}Zn was non-conservative whilst that of ^{51}Cr and ^{124}Sb was conservative (Figure 5.11a). Laboratory experiments with environmental materials showed that about half of the ^{54}Mn and one-third of the ^{65}Zn could be desorbed by a slow reaction (1 week) from suspended river sediments. This rate is slow compared with typical desorption reactions and it suggests that the release mechanism may be complex. In general, remobilization from fluvial sediments can occur for radionuclides of heavy metal elements for which there is good evidence of non-conservative behaviour in estuaries, especially those with high organic loads. For example, in the Scheldt Estuary, Netherlands and Belgium, Mn, Cd, Zn have been shown to be released from contaminated fluvial sediments (Duinker *et al.*, 1982). For the first two metals, redox reactions form species of different solubility in the almost anaerobic zone generated in the upper estuary by the oxygen demand of the organic matter (sewage), whereas Zn may be being released directly by degradation of Zn bearing organic particles.

However, in the Elbe Estuary, Germany, Cd, Pb, Cu, Cr, Zn show conservative behaviour, in keeping with a sediment and water mixing model (Förstner and Wittmann, 1979).

Desorption of Cs from fluvial sediments has been proposed for many estuaries, especially in Europe and America, using a variety of evidence, but very few of these examples provide incontrovertible evidence for caesium desorption. Since the experimentally determined K_d for Cs is around 10^3 l kg^{-1} in seawater and 10^5 l kg^{-1} in freshwater, it is to be expected that desorption should occur as a result of exchange with seawater ions, where fluvial sources of Cs dominate. However, Cs is known to occupy adsorption sites on sheet silicate minerals (clay minerals/micas) with different degrees of exchangeability, at least in freshwater sediments (Evans *et al.*, 1983). Thus, it is important to consider the availability of labile Cs with the appropriate reaction kinetics when assessing the likelihood of desorption. For example, Zucker *et al.* (1984) considered that labile, reactor-derived, ^{134}Cs was being desorbed from sediments in the James River Estuary, on the evidence of the ^{134}Cs : ^{60}Co ratio in sediment deposits, whereas fallout ^{137}Cs, fixed in interlattice sites, was not desorbed. Desorption of ^{137}Cs (which may have had a reactor source) also was inferred by Linsalata *et al.* (1985), from solution activities in the Hudson River Estuary, US, which increased non-linearly with salinity, although the seawater end-member concentration was not determined.

Evidence for desorption based on particulate phase activities alone has to take into account the possibility that the observed trends could be produced as a result of mixing of sediment from different sources. Much of the evidence from US estuaries comes from variations in ratios of Cs : Pu radionuclides in bottom sediments, e.g. Columbia River (Beasley and Jennings, 1984), Hudson River (Olsen *et al.*, 1981) Savannah River (Hayes and Sackett, 1987). At least in the Savannah River, Olsen *et al.* (1989) consider that the bottom sediment radionuclide ratio is evidence for dilution of the fluvial sediment by Pu-bearing marine sediment rather than for Cs desorption. This is supported by their experimental determination that the fallout Cs in the fluvial sediments is non-exchangeable, and from a consideration of the Pu isotope ratios in the inputs. Such an explanation is compatible with the landward transport of marine sediment by the non-tidal circulation of partially mixed estuaries. Desorption of ^{137}Cs from bottom sediments, together with ^{106}Ru and ^{144}Ce, has been described also from the Ulhas River Estuary (Trombay),

India (Patel *et al.*, 1978). There, levels in a surface sediment layer were found to decrease with a half removal time, ignoring radioactive decay, of 2 y following decreases in discharges from the nuclear installation. Since the environmental half-lives for ^{106}Ru and ^{144}Ce were different, it was argued that dilution by new sediment could not explain all of this decrease. However, a fuller study, based on core total inventories, would be needed to provide absolute evidence for desorption. The evidence from the Loire, France (Jeandel *et al.*, 1981) also is ambiguous in that the suspended particulate phase seems to show non-conservative behaviour whilst the solution phase remains conservative. This does not support the case for desorption and the data might be explained by dilution with uncontaminated sediment from an estuarine source.

Desorption of plutonium from marine sediments has been extensively investigated in the Esk Estuary, UK, where the discharges to the sea from the Sellafield reprocessing plant provide marine sediment with higher activity concentrations than on the fallout-contaminated fluvial sediment input (e.g. Assinder *et al.*, 1984; Eakins *et al.*, 1985). In particular, plutonium in solution shows enhanced levels in the estuary due to desorption from the sediments of marine origin, as compared to the range of levels predicted by dilution curves (Figure 5.11b; note that the seawater activities vary during the tidal input).

Following a recent reduction in the levels of Pu discharged from Sellafield, the levels of Pu present in solution in the estuary, after desorption has occurred, can be higher than in the seawater entering it. The behaviour of Pu contrasts with the more conservative behaviour of ^{137}Cs and ^{106}Ru from the discharges. Desorption occurs because the tidal excursion on a spring tide in this shallow, 13 km long estuary, brings low-salinity water into contact with sediment deposited from full-salinity seawater. It is assumed that desorption follows resuspension of these sediments into the low-salinity water, as well as from the dispersion of suspended marine sediment into low-salinity water during the mixing of salt and freshwaters. Laboratory experimentation with Esk sediment has shown that only a small labile fraction (< 5 per cent) of the plutonium on the marine sediment is involved in a rapid (< 1 h) exchange reaction in the freshwater (Hamilton-Taylor *et al.*, 1987). The experiments have also shown that it is primarily Pu(IV) species which are involved in the sorption reactions. However, the Pu behaviour is further complicated by redox reactions, between Pu(IV) and Pu(V), both on the sediment surface and in solution (Mudge *et al.*, 1988).

Estuarine scavenging of Pu has been proposed for the Connecticut River, Delaware Bay and Chesapeake Bay Estuaries, US (Figure 5.11c) (Sholkovitz and Mann, 1987), from a study of the distribution of solution and particulate phase activity concentrations. The sources of the solution phase Pu for these estuaries are fallout Pu in the marine and fluvial inputs, with the highest activities in the former. The origin of the sediment is not clear but is likely to be older, uncontaminated or lightly contaminated estuarine sediment and, maybe, fluvial sediment. In these examples, the loss of Pu to sediment appears to be inversely related to the flushing time in the estuary, being greatest in Chesapeake Bay with a 6–12 month residence time. A similar non-conservative behaviour of Pu was observed in estuaries in France, e.g. Seine, Loire and Gironde estuaries (Jeandel et al., 1980, 1981). At the time, coagulation was considered as one of the possible causes but, in the absence of a high dissolved organic carbon content in the river, there seems no reason to consider it as anything other than an adsorption reaction. The dominant sources of the Pu in solution in these French estuaries appear to be marine, with a fallout origin alone in the Gironde and fallout plus industrial discharges in the others. In the Gironde, at least, the sediment particles which are scavenging the Pu are considered to be of fluvial origin (Elbaz-Poulichet *et al.*, 1982). In the Seine and Gironde estuaries, the particulate fractions show high activity concentrations persisting down to low salinities and this may be a residual grain size effect (not removed by the normalization procedure) combined with scavenging and mixing effects, although other explanations were also considered (Jeandel *et al.*, 1981).

The difference in behaviour of marine Pu inputs in these estuaries and in the Esk Estuary, where desorption occurs, may be due to differences in the speciation of the Pu from the fallout and discharge sources (Sholkovitz and Mann, 1987) and/or to the different characteristics of the estuarine systems, i.e. no significant uncontaminated sediment inputs are available in the Esk. Interestingly, it has been suggested that both reactions may be occurring in the Savannah River Estuary, US (Olsen *et al.*, 1989), where the Pu levels in the solution phase suggest that desorption is occurring at very low salinities and adsorption at higher brackish salinities. It is considered that the scavenging leads to a threefold increase in particulate $^{239,240}Pu$ activity concentrations. Since the suspended sediments are depleted in ^{7}Be whilst enriched in Pu, it is deduced that the frequency of the resuspension events and scavenging process is slow relative to the ^{7}Be half-life (53 d).

Natural radionuclides, also, can be involved in estuarine sorption reactions. Highly particle-reactive radionuclides should show adsorption. This is clearly the fate of the direct atmospheric input of excess ^{210}Pb to estuaries (i.e. unsupported by ^{226}Ra). Also, scavenging of dissolved excess ^{210}Pb from marine inputs has been suggested for the Savannah River Estuary (Olsen *et al.*, 1989).

Thorium is also largely present on particles in aquatic systems. Adsorption of thorium isotopes onto sediments occurs in the Ribble Estuary, UK, from nuclear fuel fabrication plant discharges direct to the estuary (Hunt, 1977–92). Uranium, however, mainly behaves conservatively in estuaries (e.g. Toole *et al.*, 1987). Exceptions to this exist where there is a significant fluvial input of anthropogenic U from phosphate processing, e.g. Charente Estuary (Martin *et al.*, 1978), where loss of solution phase U occurs in the estuary by co-precipitation with phosphate. In contrast, radium (^{228}Ra and/or ^{226}Ra) desorption from river sediments has been shown from several estuaries including Hudson (Li and Chan, 1979), Scheldt (Kirchmann *et al.*, 1985), Mississippi (Key *et al.*, 1985) and Amazon (Moore and Scott, 1986). Differences in the clay mineralogy were considered to be the reason for the considerably higher amounts desorbed in the Mississippi. A reduction of

Ra concentrations at higher salinities is the result of mixing of the enriched estuarine water with seawater which has a low Ra content due to the biological removal of Ra by plankton.

Colloidal Fe and humic acids, present in the operationally defined 'solution' phase of freshwaters, are known to flocculate or coagulate in the low-salinity zone of estuaries (e.g. Sholkovitz *et al.*, 1978). Experiments with freshwater with a high dissolved organic carbon content showed that dissolved Pu was lost to the coagulated Fe humic acid colloids when mixed with seawater (Shen *et al.*, 1983). This is likely to be due to the Pu being associated with the original organic colloids in solution. This mechanism is considered to be involved in the loss of Pu from solution in the organic rich river water in the Mullica River Estuary, US (Sholkovitz and Mann, 1987).

Non-conservative behaviour of dissolved constituents in estuaries can occur through biological uptake. This will affect particularly the major nutrients, such as C, N, P, K. In addition, plankton blooms potentially could have an effect on micronutrients such as trace metals. As well as being involved in metabolism, passive adsorption onto cell surfaces could occur (see Wolff, 1980, for a summary). Such studies have not been carried out in estuaries.

5.3.6 DIAGENESIS

Diagenetic, or post-depositional processes modify the distribution of radionuclides and their partition between solution and particulate phases in estuarine deposits.

Mixing of sediment by bioturbation in estuaries is due to a variety of sediment-inhabiting organisms, e.g. annelids, arthropods, molluscs and echinoderms. Although the fauna is characteristically species-poor, because of the variable salinity, it can have high population densities, with several species having densities in the range of 10^3–10^4 m^{-2}. The depth of this bioturbated zone in estuaries varies between a few centimetres (e.g. *Corophium*) to 10s of centimetres for larger annelids and molluscs. The intensity of sediment disturbance depends on the relative rates of bioturbation and sedimentation, but it can be sufficient to totally destroy the original sedimentary structure and smooth out variations in radionuclide fluxes to the sediment.

Advective movements of porewater fluids will also take place, due to consolidation, surface infiltration, lateral recharge and groundwater flow. In the Esk Estuary, UK, percolation down open burrows resulted in the rapid labelling of the upper 5 cm of sediment with Chernobyl ^{134}Cs from rainwater and with Sellafield-derived ^{95}Zr from seawater (Kelly and Emptage, 1991). Labelling due to percolation to greater depths down desiccation cracks and by lateral recharge have also been used to explain the presence of short-lived radionuclides (^{95}Nb) at depth (70 cm) in Esk sediments (Carr and Blackley, 1985). This mechanism can result in the extension of the zone of contaminated sediments.

Changes in the porewater chemistry can result in redistribution of radioactivity between the particulate and solution phases. The advective porewater transport described above could change the salinity in the porewater, leading to sorption reactions. Also, changes in redox state are a feature of sediment deposits, from oxic to anoxic conditions, due to biogeochemical reactions. This is also true of subaerially exposed intertidal sediments, especially muds. It is well known that such elements as Fe and Mn are mobilized from sediments by these processes and redistributed by diffusion within the sediment body or lost to overlying waters, e.g. in the Mississippi delta half the Mn has been lost from the sediments (Presley and Trefry, 1980). Similar fluxes of Mn have been described from other estuaries, e.g. Tees and Conway Estuaries, UK, (Elderfield and Hepworth, 1975). In the last two instances it was suggested that Cu, Zn, Co, Ni, Pb and Fe were also released. Conversely, it is considered that anoxic bottom sediments scavenge U from solution in surface waters by precipitation of reduced U phases, e.g. Narbada Estuary, India (Borole *et al.*, 1982) and Ogeechee Estuary, US (Maeda and Windom, 1982).

It is to be expected that radionuclides of the metallic elements listed above, as well as others with more than one redox state, can similarly undergo redox remobilization. Although radioactive tracers have been used in studies of the anaerobic bacteria sulphate reduction process which underlies the redox changes (^{35}S and ^{55}Fe) (King, 1983), detailed studies of radionuclide diagenesis have not been carried out in estuarine areas. However, studies in other aquatic environments have shown that Cs remobilization takes place by exchange with anaerobically produced NH_4 (Comans *et al.*, 1989). In contrast, similar studies have been unable to demonstrate the occurrence of remobilization of Pu (from both fallout and nuclear industry discharges) and americium by redox-related reactions (Buesseler and Sholkovitz, 1987; Malcolm *et al.*, 1990).

5.3.7 RADIONUCLIDE BUDGETS AND INVENTORIES

The interaction between the radionuclide sources and the system of hydrodynamic, sediment dynamic and geochemical processes in an estuary determines the balance between the inputs and outputs. Several estimates have been made of Pu budgets for individual estuaries. These emphasize the important role of estuaries as transient or permanent stores of particle-reactive radionuclides.

A budget for Pu over a single tidal cycle for a part of the Esk Estuary (Kelly *et al.*, 1991) shows how the particulate phase behaviour of a highly particle-reactive radionuclide like Pu can dominate the budget when there is an adequate supply of sediment (Figure 5.12a). The only significant Pu source was the Sellafield reprocessing plant marine discharges, mainly introduced to the estuary via fresh marine sediment, but also partly via eroded older contaminated estuarine sediment. In this well-mixed estuary, tidal advection produced relatively large fluxes of $^{239,240}Pu$ (and other radionuclides) in and out of the estuary over the tidal cycle (input 283 MBq per tide). However, overall, there was a net input of sediment (18 t per tide) and of Pu (85 MBq per tide) to the estuarine sediment deposits. The solution phase was a small component of the budget, although it showed a net output to the sea, due

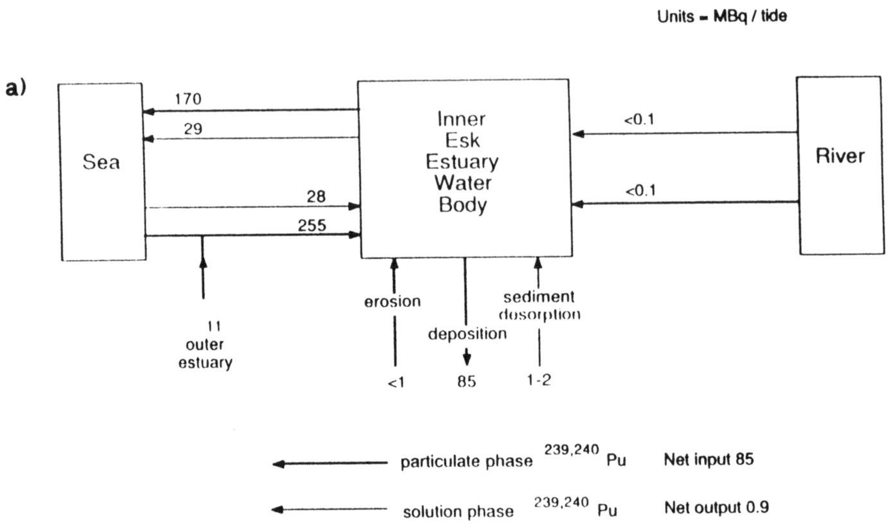
Units = MBq / tide
a)
Sea
170
29
28
255
11
outer
estuary
Inner
Esk
Estuary
Water
Body
<0.1
<0.1
River
erosion
deposition
sediment
desorption
<1
85
1-2
particulate phase 239,240 Pu
Net input 85
solution phase 239,240 Pu
Net output 0.9

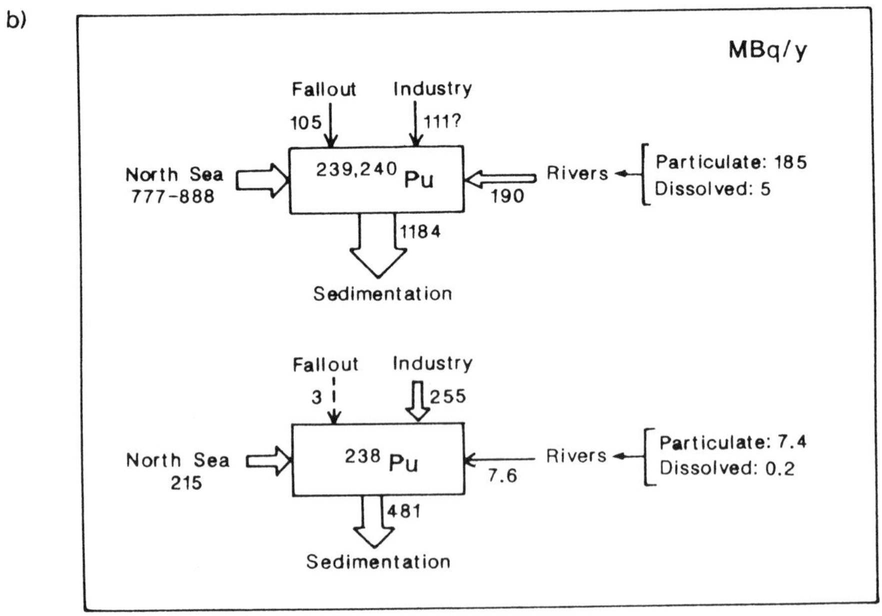
b)
MBq/y
Fallout
Industry
105
111?
North Sea
777-888
239,240 Pu
Rivers
190
Particulate: 185
Dissolved: 5
1184
Sedimentation
Fallout
Industry
3
255
North Sea
215
238 Pu
Rivers
7.6
Particulate: 7.4
Dissolved: 0.2
481
Sedimentation

to the desorption of Pu from suspended sediment in the estuary. The extent to which these data indicate the state of balance of the Pu budget over a longer period is not known. Not only is it a question of the sediment budget, but also of the activity concentrations of the sediment inputs, since declining discharges could mean that the higher activity concentration of the eroding older contaminated sediment in the estuary could result in net output of activity, as has been claimed (e.g. Burton and Yarnold, 1988). However, changes in the discharges are likely to be buffered by the availability of older contaminated sediment on the floor of the Irish Sea.

From sedimentation rate measurements, the annual net sediment input to the main depositional areas in the whole Esk Estuary, i.e. intertidal banks, is estimated at 15.5 kt y^{-1}. This gives a net input to these areas of about 40 GBq y^{-1} of $^{239,240}Pu$, if the activity concentration of the input is assumed to be the same as measured in the tidal budget study.

A mean annual Pu isotope budget has been estimated for the Western Scheldt (Duursma *et al.*, 1985) (Figure 5.12b). Four different sources supply the Pu: catchment fallout (river input), direct fallout, estuarine and marine nuclear industry discharges. Their balance shows that direct fallout to the estuary is not a negligible term. However, the budget emphasizes the important role of nuclear industry discharges, even when the discharge points are remote, as is this case where the marine sources are the discharges on the west coasts of France and UK. The marine input given is the net value needed to balance the budget and an estimate of the gross input would be higher, including a flux equal to the mean tidally averaged advective and dispersive transport to the sea of the solution and, perhaps, the particulate phase Pu. Overall, the budget again shows the importance of the estuarine sediments as a sink for Pu, with a deposition of 1.2 GBq y^{-1} of $^{239,240}Pu$.

A similar approach was used for determining the Pu budget for the Hudson River Estuary by Olsen *et al.* (1984–85) who considered that about 0.8 GBq y^{-1} of fallout $^{239,240}Pu$ was delivered in the particulate phase by the river and about the same amount from the sea, at a higher activity concentration on a smaller amount of sediment, i.e. giving a total deposition to the sediment deposits of about 1.5 GBq y^{-1}. Since two-thirds of the total sediment input (1.5 Mt y^{-1}) is removed by dredging, there is a major loss of Pu from the estuary, although a net gain occurs in non-dredged areas. Any sediment scavenging of Pu from solution, as described from other US east coast estuaries (Sholkovitz and Mann, 1987), is probably only a small fraction of the Pu delivered as sediment labelled with fallout.

The inventories of artificial radionuclides stored in the estuarine sediments are the result of the 50 y of inputs from fallout and nuclear industry liquid discharges (Table 5.4). The inventories are relatively low in estuaries mainly contaminated

Figure 5.12 Radionuclide budgets for estuaries. (a) $^{239,240}Pu$ budget for a single spring tide (12.5 h), Esk Estuary, UK. From Kelly *et al.*, 1991. (b) Estimated annual ^{238}Pu and $^{239,240}Pu$ budgets, Western Scheldt Estuary, Netherlands. From Duursma *et al.*, 1985; reproduced by permission of the Commission of the European Communities.

Table 5.4 Estimated inventories in estuarine sediments for artificial radionuclides from fallout and nuclear industry liquid discharges

Estuary	Area (km^2)	^{60}Co (GBq) (fallout %)	^{137}Cs (GBq) (fallout %)	$^{239,240}Pu$ (GBq) (fallout%)	^{241}Am (GBq) (fallout %)
Hudson, US, 1975	240	8 (0%)	925 (76%)	41 (100%)	
Columbia, US, 1977/78	285	555 (0%)	851 (?)	30 (95%)	9 (0%)
James, US, 1981	380	95 (0%)	766 (83%)		
Esk, UK, 1989	5.6	?	4500 (<1%)	380 (<1%)	490 (0%)

References: Hudson (Olsen *et al.*, 1981); Columbia (Beasley and Jennings, 1984); James (Zucker *et al.*, 1984); Esk (Burton, 1988; Kelly and Emptage, 1991).

by nuclear weapon test fallout, ~2 GBq km^{-2} for ^{137}Cs and 0.1–0.2 GBq km^{-2} for $^{239,240}Pu$, but considerably higher inventories are found where discharges are high, with ~800 GBq km^{-2} ^{137}Cs and ~70 GBq km^{-2} for $^{239,240}Pu$ in a total of 1.2 Mm^3 of contaminated sediment in the Esk Estuary (Burton, 1988; Kelly and Emptage, 1991).

The inventory for $^{239,240}Pu$ in the Esk Estuary, together with the estimated annual sediment input given above, would indicate a residence time of 9.5 y for the Pu-contaminated sediments if the estuary was in a steady state. In reality this is an underestimate, and the estuary is still gaining activity, at least for the actinides. The residence time for contaminated sediments in the Hudson Estuary is greatly shortened by dredging, with two-thirds of the inventory in Table 5.4 having been removed for offshore dumping (Olsen *et al.*, 1981).

5.3.8 OTHER COASTAL INTERTIDAL ENVIRONMENTS

A wide variety of other coastal environments exist besides estuaries, although none exhibit the full range of physical and geochemical processes seen in estuaries. In particular, the particle–solution reactions in the water column associated with the mixing of fresh and saline water are unimportant. The behaviour of radionuclides in these environments can be considered as variants of their behaviour in estuaries and, consequently, a detailed discussion is not provided. The main environments concerned include beaches, embayments, lagoons and barrier saltmarshes.

Beaches are essentially high-energy environments whose physical processes are dominated by wave activity, with a wave-generated circulation system. The sediments are characteristically coarse-grained sands and gravels, derived from the seabed offshore, from coast erosion and, via longshore transport, from rivers. Beaches are themselves the source of sediments for aeolian dune systems built behind the beach. Beaches are often transient stores for sediments and they typically undergo a cyclical development which is often seasonal, with removal during storms and accumulation during quiet weather periods. Longer-term storage occurs where there is construction of spits and barrier island complexes. The coarse grain size means that radionuclide activity concentrations normally will be relatively low on beaches. However, their total inventories may still be significant, e.g. along 20 km of coast close to the discharge point from Sellafield, UK, the beaches (7 km^2) had inventories of 1800 GBq $^{239,240}Pu$, and 3690 GBq ^{241}Am (1983/84) (Eakins *et al.*, 1990). Also, where activity is associated with organic particles this may end up on beaches in the flotsam deposited at the tide-line, as was found to have occurred after an accidental release of activity from Sellafield (DOE, 1984), which contaminated organic objects in the flotsam with up to 69 kBq ^{106}Ru.

Embayments, characterized by relatively large intertidal areas, are typical of shallow coastal seas and/or areas with large tidal ranges. The sediment sources are principally from offshore or longshore transport. The tidal energy and degree of exposure to wave activity determine the distribution of sediment grain size and, thus, the relative activity concentrations in the deposits. Typically, however, there will be a zonation from an outer zone of sand banks, succeeded by mud banks and, finally, salt marshes.

Lagoons are morphologically similar to some estuaries, in being partially enclosed from the sea, but by definition lack the freshwater input. They are developed by the construction of spits/barrier islands, as on the coasts of the eastern US and south-east North Sea. Barrier salt marshes are similar in origin but lack the open water lagoon. Lagoons can also originate from coral reef construction. They are essentially low-energy environments, with weak circulations due to tidal discharges through the restricted inlets and density currents generated by the development of hypersalinity (in warm climates). Consequently, sediments are characteristically muds, both intertidal and subtidal, except for deltas of coarser sands formed at tidal inlets and where storm washover occurs over the barrier with the open sea. In coral reef areas, the sediments will be characteristically composed of carbonate minerals unlike the predominantly siliceous minerals elsewhere. With the fine grain size will be associated relatively high radionuclide activity concentrations. Also, in the lagoon environment the potential for rapid reworking of the sediment is low and sediment residence times will be on a geological timescale, determined by the evolution of the coastline. Although coral reef lagoons are unusual environments they have a particularly significant place in the story of the radioactive contamination of coastal environments, as coral islands in the Pacific Ocean have been a favourite site for nuclear weapons testing since the early days of the nuclear industry.

5.3.9 THE FUTURE OF RADIOACTIVE CONTAMINATED COASTAL ENVIRONMENTS

The future levels of radioactivity in coastal areas will depend not only on the future releases of radioactive materials to the environment but also on the behaviour of the already-contaminated areas, which can act as secondary sources of solution and particulate phase radionuclides for coastal environments. These releases can be further exacerbated by human activity, insofar as changes in land use, increase or decrease in agricultural activity, mining or construction, will affect sediment yield from the catchment and fluvial sediment input to the estuary. In addition, the impact of contaminated estuarine and coastal areas will be influenced by the future management of these areas, with major changes in use potentially affecting radiation exposure pathways. The possible changes include:

1. reclamation for agriculture, resulting in desorption and plant uptake of radionuclides;
2. reclamation for development, which could lead to dispersal of contaminated sediments to land during construction work, both inadvertently by wind and mechanical transport and deliberately via disposal of excavation spoil;
3. construction of training walls in the channel, which stabilize the channel and inhibit natural erosion cycles and increase the residence time of contaminated sediments (and encourage reclamation);
4. construction of tidal barriers and barrages, the latter, especially, encouraging permanent sediment deposition downstream and increase of sediment residence times and inventories;
5. construction of any structure in a coastal area, which will result in local sediment deposition or erosion affecting radionuclide storage and mobility;
6. the prospect of a rise in sea level caused by the greenhouse effect, which will encourage net sediment accumulation and radionuclide storage.

5.4 COASTAL, SEMI-ENCLOSED BASINS, SHELF AND CONTINENTAL MARGINS

5.4.1 INTRODUCTION

Coastal zones and shelf seas are considered here to represent the region from the low-water mark along open coasts to the shelf/slope oceanic boundary. This section also considers semi-enclosed basins representing distinct marginal zones having limited exchange with adjacent coastal and shelf regions. General descriptions of the physical, chemical and biological characteristics of such environments can be found in a number of texts such as Dyer (1986); the main processes of interest for the understanding of radionuclides are summarized in Section 5.1.

The sources of artificial radionuclides in the environment are described in the General Introduction. In regions remote from nuclear installations, inputs of artificial radionuclides in the marine environment have been dominated largely by fallout from nuclear weapon tests performed in the atmosphere. The coastal marine environment can also receive direct inputs of artificial radionuclides arising from the discharges of nuclear facilities: nuclear power plants, nuclear reprocessing plants and industrial users of radioactivity. The largest quantities of low-level radioactive waste discharges have resulted from nuclear reprocessing plants, either by direct discharge of the liquid effluents into the marine environment (Sellafield,

La Hague, Bombay) or via rivers (Hanford and the Columbia river in the Pacific; Marcoule and the Rhône river in the Mediterranean Sea). Of these sources, the Sellafield reprocessing plant has had the greatest impact and the largest number of studies carried out (see General Introduction and review in IAEA, 1985a). In contrast, routine releases from nuclear power plants and industrial users of radioactivity have had a much lower impact. The Chernobyl accident in 1986 produced a significant input of artificial radionuclides in the marine environment, both by direct fallout, especially on the Baltic, Mediterranean and Black Sea, and by indirect inputs, for example via the Dniepr into the Black Sea and thence into the Mediterranean Sea (Figure 5.13). As regards the open ocean, in addition to nuclear

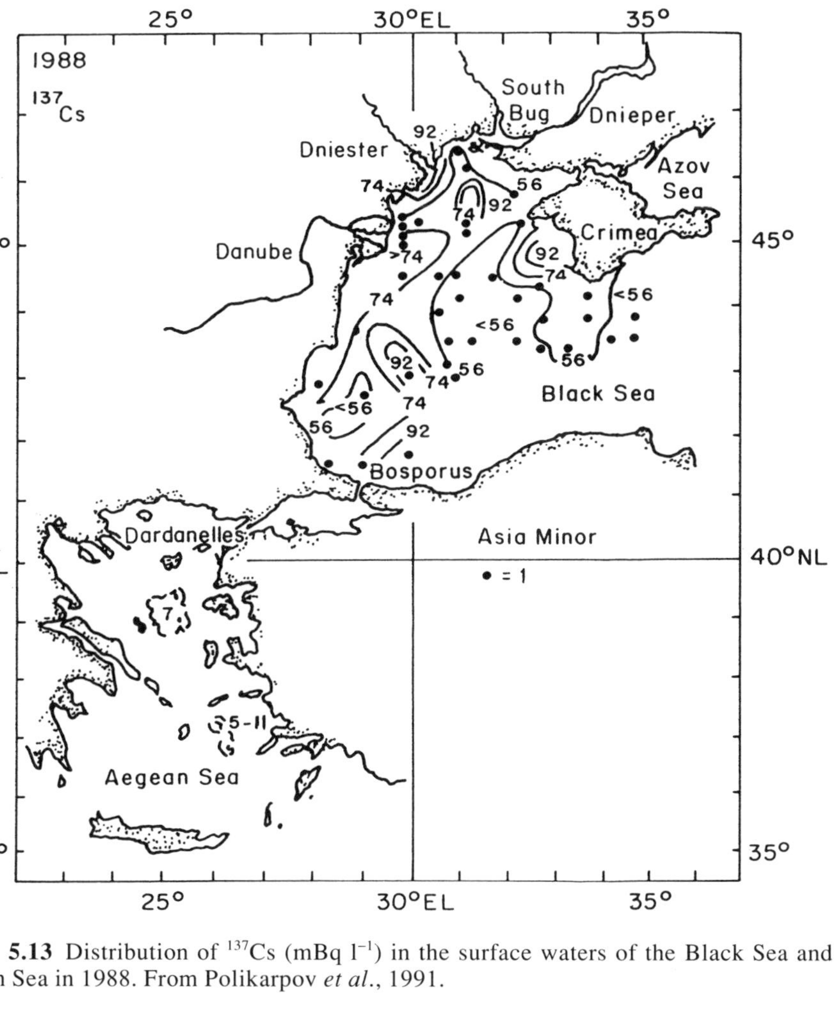

Figure 5.13 Distribution of ^{137}Cs (mBq l^{-1}) in the surface waters of the Black Sea and Aegean Sea in 1988. From Polikarpov *et al.*, 1991.

weapons test fallout, one must address the inputs due to the disposal of low-level radioactive wastes in the deep ocean, and be aware of the accidental releases from satellites re-entering the atmosphere and from nuclear-propelled submarine failures (IAEA, 1992a).

The physical and chemical form of radionuclides in the source term affect their initial behaviour and distribution in the receiving environment. It may be difficult to predict the chemical form of any specific radionuclide in view of the many processes which take place at a major reprocessing site, and the differences in technology which may occur with time and between sites. Of particular importance is the occurrence of particulate forms. In one study over 95 per cent of the Pu and Am in the Sellafield effluent was associated with a particulate fraction, comprising a ferric hydroxide floc, with a significant quantity present as discrete 'hot' particles, defined as clusters of alpha-tracks revealed using the CR-39 dielectric detector (Pentreath *et al.*, 1986). Such 'hot' particles, unequivocally having an effluent origin, have been observed in environmental samples and may persist for several months following release (Hamilton, 1985). It cannot be assumed that radionuclides from different sources will behave similarly. For example, ^{60}Co released from an experimental facility at Winfrith, UK, behaves differently from ^{60}Co released from Sellafield. Most of the Pu discharged from Sellafield is in the reduced form (III+IV), but there is a progressive change to the oxidized form on mixing with seawater. The K_d values of the reduced and oxidized forms differ by about two orders of magnitude. Studies carried out in the late 1970s showed that between 70 and 90 per cent of the plutonium present in the filtrate fraction was in an oxidized form, either Pu(V) or Pu(VI), while that adsorbed to the particulate fraction was almost entirely in a reduced form, Pu(III) or Pu(IV) (Nelson and Lovett, 1978). Little variation in oxidation state distribution was evident over a concentration range of 0.05–50 Bq m^{-3}. In contrast to coastal zones, it has been reported that the two oxidation state groups are present in almost equal concentrations in the open ocean. Little variation with depth is observed except near the bottom of the water column where the distributions change rapidly and become similar to those found in shallow waters. It is now generally accepted that the reduced category is mainly Pu(IV), very likely as $Pu(OH)_4$ adsorbed on negatively charged suspended sedimentary material, while the oxidized category is predominantly Pu(V) probably as PuO_2^+ and its complexes.

Of special interest is the comparison of the role of colloidal or pseudo-colloidal solutions in freshwater and seawater systems. The formation of colloidal solutions is always observed in solutions in which insoluble hydrolysis products can be formed. Pseudo-colloidal solutions consist of adsorption formations of various dimensions on foreign colloidal particles present in the solution. Compared to the freshwater environment, relatively few data have been reported on the association of radionuclides with colloids in seawater. One of the earliest studies to link the enhanced mobility of Pu and Am to colloidal aggregates was carried out in the lagoon waters of Bikini Atoll by Nevessi and Schell (1975) who reported that almost all of the Am in the soluble phase was very probably in the colloidal form.

Later work showed that about 12 ± 8 per cent (standard deviation) of the Pu in lagoon samples was present in a colloidal state. The presence of colloidal organic carbon (COC) is known to have a profound effect on the affinity of a number of radionuclides for the particulate phase (see Section 5.2.4.2). For example, the K_d coefficients of both Pu and Am have been shown to decrease significantly at COC concentrations in excess of about 1–10 mg l^{-1}. In open waters, however, COC concentrations are unlikely to reach these levels and the fraction of Pu and Am associated with colloids or pseudo-colloids is considerably smaller. Moreover, recent studies based on ultrafiltration techniques suggest that in saline environments colloidal Pu and Am appear in the < 10 000 Dalton size fraction, in sharp contrast to the freshwater environment. There is also evidence that the colloidal fraction of both elements is in a chemically reduced form, i.e. Pu(IV) and Am(III). Clearly, differences in chemical form may have a significant impact on the initial and subsequent behaviour of radionuclides in coastal waters, which is likely to be reflected in resulting pathways.

The behaviour of most radionuclides is linked to particle dynamics because of their affinity with particle surfaces. Thus, for radionuclides with a high K_d, such as Pu and Am, their fate will largely be controlled by sedimentation processes (sediment accumulation, resuspension and transport). In the Irish Sea, for instance, the distribution of plutonium broadly reflects the distribution of settled sediments, although the influence of local water movements is also important (Figure 5.14). Radionuclides with a low K_d (Tc, ^{3}H) will be controlled mainly by water mass circulation. Nuclides which have an intermediate K_d such as Cs will be predominantly influenced by the water circulation but also affected by sedimentation processes.

Surveys of radiocaesium concentrations in filtered waters from throughout the Irish Sea have been conducted on an annual basis for many years (Mitchell, 1967–77; Hunt, 1977–92). The extensive data available show that general levels reached a peak in the years between 1975 and 1978 and have been declining more or less steadily ever since. In Figure 5.15 the distribution of ^{137}Cs in the waters of the Irish Sea in the years 1974, 1976, 1981 and 1986 is reproduced and gives some indication of the temporal evolution of these levels as well as the decrease in concentration with increasing distance from Sellafield. Concentrations in the waters of the Irish Sea attained 5–50 Bq kg^{-1} and were more than three orders of magnitude higher than representative fallout levels at similar latitudes. Data on the distribution of other radionuclides in the waters of the Irish Sea are not as extensive, though the distributions of $^{239,240}Pu$ and ^{241}Am in filtered waters have been determined in a number of surveys. For both elements the fall-off in concentration with increasing distance from Sellafield is noticeably more rapid than for ^{137}Cs. The geographical distribution of Sellafield-derived radioactivity in the Irish Sea is in close accord with the pattern of water circulation once account is taken of the affinity of certain radionuclides for sedimentary material and the distribution of the fine sediments themselves. For example, the influx of Atlantic seawater from the south is clearly evident from the shape of the isolines of ^{137}Cs concentrations in seawater depicted

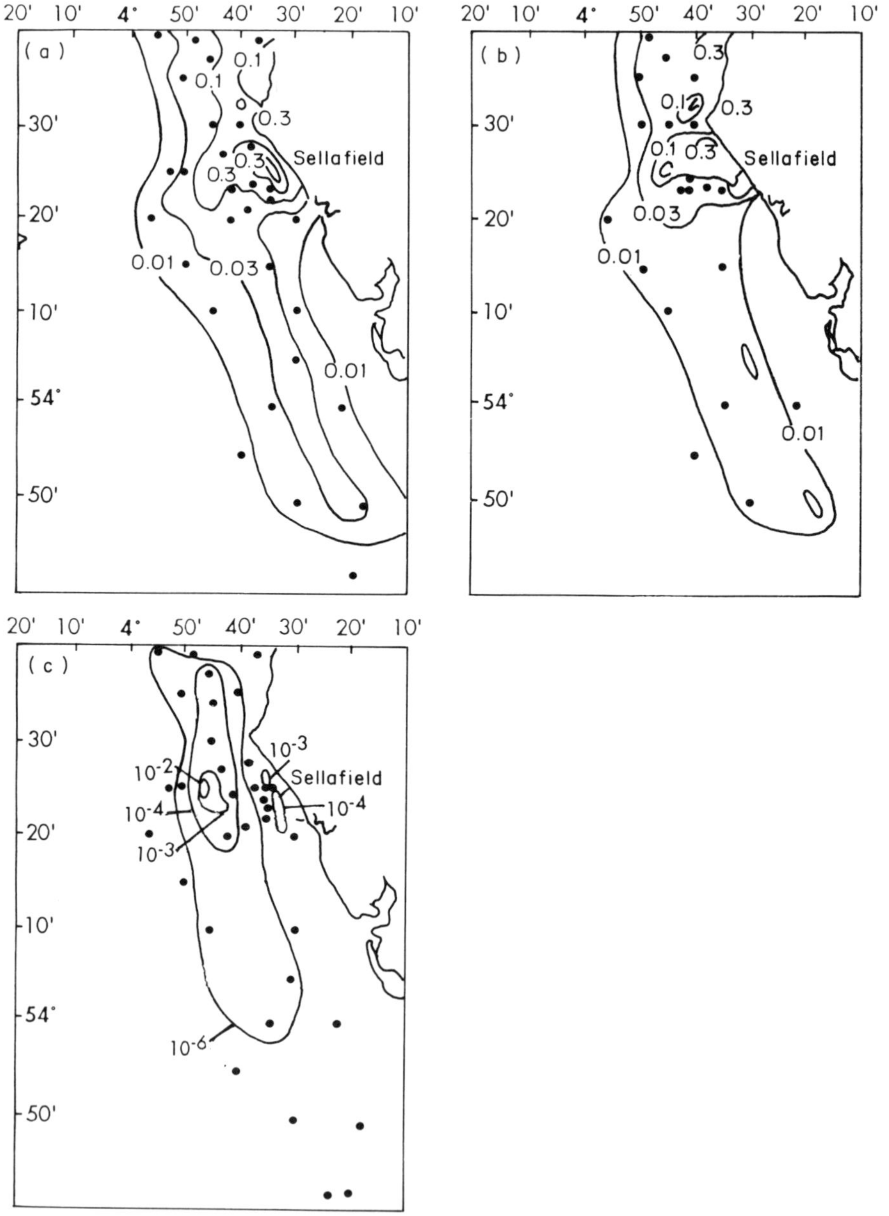

Figure 5.14 Inventory of $^{239,240}Pu$ in the Irish Sea ($MBq\ m^{-2}$) adapted from Woodhead, 1988. (a) Total core; (b) at depth > 5 cm; (c) at depth > 71 cm.

in Figure 5.15. It has been shown that the caesium discharged from Sellafield mixes with the seawater transported through the Irish Sea in a northerly direction, exits through the north channel and follows a well-defined path around the west and north coasts of Scotland where it enters the North Sea.

5.4.2 BIOMEDIATED PATHWAYS

5.4.2.1 Biomediation in the water column

The relative importance of biologically mediated removal, transformation and regeneration mechanisms of radioactive materials is a function of the production of oceanic biomass. At any given location, most of the oceanic biomass resides in the upper few hundred metres. Thus, the zone of most active biogenic particle production, particularly near oceanic margins, is in the surface layers where primary production occurs. Therefore, to understand fluxes and transfer of radionuclides it is important to evaluate the biologically mediated processes which regulate their biochemistry of uptake, removal and regeneration. This mechanism is of particular importance in upper ocean and deep marginal basins such as the Mediterranean Sea. The effect on the behaviour of radionuclides from global fallout has been studied over a period of 20 years but the Chernobyl accident in 1986 provided additional information that helped to quantify the effect of these processes. Three such studies, involving the use of sediment traps, have been reported for the North Sea, the Black Sea and the Mediterranean.

Primary production can occur on the sediment surface in coastal waters, in contrast to the open ocean, since in shallow waters the photic zone reaches the seabed. Films of microalgae, such as diatoms and other photoautotrophic microbes, develop on the surface sediment particles as light intensities increase during the spring. This benthic primary production has a cyclical pattern synchronized to that in the plankton with benthic microbes rapidly oxidizing particulate matter sedimented from the water column and returning nutrients, and in some cases sorbed radionuclides, in soluble form, to the water column. Indeed during the winter this process resupplies the water column with the nutrients which prime the plankton production cycle for the next spring.

The importance of scavenging and removal of surface-introduced radionuclides by incorporation into plankton and their particulate detrital products (e.g. faecal pellets, moults, eggs, carcasses and algal flocs) has been derived from both laboratory and field studies. In the north-western Mediterranean during the mid-1970s, freshly collected euphausiid faecal pellets contained on average 5.3 Bq ^{239,240}Pu kg^{-1}. Applying these relatively high concentrations to estimates of zooplankton faecal pellet production rates, plutonium fluxes via sinking faecal pellets were derived. This in turn led to the computation of an upper mixed layer residence time due only to sinking faecal pellets of 3.6 y for ^{239,240}Pu in this part of the Mediterranean. Only recently

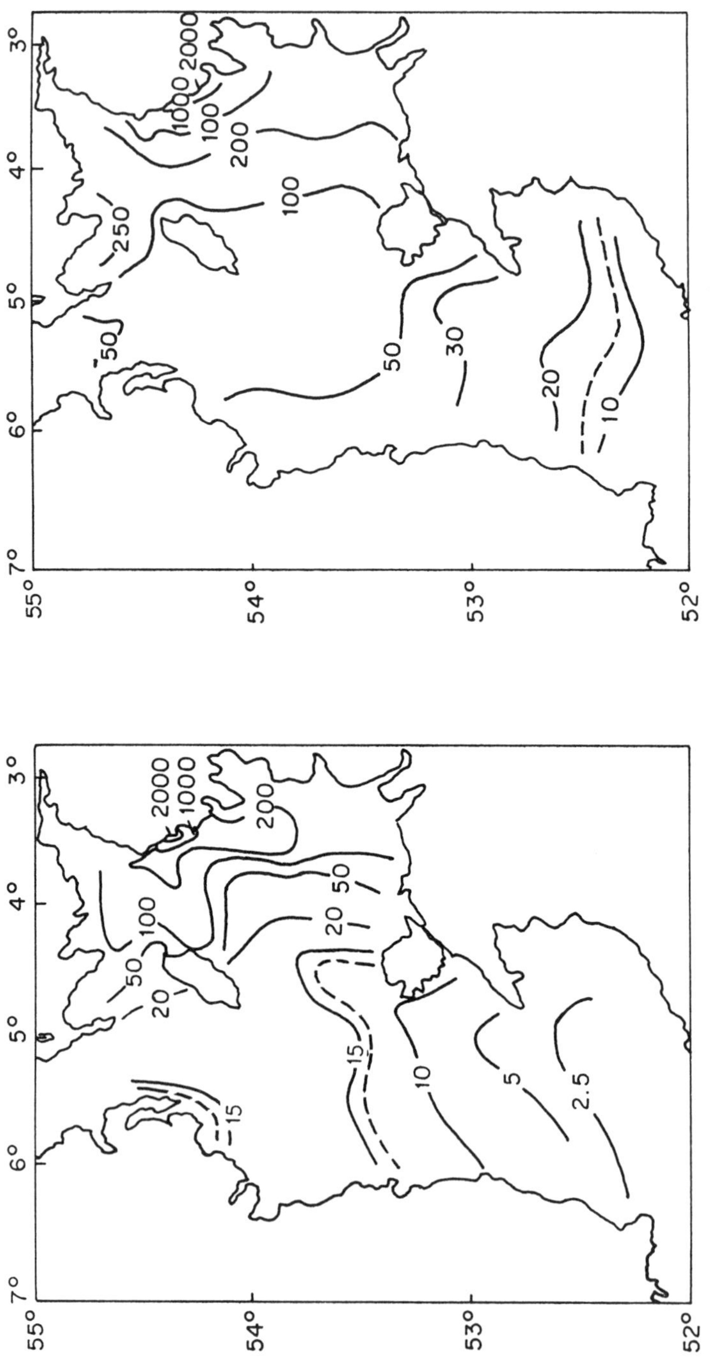

(b)
55°
54°
53°
52°
7°
6°
5°
4°
3°
2000
1000
100
200
100
250
50
50
30
20
10
(a)
2000
1000
200
100
50
20
50
20
15
15
10
5
2.5

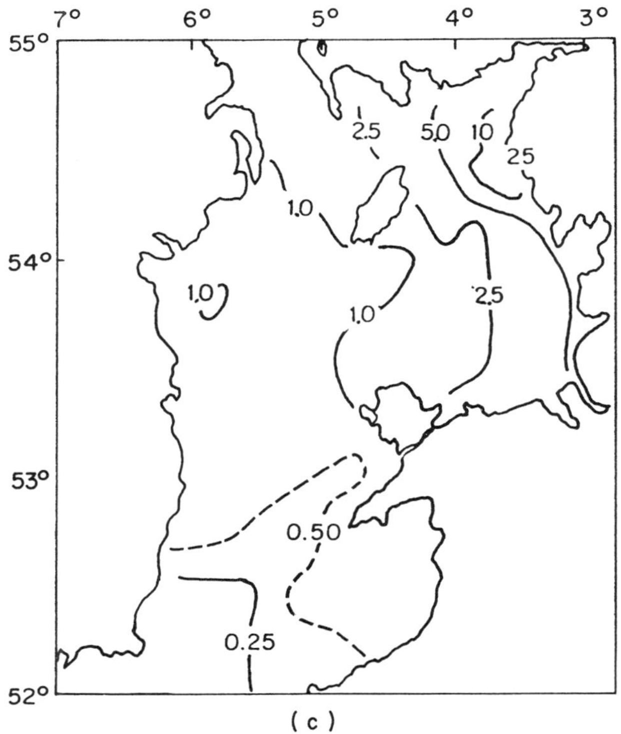

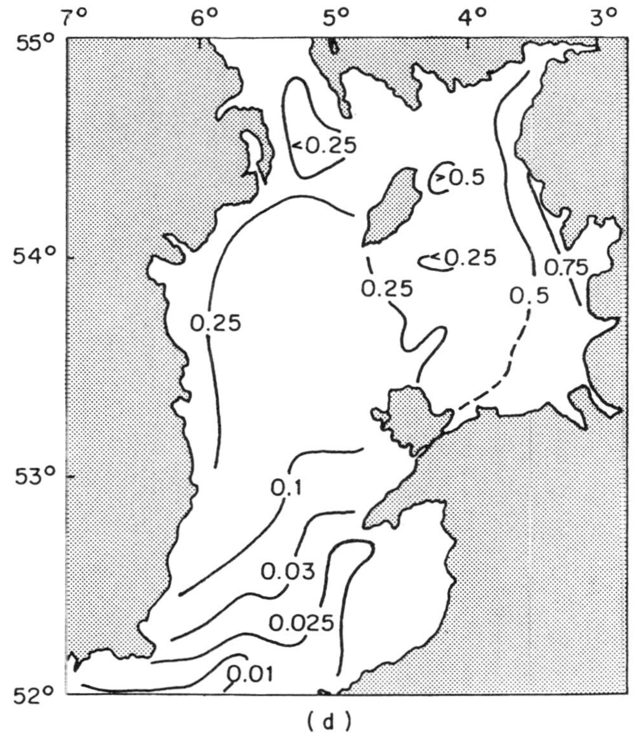

Figure 5.15 Distribution of ^{137}Cs in filtered surface seawater in the Irish Sea (a) July 1974 (pCi l^{-1}); (b) January 1976 (pCi kg^{-1}); (c) March 1981 (Bq kg^{-1}); (d) April 1986 (Bq kg^{-1}). The broken line represents the approximate position of the 0.5 Bq kg^{-1} isopleth in each case. Adapted from Mitchell, 1967–77.

have direct measurements of the vertical flux of $^{239,240}Pu$ been made by sediment traps moored in the Gulf of Lyons (Fowler *et al.*, 1990a). These plutonium fluxes, which were measured in an area where particulate mass fluxes are relatively high, resulted in a relatively short $^{239,240}Pu$ upper water layer residence time of 2.5y. The similarity in the two residence time estimates is noteworthy; however, characterization of the sediment trap material indicated that its composition was only partially biogenic. In this region, lateral transport of lithogenic material in the form of nepheloid layers also contributes to the vertical particulate flux, and that of associated radionuclides.

The most convincing data supporting the involvement of biota in the transfer and transport of artificial radionuclides in enclosed seas comes from experiments carried out prior to and following the Chernobyl accident (Fowler *et al.*, 1987, 1990b). In the Mediterranean, high-resolution, time-series, sediment trap measurements demonstrated that fission products arriving at the sea surface as a single pulse were rapidly transported to 200 m in approximately 7 d (Table 5.5). Examination of samples containing the bulk of the radioactivity indicated they were composed mainly of copepod fecal pellets, e.g. 70 per cent of the sample by mass. Fresh pellets collected from copepod grazing above the traps were found to contain similar radionuclide concentrations and ratios as those in the trap material, confirming that such pellets were responsible for removing the radioactivity to depth (Table 5.6). Furthermore, isotopic ratios in pellets, water and copepods indicated that particle-reactive radionuclides like ^{144}Ce, ^{141}Ce, ^{106}Ru, ^{103}Ru were scavenged to a far greater extent by sinking faecal pellets than the caesium nuclides (Table 5.6). This and other Chernobyl studies in the Black Sea have clearly demonstrated the rapidity (18–200 m d^{-1}) by which sinking biogenic particles can transfer surface-introduced particle-reactive

Table 5.5 Chernobyl fallout radionuclides in particles collected at 200 m depth in a 2200 m water column and in copepod faecal pellets from surface waters. (from Fowler *et al.*, 1987, 1990b)

	Sediment trap samples						Faecal pellets
	13–20 April 1986	20–26 April 1986	26 April –2 May 1986	2–8 May 1986	8–15 May 1986	15–21 May 1986	6 May 1986
^{95}Zr	< 0.07	< 0.2	< 0.3	< 0.2	24.5±1.4	< 0.2	1.4±0.8
^{95}Nb	< 0.03	< 0.1	< 0.2	< 0.1	31.8±1.1	< 0.2	< 0.2
^{103}Ru	< 0.06	< 0.1	< 0.2	3.7±0.2	23.6±1.0	14.0±0.4	16.0±1.9
^{106}Ru	< 0.2	< 0.4	< 0.8	1.1±0.5	5.4±1.8	3.5±0.7	5.8±2.9
^{134}Cs	< 0.05	< 0.05	< 0.05	0.41±0.05	2.1±0.2	1.9±0.1	3.4±0.6
^{137}Cs	< 0.05	< 0.05	0.15±0.08	0.85±0.08	3.8±0.3	4.0±0.1	6.3±1.0
^{141}Ce	< 0.2	< 0.2	< 0.3	1.3±0.7	12.6±0.6	1.1±0.5	0.9±0.4
^{144}Ce	< 0.06	< 0.3	< 0.3	< 0.2	13.6±0.7	< 0.4	2.5±1.3
$^{239+240}Pu$	5.43	2.00	3.00	3.22	9.70	4.71	7.4
^{241}Am	0.87	0.68	1.51	1.05	3.63	2.83	0.63

Values are decay-corrected for midpoint of sampling period.

Table 5.6 Selected radionuclide activity ratios of Chernobyl-derived radionuclides in different samples from the north-western Mediterranean (from Fowler *et al.*, 1987, 1990b)

Sample	Collection date (1986)	Activity ratio $^{137}Cs/^{141}Cs$	$^{103}Ru/^{141}Ce$	$^{103}Ru/^{137}Cs$	$^{241}Am/^{239+240}Pu$
Air	3 May	58	110	1.9	0.13
Unfiltered seawater	7 May	11	8.0	0.73	—
Zooplankton	6 May	1.7	14	8.2	0.22
Fresh faecal pellets	6 May	7.0	18	2.5	0.09
Trapped particles at 200 m	11 May	0.27	1.8	6.8	0.37

radionuclides to depth (Buesseler *et al.*, 1990). Whereas these studies underscore the importance of micro-crustacean faecal pellets in radionuclide flux, pellet deposition from salps, a gelatinous macroplankton species common in the Mediterranean, would also be an effective mechanism since Mediterranean salps produce pellets highly enriched in natural radionuclides and at rates greater than those of planktonic crustaceans (Krishnaswami *et al.*, 1985).

The effectiveness of this biological packaging mechanism in removing surface-introduced radionuclides and transporting them to depth should not be underestimated. For example, based on total wet and dry fallout deposition data obtained near the Mediterranean trap site, it was calculated that 50 and 75 per cent of the total Ce and Pu deposited on the sea surface had fluxed through 200 m in the form of large particles approximately 1 month after the accident. In contrast to these particle-reactive radionuclides, only a small fraction (0.2 per cent) of the ^{137}Cs deposited in the region was carried downward by the same particles. What is not evident is to what depth the biogenic particles transported these radionuclides. There is some evidence from deep sediment cores that even up to 1 year after the accident, Chernobyl radionuclides had not reached the sediments in certain areas of the north-western Mediterranean (Thommeret and Huynh Ngoc, 1990). This may result from a gradual release of radionuclides from particles through desorption or decomposition processes as the particle sinks into deeper waters, as has been demonstrated for Chernobyl radionucldes in the Black Sea.

5.4.2.2 Biomediation in the seabed

Benthic infauna have evolved a number of feeding strategies to exploit the organic content of the suspended particles over the seabed. The major nutrient sources are abundant in the top few centimetres, whereas deeper sediment represents a relatively poorer food source. The benthos can influence the behaviour of radionuclides in a number of ways. Filter-feeding forms increase the fluxes of particulate bound radionuclides across the sediment–water interface. Other surface-feeding forms may deposit a proportion of this material at depth in the form of faecal material in a manner of a conveyer belt. Conversely, deeper burrowing forms evacuate low-activity sediment onto the surface. The presence of burrow systems, opening at the sediment–water interface, results in increased ventilation of the sediment by irrigation. This in turn influences the chemical cycling of redox-sensitive elements such as Fe, Mn and Pu (see Section 5.4.3). Bioturbation has been shown to be the dominant mechanism by which Sellafield-derived radionuclides are transported below the sediment–water interface (Kershaw *et al.*, 1992; Section 5.4.4.1). The representation of bioturbation, a complex interaction of many different processes, by a single coefficient has been criticized by many authors and other models have been suggested (Berner, 1980; Jumars *et al.*, 1981; Aller, 1983; Wheatcroft *et al.*, 1990). However, these models lack general applicability and the commonest approach has remained that of modelling bioturbation as a diffusive process. The diffusive model assumes that processes affecting bioturbation are temporally and spatially continuous. They are clearly not, as inputs to the sediment are often episodic, especially anthropogenic inputs, or vary over a seasonal cycle and often occur in limited areas. Further, the infauna are not evenly distributed either in the horizontal or vertical planes, patchy distributions being common, and abundance and biomass vary with seasonal and other temporal cycles. The diffusive model also assumes that sediment particles are mixed randomly and isotropically over local distances, that is over dimensions similar to that of the grain sizes themselves. However, again, these are not good assumptions as most mixing is over non-local distances and usually most strongly in the vertical direction.

5.4.3 DIAGENESIS

The extent to which sediments retain contaminants depends, in part, on the nature of the contaminant–particle interaction and on the effect of diagenetic processes. Diagenetic reactions encompass all of the chemical and physical reactions which occur in sediments that bring about significant changes in the chemistry of both the solid (sediment) and liquid (interstitial water) phases. Assessment of the relative mobility of different radionuclides has often been made by comparing only the solid phase distribution of the elements in sediment profiles. Few studies have, however, used the most sensitive method for examining radionuclide mobility directly, that is, by the study of the composition of interstitial waters. This is no doubt due to the significant analytical/technical problems that are inherent in such studies. It is necessary to ensure that measurements are representative of chemical conditions within the seabed (temperature, pressure and oxidation changes) and that sufficient sample is collected to determine the concentrations, chemical associations (e.g. colloids) and oxidation states.

The diagenetic changes which occur in shelf sediments are driven by the bacterially mediated oxidation of the organic matter which is deposited in the sediments. The major chemical change which occurs in most sediments is the change from oxic to anoxic conditions. The significant drop in the redox potential of the system may, of course, have a direct effect on radionuclides which can occur in a number of different valence states, for example plutonium.

The major diagenetic processes that occur in sediments are the reduction of oxygen, manganese, nitrate, iron and sulphate and finally fermentation. These processes tend to occur in the order of decreasing free energy yield and this results in zones of activity arranged horizontally with depth in the sediments, with oxygen as the first electron acceptor in the sequence.

As a result of diagenesis, significant changes occur in the chemistry of particles, especially their surfaces. Radionuclides adsorb to particle surfaces. Some, like caesium, may also be absorbed into the mineral lattice of sediment particles, readily exchanging with potassium. The nature of the interaction will of course determine the subsequent behaviour of any particular radionuclide. For instance, a radionuclide adsorbed to the iron and manganese surface coatings of sediment grains will be subject to mobilization, as the nature of the surface changes, on transition from an oxic to an anoxic regime. The metal oxides will be solubilized and the radionuclide released to react with different mineral substrates. It will be the kinetics of the reaction in competition with the rate of transport which will determine the extent to which any radionuclide will be relocated or lost from the sediment. Diagenetic processes also bring about significant changes in the composition of the interstitial water, with marked increases in the concentrations of total carbon dioxide, phosphate, ammonium, ferrous ion, manganese and dissolved organic matter. The speciation of dissolved radionuclides may alter in such a way as to increase mobility. For instance, interaction of radionuclides with dissolved organic matter may enhance solubility. It will be in fine-grained sediments that diagenetic processes are best expressed, and these are also the sediments which contain the highest concentrations of radionuclides.

Two shelf areas have received detailed study of both sediments and the contained interstitial water: (a) the north-west Atlantic shelf (Sholkovitz and Mann, 1984); and (b) the north-east Irish Sea (Malcolm *et al.*, 1990). The north-east Irish Sea study site has a distinct analytical advantage in that the concentrations of artificial radionuclides in the sediments are relatively high due to the proximity of a major source: the discharges from the Sellafield reprocessing plant. In such studies, however, relatively large volumes of interstitial water have to be extracted under controlled conditions, which makes the sample collection operation very demanding technically. The higher concentrations present in the north-east Irish Sea enabled the analytical separation of the higher and lower valence states of plutonium adding a useful dimension to the study in terms of particle reactivity—the reduced form of plutonium adsorbs to particles more readily than the oxidized form (Section 5.4.1).

The data shown in Figure 5.16 derive from a site near to the Sellafield discharge point in the Irish Sea. The two artificial radionuclides ^{239,240}Pu and ^{241}Am show major subsurface peaks of concentration which are not related to the solid phase

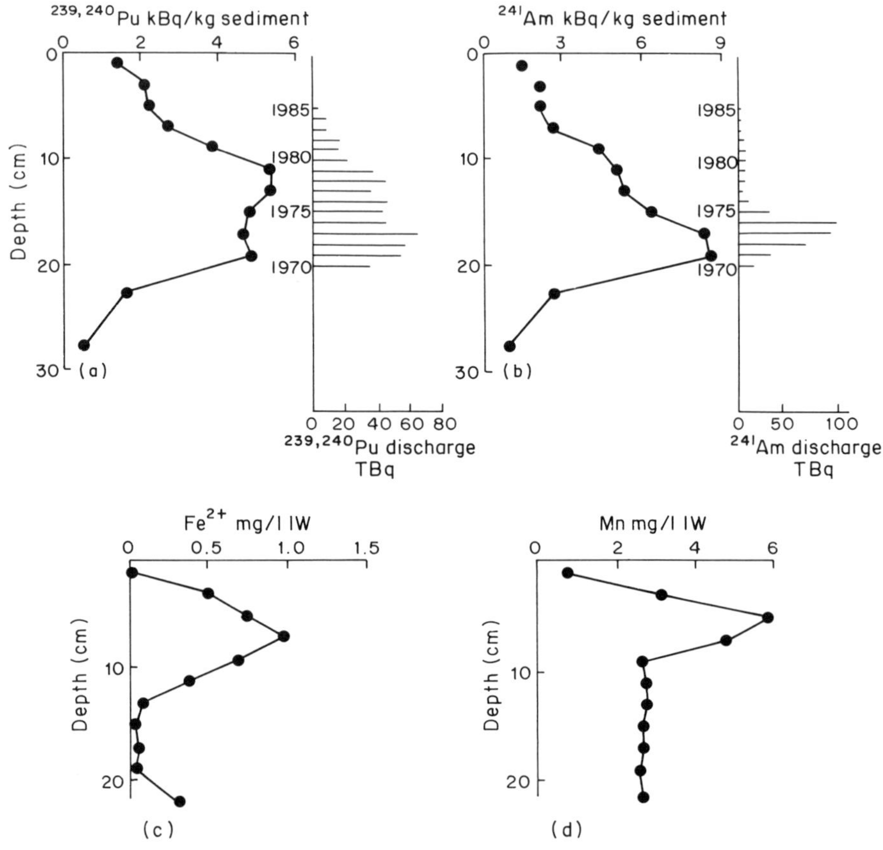

Figure 5.16 Solid-phase profiles of (a) $^{239,240}Pu$ and (b) ^{241}Am, ($kBq\ kg^{-1}$), from a sediment core close to the Sellafield discharge pipe. The concentrations profiles show a close correlation with the discharge histories (TBq) and do not appear to be influenced by iron (c) or manganese (d) cycling. From Malcolm *et al.*, 1990.

distributions of iron, manganese or organic carbon but may reflect the input due to the discharge from Sellafield (the discharge history is shown on Figures 5.16a and 5.16b). This would imply that diagenetic changes in these sediments are not able to remobilize significant amounts of plutonium or americium in timescales of several years. Examination of the interstitial water data again shows little evidence of any correlation between the artificial radionuclide concentrations and the major indicators of diagenetic processes in these sediments (Figures 5.16c and 5.16d). Only the interstitial water profiles of iron and manganese are shown but nitrate, phosphate and dissolved organic carbon were also measured. Association of the transuranic elements with dissolved organic matter has been suggested as an important factor in determining the behaviour of these elements in freshwaters. However, there is only limited evidence for this in marine interstitial waters. Unpublished studies in the north-east Irish Sea sediments using ultrafiltration techniques suggest that some plutonium and some americium is associated with material of molecular weight greater than 10 000 Daltons (Malcolm, unpublished data; Section 5.4.1). The rate of diffusion of high molecular weight complexes is likely to be very slow, so redistribution of the plutonium and americium is likely to be of limited importance.

In the surface sediments at this site between 27 and 69 per cent of the plutonium is present in the oxidized form. This contrasts with the underlying sediments where all of the plutonium is in the reduced form. The apparent distribution coefficient is within the range expected for reduced plutonium in non-carbonate marine sediments, and indicates the very strong particle association of this radionuclide. Close inspection of the distribution coefficient profile suggests that mobilization may be taking place but the signal is not sufficient to suggest major relocation of plutonium or americium within the sediments.

The studies from the north-west Atlantic shelf also suggest that diagenesis does not have a significant effect on the mobility of plutonium; interstitial water plutonium was found to be in equilibrium with the solid phase plutonium concentrations. The apparent distribution coefficient varied with the carbonate content of the solid phase.

The behaviour of caesium in marine sediments has not been well studied in interstitial waters but it has generally been assumed that caesium is more mobile than elements like plutonium. There was some evidence for this in studies reporting the solid phase concentrations of the elements, suggesting that the greater penetration of caesium into sediments was due to its greater mobility. The part that diagenesis plays in this process has not been defined. One study in Buzzards Bay (Sholkovitz and Mann, 1984) suggests that caesium in marine sediments behaves in the same way as caesium in freshwater sediments and that ion exchange reactions are important. Notable in this connection is the increasing mobilization of caesium with increasing concentrations of ammonium, a product of diagenetic reactions in anoxic sediments. While the Buzzards Bay data strongly suggest mobilization, it was not possible to quantify the process.

Natural series radionuclides are used extensively to study the rates of particle transport in marine sediments. An assumption underlying many of these studies is that the nuclides being used are strongly particle associated and not subject to diagenetic remobilization. The assumption appears to be justified in the use of thorium isotopes, though there has been little study of thorium mobility in marine sediments. Interstitial water data from Buzzards Bay (Cochran *et al.*, 1986) for thorium shows depth-invariant low concentrations of ^{232}Th and ^{238}Th in some nearshore sediments, whereas ^{234}Th did show remobilization associated with the redox-driven dissolution of iron and manganese. The implications for the use of ^{234}Th as a tracer have not yet been fully considered.

In contrast to thorium, the behaviour of uranium in sediments is more widely influenced by diagenetic processes. Uranium, like plutonium, is present in a number of redox states in the environment. Reduction of U(VI) and U(IV) occurs together with the reduction of iron and manganese. This leads to the removal of uranium from interstitial waters at the same depth as iron and manganese are dissolved. At greater depth in the sediment, uranium is released from the sediments while the organic matter with which it is associated is oxidized and the interstitial waters alkalinity increases accordingly.

The use of ^{210}Pb for estimating sedimentation rates and sediment mixing also relies on the assumption that it is predominantly particle-associated and non-mobile. This is likely to be true in sediments which are anoxic below a very narrow surface oxidized layer. This includes the sediments in California borderland basins where the use of ^{210}Pb was first established. However, in the suboxic layer of sediments, where there is mobilization of Fe and Mn, Pb may be mobilized. The implications of this for the use of Pb in measuring various rates has not been sufficiently studied.

More studies of the behaviour of radionuclides in interstitial waters and sediments are required, as they will provide increasingly important tools for the study of sediment biogeochemical processes.

5.4.4 TRANSPORT AND ENHANCED BOUNDARY SCAVENGING

In general, the effect of discharging effluent from a point source into a high-energy coastal environment will be to dilute and disperse. However, locally enhanced concentrations can occur and the degree to which these are maintained, increase or even decrease will depend on the discharge history and the timescales of the various geophysical, chemical and biological processes responsible for nuclide migration.

5.4.4.1 Coastal mechanisms

A considerable volume of work on the mechanisms controlling the behaviour and transfer of radionuclides in coastal regions of the open sea has been generated by the discharges from the Sellafield nuclear fuel reprocessing plant (Kershaw *et al.*, 1992). Some of these studies have been referred to in an earlier section (5.4.1).

The history of Sellafield releases is provided in the General Introduction, and the resultant distribution of Cs in seawater with time is described in Section 5.4.1. Although plutonium from Sellafield is found throughout the Irish Sea in both reduced and oxidized states, research based on selective precipitation has shown that more than 90 per cent of the ^{239,240}Pu in filtered surface waters exists in the oxidized forms and suggests that plutonium may be transported throughout the Irish Sea as a soluble species by residual currents. It should not, however, be inferred that this is the dominant mechanism of transport. In inshore waters the situation is greatly modified. Whilst the total concentration is similar to that observed in corresponding offshore samples, the proportion of particle-associated to soluble plutonium is much higher. This is fully consistent with the fact that plutonium becomes rapidly reduced in the presence of scavenging particulates. The reduced state is considered to be Pu(IV) and is mainly, though not exclusively, associated with the particulate phase. In inshore waters, tides and currents cause large quantities of sediment to undergo a continuous cycle of resuspension and deposition, with the result that scavenging is enhanced.

The speciation of plutonium and americium in seawater and the mechanism by which they can, in certain circumstances, form colloids, is fundamental to an understanding of how transuranic nuclides become dispersed in the marine environment and warrants further study.

The bulk of Sellafield-derived transuranic radionuclides now reside, in association with deposited sediments, relatively close to the outfall and, in particular, in the mud patch extending south from St Bees Head as well as in the local estuaries. A smaller fraction, composed of plutonium adsorbed onto fines in suspension and plutonium in a soluble or colloidal form, is dispersed throughout the Irish Sea with an estimated 3–17 per cent of each year's discharge being removed through the North Channel (Pentreath *et al.*, 1986). Radionuclide measurements were made at three selected sites along the eastern shore of the Irish Sea in the period 1968–87. The variation in the concentration of ^{137}Cs in *Fucus vesiculosus* samples closely reflects that of the ^{137}Cs discharges from Sellafield, which were at their peak in the mid-1970s. There is also evidence from these and other data which suggests that the decline in the concentrations of ^{137}Cs and other fission and activation products in environmental materials is most rapid close to the Sellafield outfall. Examples of the temporal evolution in anthropogenic radionuclide concentrations in algal indicators can be cited to confirm that a general decline in these concentrations is taking place in the biological compartment throughout the Irish Sea.

Substantial reductions in the concentrations of a number of radiologically important radionuclides in fish and shell-fish sampled in the eastern region of the Irish Sea have been observed since the mid-1970s. Concentrations in both fish and shell-fish diminish with distance from Sellafield, the rate of reduction being least for nuclides which are relatively mobile in seawater. A similar pattern is observed for seawater and algae. Variations in concentrations between fish species sampled in a given area are comparatively small and can be explained in terms of residence time in the area, as well as feeding habits. On the other hand, substantial variations are observed between shell-fish species. For example, molluscs tend to concentrate the transuranics to a considerably greater extent than crustaceans, which in turn accumulate them more than fish (IAEA, 1985b).

The principal feature observed in concentrations of radionuclides in surface sediments sampled close to the outfall is a general reflection of the reduction in discharges since 1978. As expected, the rate of reduction is least for nuclides which both exhibit strong sediment-seeking properties and have long half-lives (McCartney *et al.*, 1992). The concentrations of ^{238}Pu, ^{239,240}Pu and ^{241}Am in Newbiggin surface sediments sampled in the period 1977–87 confirm that these elements are subject to dispersion, albeit slow. A similar, though weaker trend, was evident in surface sediments sampled at Heysham. Further afield, the reduction in transuranic concentration is not always evident; for example, at Garlieston in SW Scotland, the concentrations of Pu and Am in silt, if anything, are still increasing, whereas those of fission products such as ^{95}Zr–^{95}Nb, ^{106}Ru, and ^{137}Cs are clearly falling.

Prior to the Chernobyl accident in 1986, the radionuclides ^{134}Cs and ^{137}Cs were regarded as near ideal tracers for continental shelf water circulation because: (a) the half-life of the latter (30 years) is long compared to typical transport timescales on the shelf; (b) the half-life of the former (2 years) is distinctly shorter than that of ^{137}Cs, which means that the ratio of these two nuclides can be used to provide an alternative determination of transport times; (c) being soluble in seawater, they are not easily scavenged from the water column by particulate matter in suspension; (d) both had been discharged from Sellafield in relatively large, known quantities for many years; and (e) neither occurs naturally, though a low background had resulted from earlier nuclear weapons tests in the atmosphere. Therefore, caesium isotopes have been widely used in marine transport studies in the north-east Atlantic, and in particular in the Irish Sea and the North Sea. For example, the ^{134}Cs/^{137}Cs ratio has been used to estimate the turnover time for water entering the Irish Sea and to trace the path of Sellafield effluent to the coast of Greenland.

The activity ratio ^{238}Pu/^{239,240}Pu in releases from nuclear fuel reprocessing plants has varied systematically and differed from that of fresh fallout and weapons-grade plutonium. It has been used to identify source terms and transfer factors. Presently, this ratio lies in the range 0.22–0.25 in marine environmental materials, including surface sediments, sampled close to Sellafield (Hunt, 1977–92) and is only slightly higher than the ratio measured in similar materials from the east coast of Ireland. Analyses of the ^{238}Pu/^{239,240}Pu ratio in sediment cores and, in particular, in the linings of *Maxmuelleria lankesteri* burrows, serve to emphasize the role of bioturbation in the incorporation of the transuranics within the deeper sediments (Kershaw *et al.*, 1984).

Attempts have been made to estimate transport rates and availability times from isotope ratios in environmental materials (e.g. Hunt, 1985; Schell *et al.*, 1989; Mackenzie and Scott, 1982) with varying degrees of success. It does appear that particle-reactive radionuclides are being transported away from the vicinity of the Sellafield discharge point in both particulate and dissolved phases. Estimates of actual rates of removal are complicated by the time-varying isotope ratios in the discharge, variations in the actual quantities of each radionuclide discharged, differential half-lives, grow-in (e.g. ^{241}Pu–^{241}Am) and the mixing of materials contaminated by recent discharges with those contaminated previously.

In most cases the present distribution of contaminated sediments and biota must be considered as representing a temporary 'snap-shot', resulting from complex interrelated processes varying both spatially and temporally. The area of muddy sediments lying in deep-water between the Isle of Man and Ireland may be considered as a probable long-term sink, but there is little evidence to suggest that sediment in any other region of the Irish Sea will act in this way (Kershaw *et al.*, 1988).

Studies, conducted on soil samples, airborne deposits, muslin screens, mosses, lake sediments and sheep faeces taken in West Cumbria, UK, since the late 1970s have shown that a small fraction of the plutonium and americium discharged into the Irish Sea from Sellafield has been transferred into the atmosphere and returned to land. The considerable volume of work undertaken has recently been reviewed (McKay and Pattenden, 1990). Bubble scavenging in the water column, together with droplet ejection from bubble bursting at the surface, may be involved in the transfer mechanism (Walker *et al.*, 1986). A similar phenomenon has been reported for heavy metals.

The soil samples were taken along transects running inland from the Cumbrian coast to about 20 km and were analysed for ^{137}Cs and Pu. The transect measurements show enhanced values of ^{239,240}Pu near the coast, between 3 and 11 times the weapon fallout baseline. These relax to the baseline value at about 5 km and further inland. Preliminary estimates have been made which suggest that 40–80 GBq of excess Pu had been deposited in a coastal strip approximately 5 km wide and 40 km long by about the year 1980. In contrast, the caesium values along the transects can be accounted for on the basis of levels in bulk seawater with little or no evidence of marine aerosol enrichment. Sea-to-land transfer has also been reported near the Cap de la Hague facility on the French Channel coast and the Dounreay reprocessing plant on the northern Scottish coast. In the latter case transport inland is rather limited and the postulated mechanism is wind-blown spume (stable foam). Fine-sediment is an important component in the sea-to-land transfer process. For this reason the relative importance of this pathway will increase, as discharges from Sellafield continue to be reduced, as a result of the legacy of sediments contaminated in the 1970s and 1980s.

5.4.4.2 Cross-shelf transport and boundary scavenging

The degree to which radionuclides released into coastal waters are transported across the continental shelf is dependent upon the chemical properties of the radionuclides and the nature of the physical environment in which dispersion is taking place. A good example is given by the dispersion of radionuclides from the Hanford reactor on the Pacific coast.

The operation of Pu-producing reactors at Hanford, Washington, beginning in 1944, introduced several artificial radionuclides to the Columbia River and subsequently to the adjacent continental shelf and slope (Figure 5.17). An estimated 300 000 Ci y^{-1} of artificial radioactivity was discharged to the river by the mid-1960s, but decreasing reactor operations and more stringent discharge controls essentially eliminated radioactive discharges to the river by 1981. Radioactive decay of short- and moderate-lived nuclides, sediment export and river flow combined to reduce

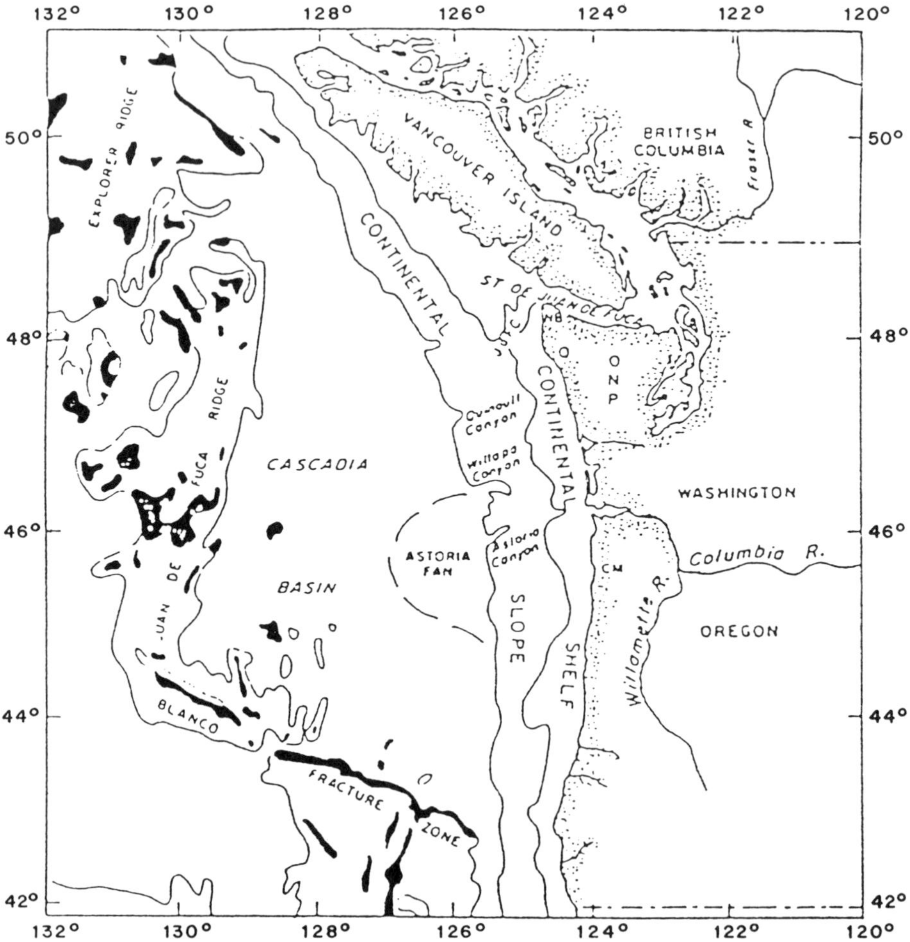

Figure 5.17 General study area off the Washington and Oregon coasts. NB = Neah Bay/Cape Flattery, Wa. ONP = Olympic National Park, Wa. JdF = Juan de Fuca submarine canyon. Puget sound is the inland body of water east of the Olympic National Park. The Hanford Nuclear Reservation is located on the Washington side of the Columbia River and just east to this chart.

artificial activities in river sediments to levels which, by 1984, could no longer easily be measured. Studies since the mid-1970s have focused on chemicals which are of concern due to their persistence and relatively toxic nature, such as the isotopes ^{241}Am and ^{137}Cs and the natural isotopes ^{210}Po and ^{210}Pb.

The artificial radionuclides have served as instructive tracers of Columbia River-derived material in the adjacent ocean. Seasonal changes in position of the Columbia

River plume were clearly evident in early studies of ^{65}Zn activity. ^{65}Zn activity in offshore sediments (Cutshall *et al.*, 1973) showed that the predominant transport of Columbia river-derived solids was along a midshelf silt deposit trending north-north-west from the river mouth towards Quinault submarine canyon (Figure 5.17). Integrated ^{65}Zn activities were at least two-thirds of the total ^{65}Zn discharged by the river, suggesting that fine-grained riverine suspended sediment was transported offshore and not trapped in the estuary. The presence of the short-lived isotopes $^{141}Ce/^{144}Ce$ and $^{95}Zr/^{95}Nb$ in sediments from Astoria Canyon was one of the first clear indications that radionuclides entering the ocean at the surface could be transported to depths of over 1 km within a few days (Osterberg *et al.*, 1963).

Pu (and ^{241}Am) activities and inventories in Washington shelf and slope sediments are 5–15 times those in sediments of the north-west Atlantic and Gulf of Mexico (Carpenter *et al.*, 1987). The Washington–Oregon coasts are in the latitude band of high atmospheric fallout of artificial nuclides, but even inventories of 2.2 ± 0.5 mCi km^{-2} estimated from terrestrial soil cores cannot account for the exceptionally high Pu activities in Washington shelf and slope sediments. Shapes of Pu activity profiles in both shelf and slope sediments show that Pu inputs to the sediments are continuing to the present—the input cannot be modelled as a pulse in the past. The continuing Pu input to the sediment is supported by a time series of Pu activities in mussels.

Inventories of Pu isotopes in bottom sediments from Hanford to the Columbia River estuary were estimated based on activities measured in 50 cores (Beasley and Jennings, 1984). $^{240}Pu/^{242}Pu$ and $^{240}Pu/^{239}Pu$ isotope data established that only about 3.5 per cent of the Pu in fine-grained sediment in the river estuary came from Hanford reactors. Thus high Pu activities in offshore sediments are not due primarily to additional Pu input from reactors at the Hanford reservation. Monthly samples of water collected near the mouth of the Columbia River between July 1978 and July 1979 were analysed for Pu and Am to estimate the quantities of these nuclides exported to the adjacent coastal zone during this year (Beasley *et al.*, 1981). From the depositional history of Pu activities recorded in a fine-grained, non-bioturbated sediment core from Youngs Bay (within the estuary), 4–8 Ci of $^{239,240}Pu$ and 1–2 Ci of ^{241}Am were estimated to have been discharged by the river to the adjacent coastal ocean since the late 1950s. Such riverine discharge of Pu and Am is small compared to total Pu sedimentary inventory of 35 Ci estimated for the Washington shelf alone.

The importance of non-riverine Pu source is seen in that, while Al contents remain relatively unchanged in moving from the estuary to and along the shelf, Pu activities of the fine-grained sediments increase from about 20 dpm kg^{-1} to 130 dpm kg^{-1}. Further evidence for major additional non-riverine Pu in shelf sediments is also seen by comparing $^{240}Pu/^{239}Pu$ isotope ratio signatures of offshore sediments with ratios in sediments accumulating in the estuary and further up rivers (Beasley *et al.*, 1982). No measurements of dissolved Pu concentration profiles have been reported for ocean waters off Oregon or Washington, but it seems reasonable that

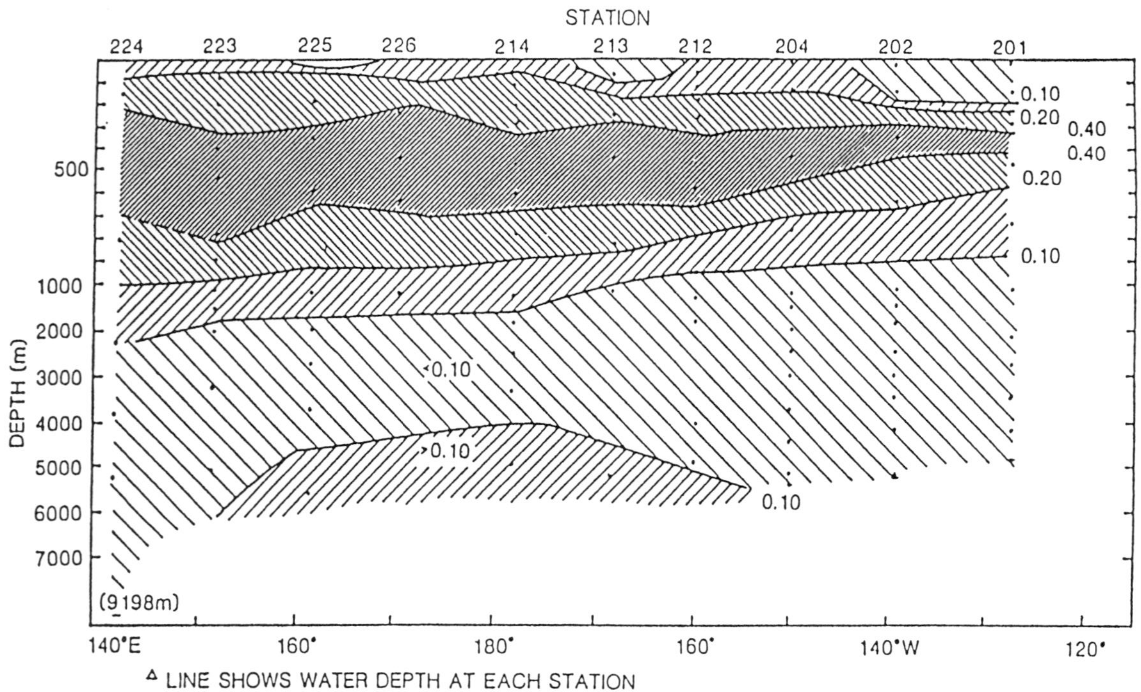

Figure 5.18 Pu concentration profiles (dpm 100 kg^{-1} seawater) in an east–west section across the Pacific at 30–35°N in 1973. From Bowen *et al.*, 1980

such profiles will resemble those found around large parts of the North Pacific in having a pronounced subsurface maximum around 400 m deep (Figure 5.18) due to dissolved Pu removal from surface waters by association with largely biogenic particles and Pu release to deeper waters after sinking and decay of most of the organic particles. This Pu remains in solution at mid-depth, due to lack of particles to scavenge it. Even when the equilibrium partition coefficient (K_d) of an element between particulate and dissolved phases is as high as 10^7, over 50 per cent of the element should still be in dissolved form when particle concentrations are ≤ 0.1 mg l^{-1}. Significant scavenging of particle-reactive, dissolved chemicals occurs when open ocean water enters the Washington coastal zone and encounters 10–100 times higher particulate concentrations due to the Columbia River discharge and the biologically productive nearshore environment. Frequent resuspension of bottom sediments by storms, bottom currents and benthic organisms also increases particle concentrations and enhances scavenging in near-bottom waters. Boundary scavenging has only been confirmed for ^{210}Pb, ^{241}Am and Pu off the Washington coast but should also be an effective mechanism for removal of other particle-reactive elements such as Be.

Waters from 200–300 m depth are now known to be upwelled onto Washington and Oregon coastal zones (Hickey, 1989). Conservative estimates based on water volumes suggest that the advective supply of these isotopes considerably exceeds their observed sedimentary inventories and fluxes. Sedimentary inventories of these chemicals on the Washington coast are therefore limited by boundary-layer scavenging reactions rather than by supply of dissolved chemicals. Particle concentrations in coastal water and published sediment–water Pu distribution coefficients combine to suggest that only 18-20 per cent of Pu flowing past should be removed to sediments, and this amount is indeed adequate to supply the observed sediment inventories (Beasley *et al.*, 1982).

Pu sedimentary phase associations have not been determined for Washington slope sediments, but in adjacent shelf sediments selective chemical leaching tests have shown that both Pu and ^{210}Pb are primarily associated with hydrous Fe–Mn oxide phases rather than with clays or organic phases. Calculations of maximum Pu flux due to biological production on the Washington shelf based on upper limits for total organic matter production and Pu concentration factor in the organic matter, and assuming all the biological matter reached the sediment and remained there, show it is unlikely that biological processes have substantially mediated the scavenging of Pu to Washington coast sediments. The key role of Pu scavenging by reductant-soluble hydrous Fe–Mn oxide phases has also been shown for soils and Lake Michigan sediments.

Because of boundary scavenging, sedimentary inventories and fluxes of excess ^{210}Pb and Pu (Beasley *et al.*, 1982; Carpenter *et al.*, 1987) off Washington are 5 to 10 times greater than off the US east coast. This enhanced boundary scavenging off the Washington coast is attributed to a combination of several reasons (Carpenter and Peterson, 1989): (a) concentrations of many dissolved chemicals are greater in mid-depth North Pacific than North Atlantic waters; (b) more upwellings, advection of deep water, and production of biogenic particles occur along the Washington coast than off other coasts; (c) the Columbia River discharges relatively large amounts of fine-grained, reactive detrital particles which include high concentrations of expandable smectite clays which have high ion exchange capacity; (d) lower O_2 concentrations and higher organic C fluxes to Washington coasts sediments may lead to greater dissolved Mn fluxes from the sediment in the overlaying water column, enhancing trace element scavenging by hydrous Mn oxides near the sea floor.

Excess $^{210}Pb/^{239,240}Pu$ inventory ratios in eight representative cores from the Washington shelf average 100 ± 19, even though absolute values of both inventories, sediment texture and accumulation rates vary much more. This surprisingly narrow ratio suggests mechanisms supplying and removing the natural isotope ^{210}Pb and the artificial isotopes Pu are similar.

Selective chemical leachings of shelf sediments have shown that both Pu and ^{210}Pb are primarily associated with hydrous Mn and Fe oxides. Pu activities and inventories in slope sediments also correlate strongly with excess ^{210}Pb, with inventory ratios increasing from 100 in shelf sediments to 300–380 in deeper slope sediments. These reasonably constant ratios, for given water depths and similar phase association, permit estimation within a factor of two of the total Pu inventories and prediction of sites of accumulation of Pu from data on the more easily measured ^{210}Pb. Geographical distributions of ^{210}Pb inventories thus lead to the prediction that Pu inventories should be 2–4 times greater in sediments from the submarine canyons off Washington than in sediments from nearby open slopes at comparable water depths.

Average $^{241}Am/^{239,240}Pu$ inventory ratios are twice as large in slope sediments as in shelf deposits (Carpenter *et al.*, 1987). This increase is attributed primarily to a greater particle affinity for dissolved Am than Pu (and hence Am deposition). The conclusion of preferential scavenging and downward transport of Am over Pu by particles is supported by data from the Mediterranean, off central and southern California, the abyssal Atlantic and Pacific, and by measurements of solid–water partition coefficients.

In contrast to Pu and ^{241}Am, ^{137}Cs inventories in both Washington shelf and slope sediments result primarily from residual Cs retention on particles discharged by the Columbia River. Rough estimates of riverine supply of ^{137}Cs-labelled particles show the river discharge is more than enough to maintain offshore sedimentary inventories, even allowing for desorption and decay of some of the riverine Cs. ^{137}Cs activities in surface slope sediments average slightly less than activities in adjacent Washington shelf sediments. The finer-grained, more clay-rich deposits of the slope do not contain more ^{137}Cs than mid-shelf silt deposits. This difference in ^{137}Cs activity between shelf and slope sediments is opposite to that of ^{210}Pb, Pu isotopes and for ^{241}Am, which are all higher in slope than shelf sediments. $^{239,240}Pu/^{137}Cs$ activity and inventory ratios in Washington slope sediments are 2–3 times ratios in shelf sediments. Higher ratios are due to a combination of preferential scavenging of dissolved Pu (predicted by the approximately 1000-fold greater sediment–water partition coefficient for Pu than for Cs) and to greater ^{137}Cs loss by decay and desorption. Partial desorption of ^{137}Cs during offshore transport of riverine-derived particulates is supported by laboratory tests of the leachability of ^{137}Cs from Columbia river sediments by artificial seawater.

5.4.4.3 Transport in marginal basins and semi-enclosed inland seas

Marginal basins and seas with limited connection to the ocean often exhibit characteristics which are atypical of more open coastal and shelf regions. Two such semi-enclosed seas are considered here: the Mediterranean and Baltic Seas. Work carried out on the Black Sea is referred to elsewhere within Section 5.4. A review of Black Sea oceanography has recently been published (Izdar and Murray, 1991).

The Mediterranean is a small-scale ocean basin with limited exchange with Atlantic water through the Straits of Gibraltar. It received direct and indirect input of Chernobyl radionuclides (Section 5.4.2) and receives input via the Rhône River of artificial radionuclides from the Marcoule reprocessing plant. In the Rhône estuary water column, the sharp decrease of ^{137}Cs from the surface to bottom waters shows an inverse correlation with the salinity profiles and illustrates the dilution of the Rhône River freshwater discharges in the marine surface layer. The predominant fraction of the Rhône sediments (about 2×10^6 t y^{-1}) accumulates up to 60 km westward of the Rhône mouth. Thus, ^{137}Cs associated with fine-grained terrigenous material released by the Rhône is deposited permanently on the continental shelf. However, more recent studies exhibit a more complex behaviour of ^{137}Cs, specially at the air–sea interface.

In 1987, a comparison of the 50 cm depth seawater concentration with the surface microlayer (top 150 μm), along a cross-section through the Rhône dilution area, revealed a ^{137}Cs enrichment factor of 4 for the fraction less than 0.4 μm by comparison with the fraction greater than 0.4 μm.

Prior to the Chernobyl accident, elevated ^{137}Cs concentrations were restricted to the coastal region influenced by the Rhône outflow. The enrichment factor of the surface microlayer is consistent with data proposed by Eakins and Lally (1984) for marine aerosols. Temporal trends were related to variations in river flow, and thus to rainfall, as well as to the dates of discharge of effluents with a low level of radioactivity.

In April and May 1986, the fallout from the Chernobyl accident added significantly to the Mediterranean marine stock of ^{137}Cs. Radionuclides such as ^{134}Cs, ^{103}Ru, ^{106}Ru, ^{110m}Ag, ^{125}Sb, ^{141}Ce and ^{144}Ce were also detected, but concentrations decreased very rapidly with time, approaching the minimum detectable limit in seawater, sediments and marine organisms. In contrast, Chernobyl-derived ^{137}Cs was still detectable, especially in sediments, two years after the date of the accident. These temporary inputs of radionuclides, which may be considered as tracers, allowed the study of various processes in the marine environment: tracing of telluric inflows in both dissolved and particulate form into seawater masses, followed by the penetration and distribution of radionuclides in the sedimentary layer, and the response of marine organisms.

Direct deposition of caesium from the atmosphere onto the seawater surface provided a source for tracing the vertical transfer of radionuclides through the water column (see Section 5.4.2.1). In the water of the Gulf of Lyons this penetration had reached, in December 1986, only the first 200 m. Twenty months after the date of the accident, a significant increase in caesium content seemed to affect the water masses as far as 400 m. The movements of water masses in this area are characterized by the occurrence of upwellings in winter, combined with the occurrence of winds such as the Mistral or Tramontane. These upwellings homogenize the hydrological characteristics of the whole water column, thus causing the vertical variations in caesium contents to disappear. However, the Liguro–Provençal basin drift along the continental shelf in this area seems to play a leading role in supplying the surface waters of the Gulf of Lyons with radionuclides originating from the Ligurian Sea. Indirect telluric inflows from Mediterranean rivers, after the leaching of continental deposits by rainwater, can also be considered as secondary coastal sources which are delayed in time (Fukai *et al.*, 1983). Following the increase in rainfall in December 1986, the ^{137}Cs distribution in surface waters of the Gulf of Lyons allowed the dispersion of river waters to be defined along the north-east to south-west line linking the Rhône prodelta to Cape Bear.

The coastal sediments also responded quickly to the injection of new radionuclides. Thus, in September 1986, three months after the Chernobyl accident, the surface layer of sediments sampled at river mouths indicated the presence of ^{137}Cs at higher concentrations than before, and the appearance of ^{134}Cs, ^{106}Ru, ^{144}Ce and ^{110m}Ag. Caesium isotopes originating from the Chernobyl fallout were detected up to 10 cm below the sediment surface showing high apparent diffusion coefficients.

This apparent mobility may limit the use of caesium as an accurate temporal marker in sediments. But caesium is a particularly useful tracer of particulate telluric inflows and sediment transport. Consequently, the measurements made on suspended matter introduced by the Rhône and collected in sediment traps show no significant difference from those made on sediments sampled at the same point. This observation reflects the predominance of telluric inflows compared to direct fallout on the surface of the water. However, the map of the Rhône prodelta down to depths exceeding 2000 m does not reveal the passage of continental shelf particles labelled with ^{137}Cs down to the bathyal region. Sediment transport must certainly have happened, in relation to the presence of a deep nepheloid layer along submarine canyons.

Following the Chernobyl accident, certain marine organisms showed a rapid response to the injection of the new radionuclides. Subsequently, the concentrations of these various radionuclides decreased substantially, reflecting their rapid removal from the water column. *Mytilus* sp. radionuclide concentrations in the wake of the Chernobyl reactor accident confirm the significance of this genus as a bioindicator for the monitoring of the Mediterranean coastal marine contamination by radionuclides (Figure 5.19, Calmet *et al.*, 1988). Figure 5.19 shows a decreasing east–west gradient emerging in May 1986 of radioactivity concentrations in *Mytilus* sp. samples. This gradient reflects the heterogeneity of Chernobyl fallout deposition in the north-western basin surface water, primarily affecting the Tyrrhenian Sea and the Gulf of Genoa. The marine radionuclide inventory thus created within the surface water was subsequently dispersed throughout the north-western basin by means of the east–westbound Liguro–Provençal current.

The Baltic Sea is the largest brackish water area in the world. As it is a semi-enclosed sea with only a limited water exchange with the world ocean via the North Sea through the Danish straits, Kattegat and Skagerrak, persistent contaminants will remain in this area for a long time. A characteristic feature of the Baltic Sea is its vertical stratification into two layers, with the surface exhibiting a low-salinity water body. The permanent halocline, at a depth of between 40 and 60 m, prevents the vertical exchange and mixing of water due to the different densities of the water types. Surface water supplied by river runoff moves into the Skagerrak with 8 to 9 PSU (practical salinity unit), while water below the halocline reflects the inflow from the North Sea with 17 to 19 PSU. The mean residence time of water in the Baltic is estimated to be 20 to 30 years.

Artificial radioactivity in the Baltic Sea mainly originates from three sources: (a) fallout from the nuclear weapons tests during the 1960s; (b) inflow of contaminated water from the North Sea; and, (c) fallout from the nuclear accident at Chernobyl. Fallout from weapons testing is a more common source of artificial radioactivity in the Baltic Sea than in other comparable shallow sea areas.

Owing to increasing discharges from reprocessing plants at Sellafield (Windscale) and La Hague during the mid-1970s, the inflow of water from the North Sea into the Baltic Sea resulted in a time-delayed increase of the activity concentration in the Baltic (IAEA, 1986) with a strong correlation between ^{137}Cs activity and salinity. A strong correlation between the annual ^{137}Cs discharge from Sellafield and the ^{137}Cs activity concentration four years later in the Danish straits could also be established (IAEA, 1986). Similar correlations cannot be ascertained for ^{90}Sr, proving that the more relevant pathway of ^{90}Sr is runoff from river systems. Figure 5.20 shows the temporal evolution of the ^{137}Cs and ^{90}Sr concentrations at a station in the western Baltic from 1970 to 1989. The 1986–90 period is enlarged sixfold in order to elucidate the Chernobyl impact and the seasonal in- and outflow dependent on the freshwater balance. The lower concentration during winter time is the result of inflow of North Sea water; the higher values during spring and summer times is the result of the southward movement of more contaminated water from the northern Baltic. The increase in the activity concentration of ^{137}Cs from 1977 to 1980 reflects the increasing discharges of this nuclide at Sellafield in previous years. The same is true for the overall decreasing values between 1980 and April 1986.

^{90}Sr was almost homogeneously distributed in the water column with highest activity concentrations in the western Baltic around 20 Bq m^{-3} and 10 Bq m^{-3} in the Bothnian bay. Plutonium isotopes were determined at very low levels between 2 and 10 mBq m^{-3} for ^{239,240}Pu, reflecting global fallout levels.

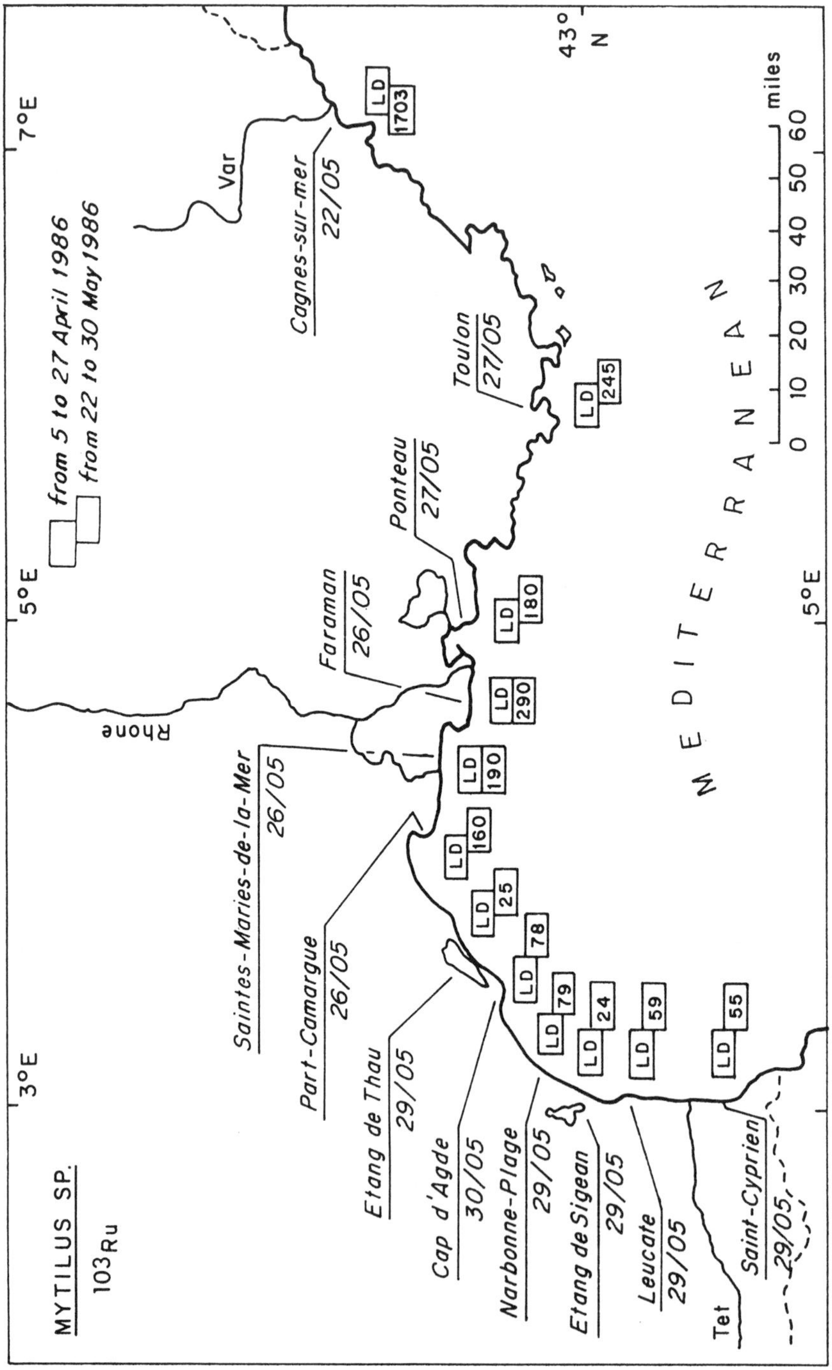
MYTILUS SP.
103Ru
from 5 to 27 April 1986
from 22 to 30 May 1986
7°E
5°E
3°E
43° N
Var
Rhone
Tet
Cagnes-sur-mer 22/05
Toulon 27/05
Ponteau 27/05
Faraman 26/05
Saintes-Maries-de-la-Mer 26/05
Part-Camargue 26/05
Etang de Thau 29/05
Cap d'Agde 30/05
Narbonne-Plage 29/05
Etang de Sigean 29/05
Leucate 29/05
Saint-Cyprien 29/05
LD 1703
LD 245
LD 180
LD 290
LD 190
LD 160
LD 25
LD 78
LD 79
LD 24
LD 59
LD 55
MEDITERRANEAN
0 10 20 30 40 50 60 miles

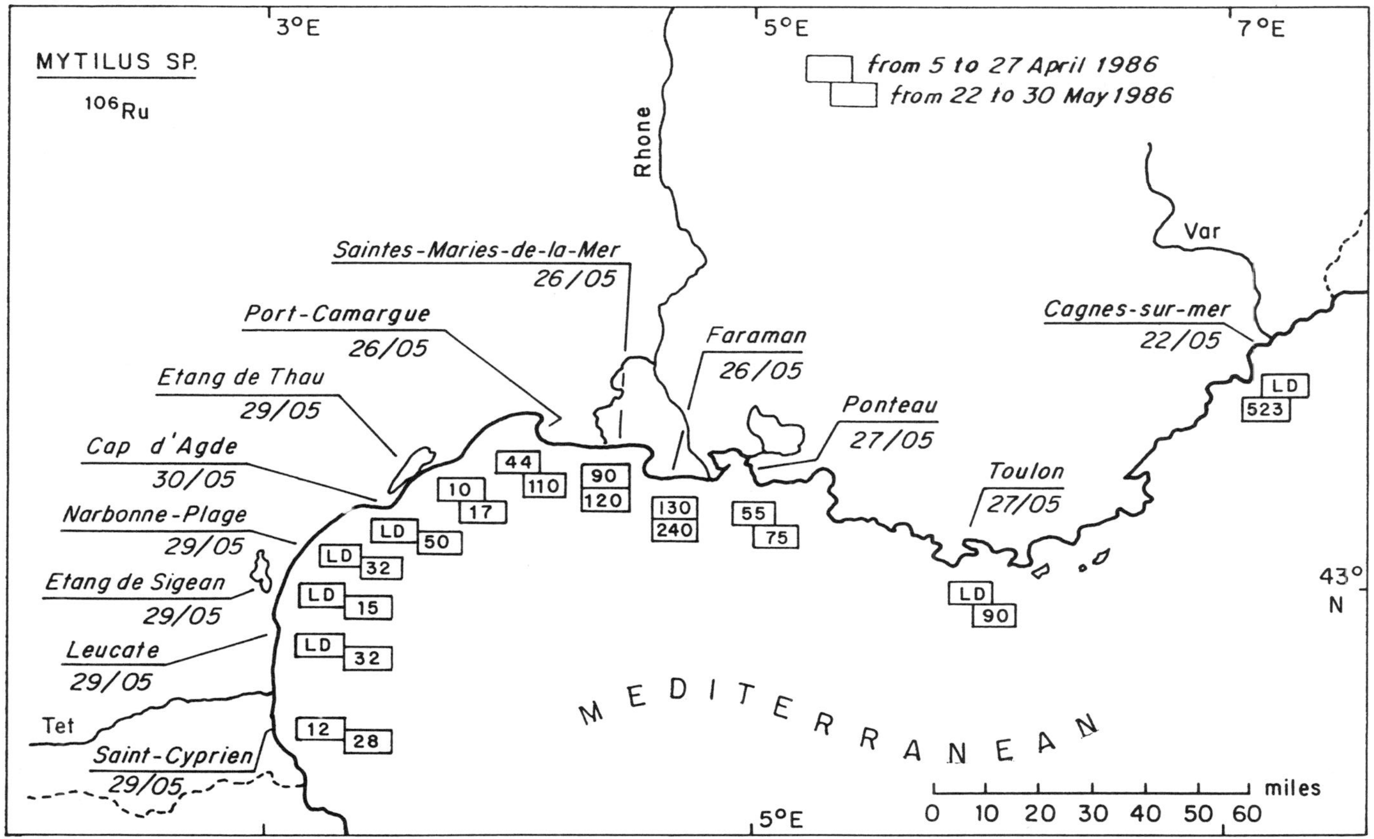

Figure 5.19 Concentration of ^{103}Ru and ^{106}Ru in *Mytilus* sp. (Bq kg^{-1} dry weight) along the French Mediterranean coast; LD = detection limit. From Calmet *et al.*, 1988.

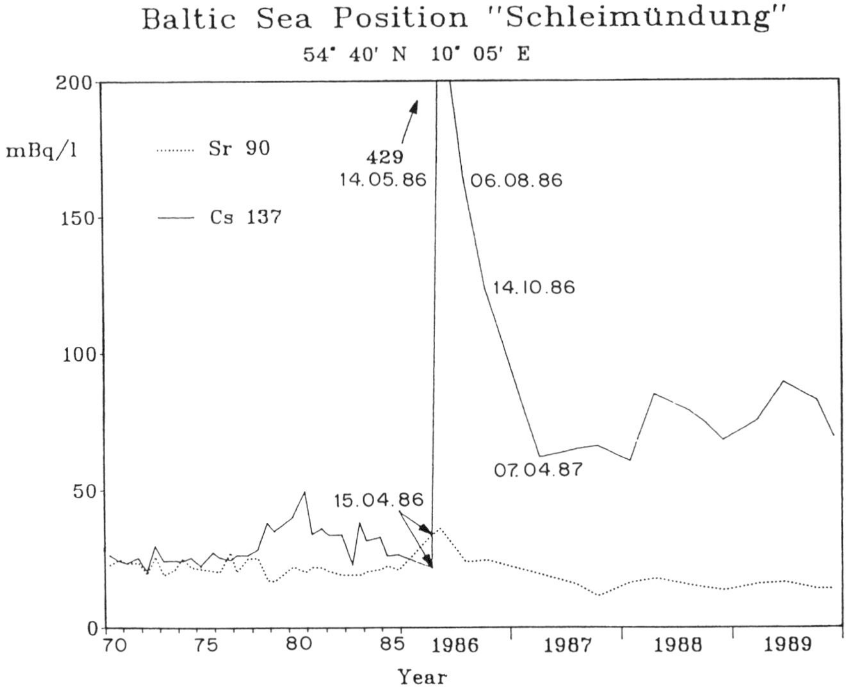

Figure 5.20 Temporal variations in the concentrations of ^{137}Cs and ^{90}Sr (mBq l^{-1}) in water from Schleimündung in the western Baltic, from 1970 to 1989. The horizontal scale has been expanded six-fold over the period 1986 to 1989. (Reproduced by permission of H. Nies.)

Prior to the Chernobyl accident, artificial radionuclides have been detected in sediments only at levels typical for global fallout. The sediments of the Baltic Sea are mostly fine-grain soft mud, especially in basins, where partly anoxic conditions are prevailing in bottom waters. This fine grained material acts as a trap for various substances dissolved in the water column. There exists an inverse correlation between Cs inventories in sediments and the corresponding water depth. This is a consequence of the decreasing accumulation rate of sedimentation with increasing depth.

The Chernobyl fallout has contaminated the sediments of the Baltic Sea to a higher degree in those areas where the contamination in the water has been the highest, namely along the northern coast of the Baltic Sea and the Bothnian Sea and the bay of Bothnia/Baltic Sea. Between 5 and 20 per cent of the Chernobyl fallout deposited at the sea surface in 1986 has been bound to sediments.

5.4.4.4 Long distance transport pathways

Waterborne radioactive contaminants from northern Europe are dispersed by the north-east Atlantic Ocean current system (Figure 5.21). For example, contaminants from

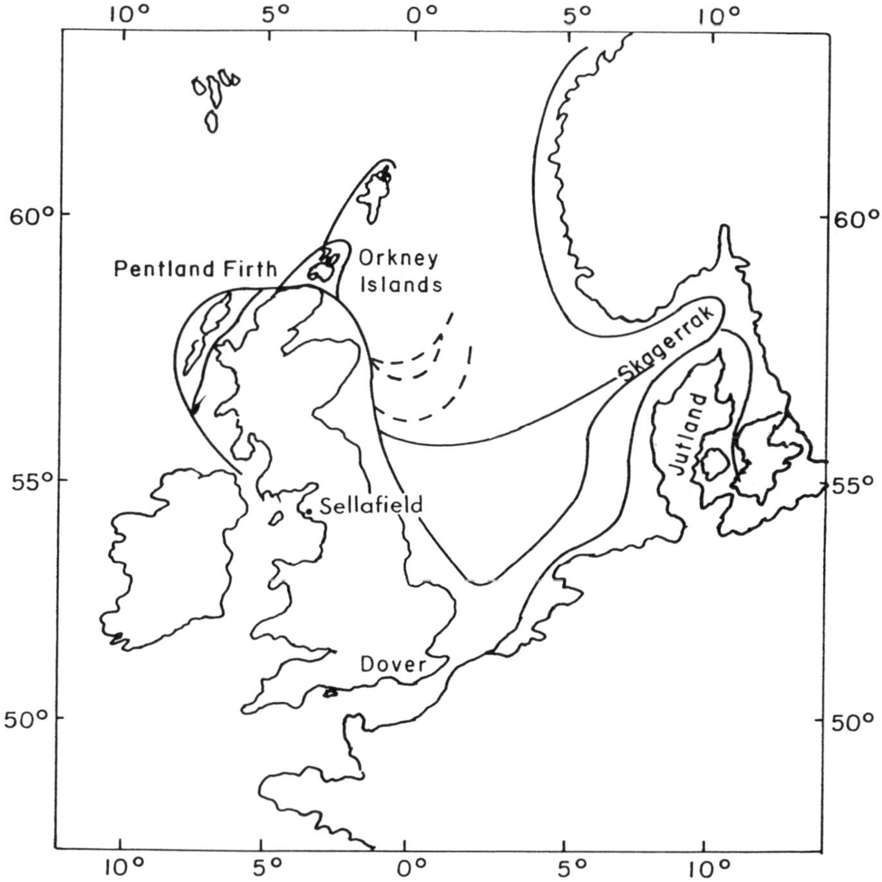

Figure 5.21 Transport routes of ^{137}Cs deduced from the measurements of the activity concentration distribution in the years 1971 to 1984. Dotted lines indicate different temporal transport routes. From Kautsky, 1988.

the Irish Sea and the Baltic Sea move via the North Sea up along the Norwegian west coast with the Norwegian Atlantic current. At about 70°N, the current branches. An eastern part runs along the Kola peninsula and moves in the direction of the Siberian north coast and Novaya Zemlya; but the major part turns northward and enters the Arctic Ocean through the Fram strait between Spitzbergen and north-east Greenland. The transfer through the Fram strait represents more than 7 million cubic metres of Atlantic water per second. From the Arctic Ocean, the water leaves through the Fram strait as a surface current: the East Greenland current running southwards along the shelf of East Greenland. The current runs clockwise around Greenland, but on its way the water is increasingly mixed into the north-east Atlantic Ocean (Figure 5.22).

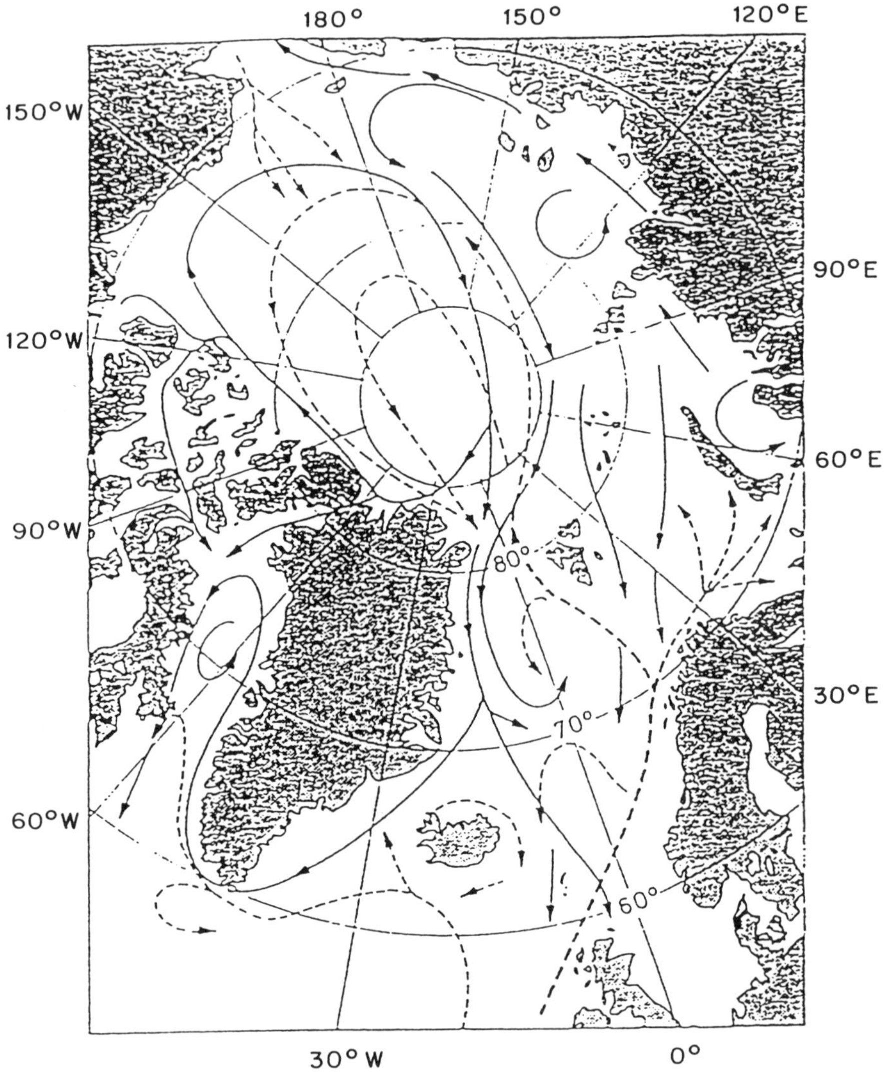

Figure 5.22 Main current directions in the northern Atlantic and Arctic Oceans; dotted lines represent relatively warm water and solid lines relatively cold water. From Rey, 1982.

Since the early 1960s seawater concentrations of ^{90}Sr and ^{137}Cs have been followed at a number of stations on the shelf in the north-east Atlantic, notably at the Faroe Islands and along the East Greenland coast. The main source of radioactive contamination at these stations has been global fallout from nuclear weapons testing in the atmosphere.

While the shelf water at the Faroe Islands is equivalent to the surface water of the north-east Atlantic Ocean, the East Greenland shelf water is almost identical to water in the East Greenland current representing the surface water of the Arctic Ocean. It is possible to estimate the mean residence time of water in the upper mixed layer of the north-east Atlantic Ocean and the Arctic Ocean to 25 years and 22 years respectively. It appears that ^{90}Sr concentrations in the Arctic Ocean are 2.1 times those in the north-east Atlantic while the ^{137}Cs levels differ by a factor of 1.7 only. Hence $^{137}Cs/^{90}Sr$ ratio is lower in the Arctic than in the Atlantic Ocean. The reason for this may be that the runoff from the Siberian rivers is enriched in ^{90}Sr and that this runoff plays a relatively more important role for the concentrations in the Arctic Ocean than runoff from rivers do for other parts of the Atlantic Ocean. Another factor could be that ^{137}Cs behaves less conservatively than ^{90}Sr, i.e. is sedimented to a larger extent than ^{90}Sr. Why the concentrations of both radionuclides are higher in the Arctic Ocean than in the north-east Atlantic is not known. The deposition of global fallout is less over the polar region than at the temperate latitudes. Although Pacific water shows two times higher concentrations than Atlantic water, inflow from the Pacific through the Bering Strait (16 per cent of the total inflow to the Arctic Ocean) is not sufficient to explain the higher levels. Two other sources may be considered. One possible explanation is run off from the Siberian rivers, in particular if these rivers have been heavily contaminated by sources other than global fallout. This has been recently suggested by T. B. Cochran following a visit to the Chelyabinsk region in the southern Urals, where it is thought that rivers and lakes may have been used for deposition of high-level radioactive waste in the late 1940s and early 1950s during the early phases of the Soviet nuclear weapons programme. In addition, recent unofficial reports have implied that substantial quantities of radioactive waste may have been dumped by the former Soviet Union in the Kara and Barents Seas. The radionuclide content and packaging of this material is not known in detail. A number of collaborative international projects have been initiated (in 1992) to seek to establish the magnitude of the input and the potential consequences in terms of environmental contamination and dose implications.

^{99}Tc is a radionuclide present in the liquid waste from nuclear reprocessing. Its half-life is 2×10^5 y and it is not readily adsorbed by sediment particles. This makes it an excellent tracer of water movement. In the order of 1 PBq of ^{99}Tc have been discharged to the sea from nuclear reprocessing in western Europe. Increased discharges began in 1970 when Sellafield commenced to release approximately 40 TBq annually. Observations of ^{99}Tc in *Fucus* collected in the Danish straits (Kattegat) in the period 1967–84 (Figure 5.23) show a significant increase in the concentrations in 1973–74 suggesting a transit time from the Irish Sea of 7 years. Measurements of water and *Fucus* concentrations have allowed the tracking of Sellafield-derived ^{99}Tc through the Norwegian and Greenland Seas and of La Hague-derived ^{99}Tc along the Channel, southern North Sea into the Kattegat.

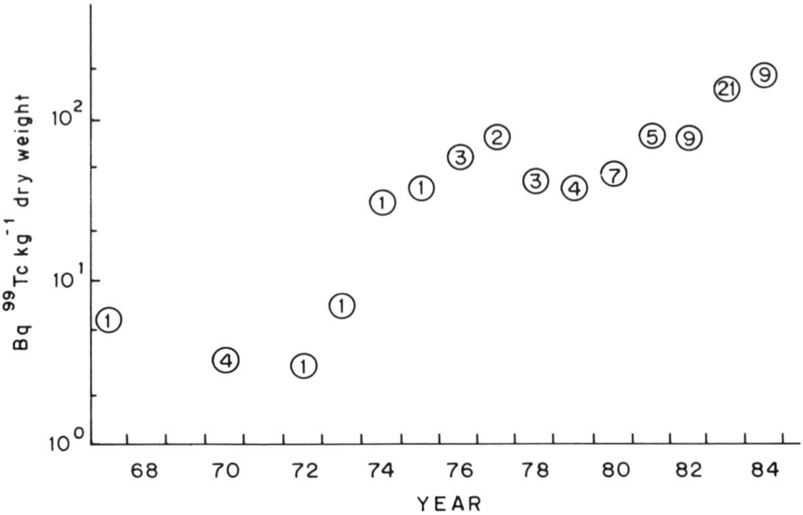

Figure 5.23. Annual mean values of ^{99}Tc (Bq kg^{-1}, dry) in *Fucus vesiculosus* collected from the Kattegat from 1967 to 1984. (The number of samples and SE are indicated). From Aarkrog *et al.*, 1987.

5.4.4.5 Radioactive tracers and transport modelling

Extensive use has been made of the more soluble radionuclides ^{137}Cs, ^{125}Sb and ^{99}Tc to develop and validate transport models in the Irish Sea, Channel, North Sea and throughout the north-western continental shelf. In some cases, models have been developed to assist in radiological dose assessments, particularly for collective dose assessment (CEC, 1990). In other cases, radionuclide data have been used to validate models developed to examine the movement of water masses, for example the Channel and North Sea (Guary *et al.*, 1988), to understand better the influence of winds and tides on water circulation, and to allow predictions to be made about the fate of other contaminants in these waters (Howorth and Kirby, 1988).

The current state-of-the-art of coastal modelling of relevance to waste inputs by sea dumping or land-based discharges, has been critically reviewed by a recent GESAMP Working Group (IAEA, 1990).

5.5 DISPOSAL IN THE DEEP OCEAN

5.5.1 INTRODUCTION

The major source of most pollutants to the open ocean is atmospheric deposition. This is also true for artificial radionuclides but direct waste dumping and land-based discharges represent additional sources.

The major processes of radionuclide transfer in the open ocean have been mentioned in previous sections. Vertical transfer of radionuclides from the surface water occurs primarily via the mediation of the zooplankton (faecal pellets, moults, carcasses). This type of transfer is described in Sub-section 5.4.2.1. Material of biogenic origin is partly degraded in the water column. Consequently, some elements are redissolved in intermediate waters where they have a longer residence time than in surface waters because scavenging particles are less numerous (see Figure 5.18). For many radionuclides and trace elements enhanced scavenging occurs at the ocean margins (Sub-section 5.4.4.2).

This section focuses principally on waste disposal.

5.5.2 RADIOACTIVE WASTE DISPOSAL

5.5.2.1 Control and Assessment

The practice of dumping packaged, low-level, radioactive waste has been carried out at a number of sites in the North Atlantic since 1946 (NEA, 1985).

This has produced significant, potential sources of a wide range of alpha-, beta- and gamma-emitting radionuclides to the bottom waters and hence via various pathways, potential sources of radiation exposure to marine fauna and ultimately to man.

Packaged waste has been disposed of in the north-east Atlantic by seven European countries at several sites (Figure 5.24). The emphasis has changed from using relatively shallow continental shelf sites in the early years to disposal on the deep ocean seafloor. The total quantities dumped in the north-east Atlantic in the period 1949 to 1982 amounted to approximately 680 TBq of alpha-emitters and 56 PBq of beta/gamma-emitters, of which 15 PBq was attributed to ^{3}H in the period 1975–82. There have been no further dumping operations since 1982. In the region of 3.7 PBq of 'mixed' activity, packaged waste was dumped by the USA in the north-west Atlantic between 1946 and 1967. About 95 per cent of the total activity, including 1.2 PBq of activation products in the 'Seawolf' reactor, was disposed of at the 2800 m site (38°30′N, 72°06′W) (NEA, 1985). In comparison, the total alpha discharge from the Sellafield reprocessing plant in the period 1957 to 1981 amounted to 1.3 PBq, with 47 PBq of ^{134}Cs, ^{137}Cs plus ^{90}Sr (NEA, 1985).

Dumping operations in the north-east Atlantic during the 1950s and early 1960s took place without international control. Since 1967, dumping operations have taken place under the auspices of the Nuclear Energy Agency (NEA, formally ENEA: European Nuclear Energy Agency) of the Organization of Economic Cooperation and Development. The OECD Council established a Multilateral Consultation and Surveillance Mechanism for Sea Dumping of Radioactive Waste in 1977 (OECD, 1983) to further the objectives of the London Dumping Convention (LDC). A periodic assessment of the continued suitability of dumping sites is required under the Mechanism, taking account of the provisions in the LDC and the International Atomic Energy Agency (IAEA) Revised Definition and Recommendations (IAEA, 1984).

Two assessments have been published for the current dumpsite at 46°N, 17°W (NEA, 1980, 1985) and a third was carried out in 1990. The later assessments were aided by the creation of the NEA Coordinated Research and Environmental Surveillance Programme (CRESP) in 1981. This was intended to increase knowledge of the processes controlling radionuclide transport to biota and humans. The programme has focused on, but not exclusively, the north-east Atlantic with site-specific studies at the current dumpsite, and the results published as CRESP reports and in a three-volume series on the oceanographic description of the site (Gurbutt and Dickson, 1983; Dickson *et al*., 1986; Nyffeler and Simmons, 1989). Fourteen countries and two international bodies have been represented on CRESP, with Working Groups covering physical oceanography, geochemistry, biology, modelling and radiological protection.

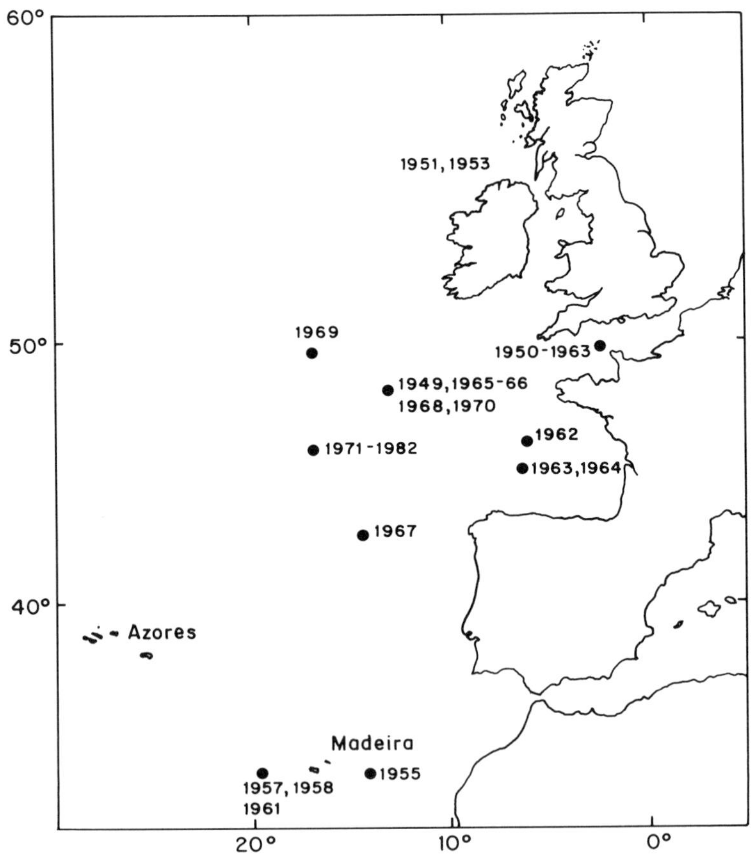

Figure 5.24 Approximate locations of dumpsites used by European nations in the north-east Atlantic from 1950 to 1982. From NEA, 1985.

To some extent CRESP can be seen as a parallel to the NEA Seabed Working Group (SWG). The SWG was set up to provide a scientific basis for assessing the feasibility of the sub-seabed disposal of high-level (or heat-generating) radioactive waste. The group considered problems associated with site-selection, emplacement, leaching, remobilization, vertical migration to the sediment–water interface, biological interactions and advective and diffusive transport in the water column (NEA, 1988). The last category is of equal relevance to low-level waste disposal. Site-specific studies were carried out at several locations, including Great Meteor East (GME) on the Madeira Abyssal Plain to the south of the current dumpsite (Figure 5.24).

Two other completed programmes are of particular relevance. The Dutch DORA programme formed an integral part of CRESP, providing detailed information on biogeochemical processes and pathways on the dumpsite (Rutgers van der Loeff and Lavaleye, 1986). The German North Atlantic Monitoring Programme (NOAMP) examined a region immediately to the north-west of the dumpsite (Figure 5.25). The principal aims were to study the pathways of suspended matter in the deep ocean and possible vertical transport out of the near-bottom layer (Mittelstaedt, 1986; Dickson *et al.*, 1986; Nyffeler and Simmons, 1989).

There are two major international programmes which have been initiated recently, and which will contribute to any further assessments of past or future disposals of radioactive waste by improving our knowledge of ocean processes. The World Ocean Circulation Experiment (WOCE) represents an attempt to describe and explain the major features of the ocean circulation. The Joint Global Ocean Flux Study (JGOFS) is designed to further our understanding of vertical fluxes, carbon cycling and benthic exchanges.

The models used in the radiological assessments of the north-east Atlantic dumpsite (46°N, 17°W) predict very low concentrations of radionuclides in seawater and sediments as a result of past dumping practices (NEA, 1980, 1985). These predictions have been supported by measurements of radionuclide concentrations in seawater, sediments and biota from on and around the dumpsite. In general, radionuclide distributions are dominated by global fallout. It follows that the biogeochemical pathways which are described below should, for the most part, be considered as potential pathways with regard to radioactive waste. In practice, it has proven to be very difficult to establish the degree of environmental contamination resulting from the dumping itself.

5.5.2.2 Advection and dispersion

The current dumpsite is located about half-way between the mid-Atlantic Ridge and the west European continental margin, on the easternmost part of the province at the southern end of the Porcupine Abyssal Plain (Gurbutt and Dickson, 1983). The area is characterized by a flat-bottom valley (> 4750 m) in the western half flanked by a relatively abrupt ridge and trough topography trending approximately

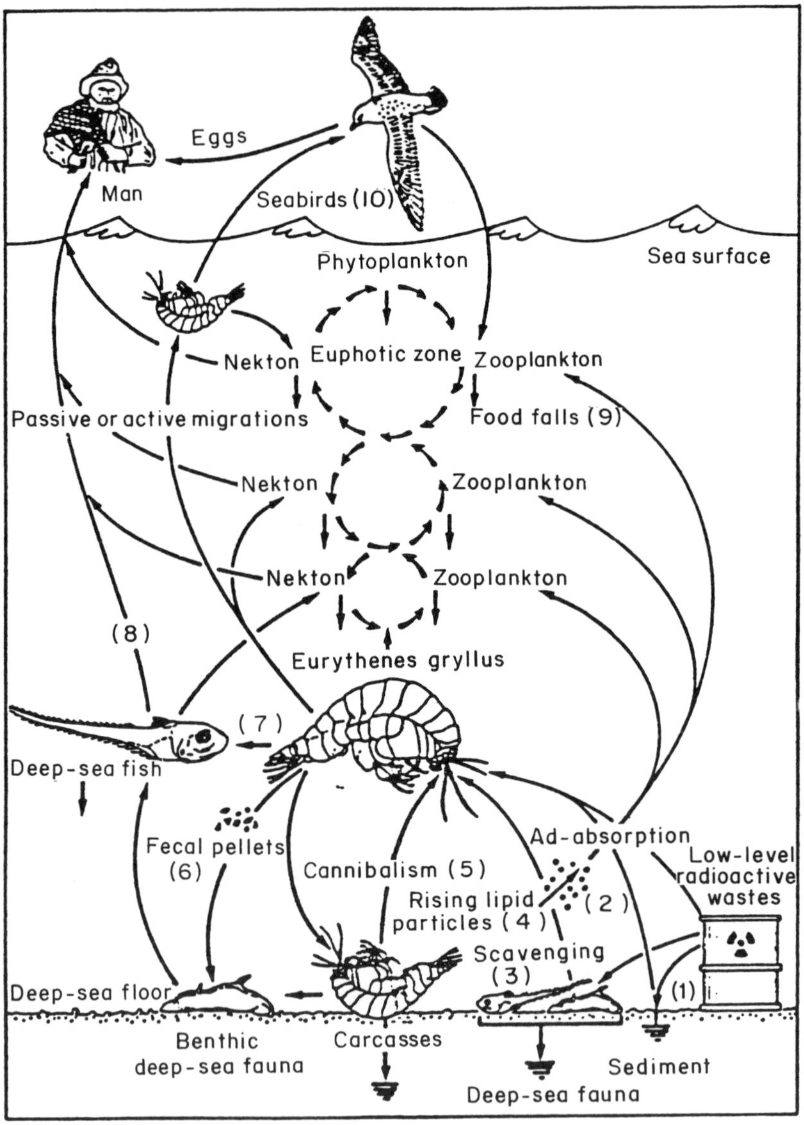

Figure 5.25 Hypothetical diagram of radionuclide transfer into different layers of the water colunm by the food-chains in which *Eurythenes gryllus* might be directly involved. From Nyfeller and Simmons, 1989.

north–south, with the crests of hills reaching to water depths of 3750 m. The entire area has been charted with a narrow beam echo sounder, 'Sea Beam', providing maps at a scale of 1 : 50 000 and 1 : 100 000 with contour intervals of 20 m (Dickson *et al.*, 1986). 'Sea Beam' has also been used in the NOAMP area.

Current velocities in the dumpsite and NOAMP regions have been examined using bottom-mounted, long-term current meter moorings. Current velocities and directions in the near-bottom region—extending to about 1000 to 1500 m above the seafloor—are strongly influenced by the bottom topography and exhibit considerable spatial variability. The mean flow in the upper water column, in the upper 1500 m, is influenced by the near-surface flow and is largely independent of the near-bottom flow. The region between depths of 1500 and 3000 m is partially influenced by the near-surface and near-bottom flows and by a thermohaline-driven circulation.

Eddy kinetic energies are greater than the mean flow and the timescales are complex. The sparsity of the current meter records, combined with the observation that the fluctuating component, resulting from eddy activity, is large compared with the mean flow, introduce uncertainties in the reliability of mean flow estimates. However, an analysis of all available records suggests an overall slow northerly mean flow over much of the region (Gurbutt and Dickson, 1983; Nyffeler and Simmons, 1989). The long-term mooring at NEADS 6 (near the foot of the Rockall Trough), whilst showing an overall north-westerly drift, does record significant periods when the residual flow was to the south-west. Although the mean current velocities are relatively low (1 to 3 cm s^{-1}) the combination of short-term (1 to 3 weeks) high bottom current benthic storms, > 10 cm s^{-1}, and the semi-diurnal tidal currents can lead to instantaneous velocities of 20 to 30 cm s^{-1} close to the bottom (Nyffeler and Simmons, 1989). This causes the resuspension of significant quantities of surficial sediments which may remain in suspension as a bottom nepheloid layer (BNL) for several days. The combination of the considerable topography (1000 m), vertical mixing and lateral advection result in the occurrence of a BNL, up to 1000 m above the seafloor, frequently showing a number of intermediate layers (Dickson *et al.*, 1986).The near-bottom eddy diffusivity on the dumpsite was estimated from measurements of excess ^{222}Rn in bottom waters, which diffuses out of the seabed following the decay of naturally occurring ^{226}Ra. The diffusivity was proportional to the local density gradient and gave a bottom mixed layer of 50 m. The occurrence of concentration maxima at greater distances above the seabed was interpreted as evidence of lateral advection from surrounding hills (Dickson *et al.*, 1986).

The pattern of water transport has also been studied using neutrally buoyant floats. Four floats released into the bottom waters of the western end of the dumpsite in 1976 were tracked for up to 17 days in a southerly to south-westerly direction, in contrast to a bottom float and two mid-depth floats which were tracked in a westerly to north-westerly direction—with mean velocities of 2 to 3 cm s^{-1} and an overall cyclonic path apparently influenced by the local topography (Gurbutt and Dickson, 1983). Fourteen SOFAR (sound fixing and ranging) floats launched in 1985 in the NOAMP area were tracked for up to 1 year at a depth of 3300 to 4000 m. These yielded a horizontal diffusivity of 10^6 cm^2 s^{-1} and a mean velocity of 1 cm s^{-1}, with a south-westerly tendency. The floats dispersed in an area of about 300 × 300 km over a year with a domain of occupation of 100 km in diameter in 180 days. Similar diffusivities have been obtained for the Iberia Abyssal Plain, although at long time scales dispersion decreases as the floats reach the sides of the basin. In contrast, floats in the Madeira Abyssal Plain have not yet filled the basin and their dispersion increases.

A major difficulty is to estimate vertical velocity of bottom waters. The general average upwelling velocity of ocean waters is about 10 m per year: such estimates are made as a compensation for the downwelling waters (waters that are saltier and cooler will sink to a level appropriate to their density). Large variations can occur in the upwelling velocity of deep waters. It has been shown that these variations can depend on local topography or meso-scale events (Armi and D'Asaro, 1980) which are quite common in the Atlantic. Such events may affect the vertical distribution of radionuclides and particles (Lambert *et al.*, 1983; 1984).

Several attempts have been made to determine waste-derived radionuclides in the bottom waters on and around the dumpsite to estimate the degree of local contamination and the effect of advective and diffusive transport. Tritium has been dumped in large quantities (> 15 PBq), has a high solubility and was mostly disposed of in vented drums. It could be anticipated that detectable levels would exist on the dumpsite. In order to check the assumptions made for the release rate and initial dispersion in the radiological assessment model (NEA, 1985) 176 water samples from the bottom 1000 m at 18 sites in the north-east Atlantic were collected for tritium analysis in 1986. There was considerable horizontal and vertical variability, with many concentrations being not significantly different from zero. Five stations produced inventories in the bottom 1000 m which were statistically significant. There were three sites on the dumpsite and two to the south and south-west (Figure 5.24). Taking a mean value from the five sites, and a 3° × 3° square around them, resulted in an inventory of 1460 TBq, which is approximately 10 per cent of the total decay-corrected inventory in 1986 from all disposals to the deep north-east Atlantic (Nyffeler and Simmons, 1989). The implied movement to the south-west was not predicted from the current meter records, but the limited float data suggest that geographically and temporally limited excursions to the south-west are possible. The inventory estimates are a gross approximation and further measurements would be necessary to attribute greater credence to these preliminary findings.

Water column profiles of $^{239,240}Pu$ and ^{137}Cs in the NOAMP and dumpsite areas showed no enhancement which could be ascribed to dumping operations, i.e. the profiles were consistent with a fallout origin and did not differ between the two regions. The approximately exponential decrease in the ^{137}Cs concentration with depth contrasted with the profiles of $^{239,240}Pu$ which exhibited marked concentration maxima at about 1000 m, as a result of biological activity and the remineralization of sinking particles. No significant enhancement of $^{239,240}Pu$ was observed in near-bottom (+ 5 m) water samples taken from throughout the north-east Atlantic basin. $^{239,240}Pu/^{238}Pu$ ratios were lower than in surface samples taken at the same position.

5.5.2.3 Biogeochemistry

Most of the seafloor at the dumpsite lies at or a little above the estimated depth of the lysocline (4700 m) and this is reflected in average Holocene sediment accumulation rates of 2 cm ky^{-1}. Deeper core profiles have yielded slightly higher accumulation rates due to the influx of ice-rafted material in the late Pleistocene. In the hillier areas there is evidence that localized slumping and episodic deposition has taken place.

The extent of bioturbation in the sediments at the dumpsite appears to be of the same order of intensity as other areas of the Atlantic. The irregular profiles of both Pu and ^{210}Pb are indicative of episodic mixing events, predominantly by Sipunculida (Rutgers van der Loeft and Lavaleye, 1986). Most of the meio-, macro- and megabenthos are deposit-feeders and occur in the upper 10 cm.

Decomposition of the most reactive organic matter takes place at the sediment surface with dissolution of some $CaCO_3$ and release of organically bound Pu, Am and Sr. These reactions take place at reduced rates with depth in the sediment. In the 10–70 cm zone there appears to be little biological or chemical change in condition, but below 70 cm Fe and Mn are reduced and associated elements may be released. Significant sulphate reduction is restricted to turbidity deposits (Rutgers van der Loeff and Lavaleye, 1986).

Concentrations of Pu and Am in bottom sediments are consistent with a fallout origin. However, $^{239,240}Pu/^{238}Pu$ ratios provide equivocal evidence of a possible dumpsite origin for some of the Pu. K_d values of Pu and Am obtained by filtering large-volume bottom-water samples were similar to those from coastal sites.

Many analyses of biota have been completed, including Holothuroidea, Actiniaria, Cephalopoda, Decapoda and *Coryphaenoides armatus* (rat tail). In general the variety and concentrations of radionuclides present are indicative of fallout origin. Some muscle samples from *Coryphaenoides* had relatively low $^{239,240}Pu/^{238}Pu$ ratios, suggesting contamination by dumpsite Pu, but the errors on the determinations were considerable. Hypothetical pathways for radionuclide transfer away from the dumpsite have been suggested based on the ubiquitous Amphipod *Eurythenes gryllus* (Figure 5.25; Nyfeller and Simmons, 1989).

The 91-box SSR model (NEA, 1985) contained representations of the perceived key physical, chemical and biological processes. Generally, pessimistic assumptions were made about the rates of processes in the model (e.g. release rate), particularly when data were lacking. Additional measurements made after completion of the 1985 SSR have allowed a comparison to be made between observed and predicted concentrations on and around the dumpsite. Clearly radionuclide concentrations were overestimated at the dumpsite, suggesting that estimates of dose to local fauna had been overestimated. The effects at greater distances is less clear but it seems likely, on balance, that doses in the far field will also have been overestimated in the 1985 SSR.

5.6 RECOMMENDATIONS

1. Greater emphasis should be given to studies of freshwater pathways, especially lakes, as these systems are clearly more sensitive to pollution.
2. The role of micro-organisms in the biogeochemical cycling of radionuclides in aquatic systems is not, generally speaking, well understood and undoubtedly warrants further research.
3. Collaborative international efforts to study hot zones such as the Urals, Ob River, Yenisei River (some of which are presently under way) should be co-ordinated and intensified.
4. Radioecological studies of polar seas (Laptev's Sea, East Siberian Sea, Kara Sea) should be organized on an international basis to minimize expense.
5. The role of the microlayer in aquatic systems and its influence on the behaviour of radioelements requires further study. In particular, improved techniques for the sampling of this layer should be developed.
6. Although the influence of colloids on the behaviour of heavy metals in freshwater and estuarine systems is well documented in the literature, comparatively little has been reported on the interaction of colloids and radioelements at trace levels in the marine environment. Current research is understood to focus on the size-distribution of radioelements as determined by hollow fibre ultrafiltration, in conjunction with chemical speciation analyses involving either selective leaching techniques or oxidation state determination using rare earth co-precipitation techniques. The results of these studies are awaited with considerable interest.
7. International intercomparisons/intercalibration of radioecological methods should be given higher priority with particular emphasis on post-Chernobyl research and studies under way in the field.

6 Behaviour and Decontamination of Artificial Radionuclides in the Urban Environment

Co-ordinator: **C. N. Hewitt**

Contributors: **R.W. Allott, M. Kelly, J. Roed and K. Andersson**

6.1 INTRODUCTION

About 70 per cent of the population of the current developed world lives in urban areas and it is therefore essential that an understanding of the pathways and behaviour of pollutants, including artificial radionuclides, in towns and cities is obtained. For various reasons, however, the urban environment has largely been excluded from terrestrial studies until recent years. Perhaps one of the greatest catalysts for research within this field was provided by the Chernobyl reactor accident. The environmental processes operating within an urban system may be similar to those observed in a rural environment. Therefore the emphasis of this chapter will not be on describing these general processes, but rather the aim is to highlight how pathways of artificial radionuclides are modified by features of the urban environment. Finally, it is shown how a knowledge of the behaviour of artificial radionuclides in the urban environment may be used to develop a strategy for decontamination.

6.2 CONTAMINATION PROCESSES

6.2.1 DRY DEPOSITION

Underwood (1987) has critically reviewed the applicability of a single deposition velocity to model plume depletion over urban complexes and found that careful interpretation is required. Since urban complexes generally have a spatial extent of many kilometres the concept of an 'averaged bulk deposition velocity' is feasible. However, in reality the roughness elements of the urban canopy are never homogeneous, thus only local deposition velocities can be determined and these will not be strictly applicable to the whole city. These fundamental problems, com-

Radioecology after Chernobyl
Edited by F.E. Warner and R.M. Harrison
 Published by John Wiley & Sons Ltd

bined with the technical difficulty in deducing the downward flux above the top of the roughness elements (buildings, etc.), has resulted in a limited application of the concentration gradient approach.

An alternative approach is to measure the accumulated flux on both outdoor and indoor sections of the urban canopy and relate that to integrated air concentrations. Initially, work concentrated on ^{137}Cs particles derived from weapons fallout, using surfaces protected from wet deposition and weathering. However, the Chernobyl accident enabled the determination of deposition velocities to numerous surfaces. Attempting to apply the deposition velocities derived by this method to the whole of an urban canopy is even more complex than using a bulk deposition velocity. Macro scale considerations include the influence of buildings on wind direction and speed, leading to greater turbulence. On the meso and micro scales the composition and texture of the individual surfaces will govern deposition. In addition, enhancement may occur at projections (e.g. window sills), discontinuities (e.g. mortar joints) or edges, negating any attempt to scale up deposition velocities from a portion of a surface, say 10 bricks, to the whole wall.

Roed (1987a) and Nicholson (1987) studied the dry deposition of Chernobyl radionuclides to various building materials and a few of their results are summarized in Table 6.1. The deposition velocities for particular radionuclides and surfaces were reasonably similar, whether measured in open spaces, near edges or in densely built-up areas. However, the deposition velocities of the majority of radionuclides to roof material (e.g. 0.027–0.09 cm s^{-1} for ^{134}Cs) were higher than those to roads (e.g. 0.0045–0.011 cm s^{-1} for ^{134}Cs). This was believed to be due to an increase in wind speeds and turbulence at roof level. There also appeared to be an order of magnitude difference between the deposition velocities to walls (e.g. 0.001–0.002 cm s^{-1} for ^{137}Cs) and roads for radiocaesium, although this was not observed in the case of ^{131}I. Deposition velocities measured for Chernobyl radioiodine were generally higher than those for radiocaesium, due to the presence of different radioiodine species in the fallout (particulate, gaseous and methyl iodide). Measurements of ^{134}Cs deposition velocities to surburban trees and bushes (0.03–0.13 cm s^{-1}) were higher on average than those to roof surfaces, while depo-

Table 6.1 Deposition velocities for ^{134}Cs, ^{137}Cs and ^{131}I to various urban surfaces

Reference	Radio-nuclide	Deposition velocities (cm s^{-1}) Roof	Wall	Road
Roed (1987a)	^{137}Cs	0.028–0.035	0.001–0.002	
	^{134}Cs	0.027–0.029		0.0045–0.011
	^{131}I	0.02–0.32	0.027–0.03	0.03–0.033
Nicholson (1987)	^{134}Cs	0.03–0.09	<0.004–<0.0005	

sition velocities to grass were nearly proportional to the mass of grass per unit area, resulting in a wide range of values (0.015–0.099 cm s^{-1} for ^{134}Cs) (Roed, 1987a).

Deposition velocities within houses have also been measured for radionuclides in Chernobyl fallout and ranged from 0.006 to 0.058 cm s^{-1} (Roed and Cannell, 1987). In addition, results from this study were used to determine the ratio of indoor to outdoor aerosol concentrations (0.25 for ^{137}Cs and 0.4 for ^{131}I). The reduction in concentration was considered to be due to the filtering effects of cracks, crevices and pores in the building structure, with the higher ratio for ^{131}I being due to a smaller particle size.

It can be seen that each surface in the urban environment will have a unique deposition velocity resulting from the nature of the surface and its location within the urban windfield. The effects of inhomogeneities appear to be minimal and therefore deposition velocities will apply to the whole surface (Roed, 1987a). A summary of urban surface deposition velocities is provided in Figure 6.1. This allows the deposition velocity to the whole urban canopy to be expressed by the following equation (Underwood, 1987):

$$V_g(\text{bulk urban}) = \sum_{i=1}^{n} A_i V_g^i \tag{6.1}$$

where A_i is the total surface area of the type i per plan area of city and V_g^i is the corresponding deposition velocity. Recent work by Roed and Jacob (1990) on Chernobyl radiocaesium produced a V_g (bulk urban) estimate of 0.03 cm s^{-1} for 0.4 μm

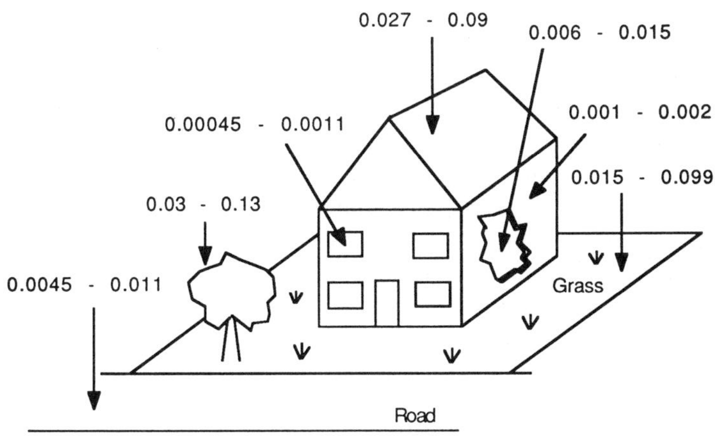

Figure 6.1 Dry deposition velocities of Chernobyl radiocaesium to various urban surfaces (cm s^{-1}). From Nicholson, 1987; Roed, 1987a, Roed and Cannell, 1987; Roed and Goddard, 1990; Roed and Jacob, 1990.

particles using this method. Despite its limitations, this 'naive' estimate of the urban deposition velocity may result in reasonable values for certain applications.

6.2.2 WET DEPOSITION

It has been well established that major conurbations (> 1 million people) cause enhancement of both precipitation, occult deposition and the frequency of thunderstorms (e.g. Changnon, 1980). In addition, scavenging ratios have been observed to decrease from upwind of a city to downwind (Gatz, 1977), but this may be due to increased air concentrations at the downwind sites. Thus, it was argued that ground level pollutants had not been efficiently scavenged by rainfall and only when the city-derived species reached cloud level, some distance downwind, would they actually be efficiently scavenged. Conversely, pollutants derived far enough upwind of the city will be subject to efficient in-cloud scavenging above an urban complex. In this situation, the apparent enhanced levels of precipitation suggest that wet deposition of particles and gases will be greater in urban areas. However, it should be realized that the observed zone of rainfall enhancement in major conurbations did not occur over the cities themselves, but developed downwind over rural regions. Therefore, since one cause of enhanced precipitation is the additional particulate matter suspended in the atmosphere, the relative position of industrial plants is of prime importance to the location of this zone. Furthermore, precipitation enhancement has been observed only for cities > 1 million people; thus, any effect that the majority of cities may have on rainfall is probably small.

Thunderstorms may have a tendency towards low wash-out efficiency due to their short duration, but their importance lies in the fact that they process large volumes of air and thus may lead to localized hot spots of deposition.

Fog frequency in central London has been shown to be greater than in the surrounding rural area (Hall, 1984), which, combined with the generally high concentration of pollutant species found in fogwater (Waldman *et al.*, 1983) suggests that this deposition pathway may be important.

Although wet deposition is more efficient at removing pollutants (particularly particles) from the atmosphere, dry deposition is pervasive through time and space. Hence, it is likely that for continuous pollutant releases both processes are of similar importance. However, for instantaneous releases of radionuclides into the atmosphere, it is of great significance whether the resulting radio-active cloud is intercepted by rainfall. This was demonstrated quite dramatically by the Chernobyl reactor accident, where large regions of north-west Britain received much higher levels of fallout due to the coincidental heavy rainfall (Clark and Smith, 1988). Hence it may be tentatively suggested that the increased frequency of rainfall in major urban complexes could alter the balance of fallout mode in favour of wet deposition, especially where particulate contaminants are concerned. For these reasons, an assessment of accident consequences needs to take into account the relative position of nuclear installations and major cities in relation to predominant wind directions.

6.2.3 MECHANICAL TRANSPORT PROCESSES

The mechanical transport of dust and pollutants in the outdoor environment is generally of greater significance in urban rather than rural areas, due to the predominance of vehicles. However, mechanical transport probably has greater impact within indoor environments, where humans provide the motive force. Consequently, the extent of outdoor contamination will determine the importance of this pathway.

Two possible sources of radionuclides to the outdoor urban environment are from nuclear fuel fabrication and reprocessing plants. Contaminated dust may be physically picked up by the tyres of heavy plant vehicles and subsequently deposited on local road surfaces. In addition, material transported by lorry may also accidentally become incorporated into street dusts. Evidence for the existence of these pathways was found by Al-Khayat (1989) with elevated levels of ^{234}Th present in dust at the entrances to the Springfield nuclear fuel processing plant (UK) and also on the roads used by the plant vehicles.

Radionuclide contamination of houses in the coastal village of Ravenglass, near Sellafield, was attributed to the mechanical transport (on the bottom of shoes) of fine-grained sediments from the Esk estuary (DOE, 1987). This conclusion was based on the significant increase in the plutonium content of house dust, from occupants who never visited the foreshore, to those who made frequent visits. Harrison (1979) observed that the particle size distributions of house dust and soil were similar and deduced that a major pathway for dust into homes was by mechanical transport of soil on shoes. This has been confirmed using a fluorescent tracer (Cannell *et al.*, 1987).

6.3 REMOVAL PATHWAYS

6.3.1 INTRODUCTION

Once deposition has occurred, the majority of pollutant species rapidly lose their individual identity, through becoming sorbed to the host soil particles. Surface deposits of urban dusts and pollutants are subject to a variety of environmental and anthropogenic removal processes, the principal transport pathways being resuspension, mechanical cleaning or hydrologically based.

6.3.2 RESUSPENSION

6.3.2.1 Resuspension in the outdoor environment

Attempts to directly assess the modifications imposed on resuspension by urban complexes have been virtually non-existent. Not only is it necessary to resolve wind-driven resuspension from a variety of hard and soft surfaces, but the influence of mechanical resuspension has to be considered. Added to this the resuspension rate is virtually indeterminate, due to the complex windfields, whilst the resuspension factor becomes even more location-specific. Thus, it is necessary to infer resuspension behaviour from a combination of separate studies.

Sehmel (1973b) has provided much of the data on vehicle-induced resuspension from ZnS tracer studies on an asphalt road. A resuspended fraction of 10^{-5}–10^{-2} was determined for a single vehicle pass through, or alongside, the freshly deposited tracer. The resuspended fraction was observed to increase with the size of the vehicle and also the square of the car speed. This was explained by an increase in turbulence, although tyre stresses may have been important. Weathering of the tracer was found to decrease the resuspended fraction, but the amount of tracer removed between experiments by wind-induced resuspension was not known and therefore the decrease represents a lower limit. Nevertheless, the observation is not unreasonable since particles may become attached to the road surface or lodged in crevices where they will be less easily resuspended.

Since the rate of vehicles driving over a contaminated surface may reach 1 vehicle s^{-1} on a major trunk road, the resuspension rate will range from 10^{-5} to 10^{-2} s^{-1}, which is relatively high when compared to wind-driven resuspension. However, the spatial extent of vehicle-induced resuspension is much more limited than the surfaces available to wind-driven resuspension.

A simple resuspension model developed by Hamilton *et al.* (1985) produced a resuspension rate of 1.7×10^{-5} vehicle^{-1}, consistent with the experimental results described above. Further predictions of the model suggest that vehicle-induced resuspension will reach equilibrium with the deposition flux after several dry days on a major trunk road, but this would never be the case for a side street with low traffic density.

Nicholson (1988a) reviewed some results on resuspension due to pedestrian movement, with a reported range of resuspended fractions from 1×10^{-5} to 7×10^{-4} per pass. This mode of resuspension is unlikely to contribute dramatically to the urban aerosol, although it may be of great significance to an individual's exposure.

Wind-induced resuspension from vegetated regions within cities will presumably differ little from rural areas, except with regard to the different wind speeds experienced. Resuspension from hard surfaces is less easy to predict: the fast drying nature of these surfaces will certainly ensure that dust and contaminants will be available to resuspension processes for longer periods of time. However, deposited species on asphalt roads become less erodable with weathering, as well as being progressively moved to sheltered positions and actually removed from the surface via runoff and cleaning. Thus, resuspension rates and factors for hard urban surfaces will have a time dependence attributable to all these processes.

Some estimates of resuspension factors have been made for urban areas and range from 5×10^{-5} up to 10^{-3} m^{-1} (Linsley, 1978). These relate to the surface contamination on paved areas, whilst J.A. Garland (personal communication, 1990) has suggested that a value of 10^{-6} m^{-1}, in relation to the initial deposition, may not be inappropriate. Unfortunately the former assumes no influence from regional resuspension, whilst the latter neglects removal of contaminants from paved surfaces via runoff.

6.3.2.2 Resuspension in the indoor environment

A number of studies were carried out in the 1960s to establish resuspension factors for indoor environments, in particular nuclear installation changing rooms. Fish *et al.* (1967) determined resuspension factors for an asphalt tile floor and found values ranging from 9.4×10^{-6} m^{-1} for 'light' work up to 1.9×10^{-4} m^{-1} for vigorous activity. Resuspension from different floor coverings was studied by Jones and Pond (1967). An increase in movement produced a corresponding increase in the resuspension factor; a value of 5×10^{-5} m^{-1} was proposed as a reasonable resuspension factor. A marked correlation between the number of personnel in a changing room and airborne contamination levels was noted by Brunskill (1967). He determined resuspension factors ranging from 10^{-3} to 2×10^{-4} m^{-1} for loose contaminants on a concrete floor. However, the resuspension factor for waxed linoleum was found to be smaller.

Resuspension factors have also been measured for carpeted surfaces and ranged from 1×10^{-5} m^{-1} for resting conditions to 2×10^{-5} m^{-1} during activity conditions. Unfortunately, no measurements were made during cleaning operations, when a large amount of resuspension would seem likely.

6.3.3 HYDROLOGICALLY INDUCED TRANSPORT

The removal of pollutants via the movement of water may occur by two distinctive mechanisms: migration due to the infiltration of water and wash-off resulting from runoff. Migration is generally considered to be a solution or colloidal phase transport process and only in certain situations, such as bioturbation, is this not the case. However, direct removal from surfaces may occur in either the solid or solution phases.

Within rural areas, infiltration is considered to be the dominant removal pathway for water, with runoff being of little significance for the well-vegetated regions of northern Europe. During urbanization, however, there is expected to be an increase in the proportion of impervious area compared to the previous rural region. This has a significant impact on the hydrological regime, with runoff increasingly dominating infiltration (Hall, 1984). Consequently, larger volumes of water are discharged into receiving streams, lag times are shortened and peak flow rates increased. Nevertheless, runoff from so-called impervious urban surfaces rarely exceeds 90 per cent and typically ranges from 34 to 83 per cent of the total rainfall (Ellis *et al.*, 1987). This deficit is not only a reflection of the road storage capacity, but is believed to be due to seepage losses at road and pavement joints, spray from vehicles and direct infiltration through the surface itself (Ellis *et al.*, 1987). Thus, in the urban environment the following processes will dominate:

1. Migration of soluble pollutants through pervious surfaces, ultimately to the groundwater.
2. Wash-off of particles and soluble pollutants from impervious surfaces into road side drains (gully pots).

3. Desorption of pollutants from particles and surfaces, followed by wash-off into gully pots or migration through surface and substrate to the groundwater.

The magnitude of these different pathways for urban pollutants will depend on the physico-chemical form of the species and initially whether the contaminants were dry or wet deposited. It has been suggested that after about a day or so dry deposited species will behave as though they had been wet deposited, presumably through moisture appearing on the surface from light rain, mist or dew (Crick and Brown, 1990). Prior to this, rainfall would be expected to remove a reasonable proportion of the dry deposited species (Section 6.4.3). The wash-off of previously deposited pollutants from road surfaces is an important removal pathway very characteristic of the urban environment and has been the subject of a great number of studies in relation to non-radioactive contaminants (e.g. Ellis, 1986). The relatively fine grain sizes and steep slopes associated with roof wash-off ensure that these surfaces do not act as a sink, although it is possible for guttering to act as a temporary trap. The remaining processes, in particular migration of soluble species through pervious surfaces and the desorption of pollutants from impervious surfaces, will be discussed under retention behaviour (Section 6.4).

Radionuclides removed from urban surfaces by hydrological processes will ultimately be discharged into natural watercourses or bodies. The reduced exposure to radionuclides in the urban environment may therefore be offset by a potential increase in exposure from aquatic pathways such as drinking water or the food chain.

6.3.4 CLEANING

The removal of dust and consequently of pollutants by cleaning is a practice widely conducted in urban environments. Studies on the effectiveness of cleaning have been confined mainly to external dusts, although one study concerned with the decontamination of internal surfaces has been reported.

It is widely accepted that street sweeping has a beneficial effect on litter control and thus provides aesthetic improvements to urban environments. However, Sartor and Gaboury (1984) addressed the question of whether street cleaning has a significant effect on the water quality of urban runoff. The removal efficiency of conventional rotary and vacuumized street sweepers was found to be dependent on the particle size of the street dust: although highly efficient at removing the large particles (> 2 mm), including litter (79 per cent), their efficiency at removing the finer, more contaminated, grain size fractions was observed to be low (15 per cent of the < 43 μm fraction).

Revitt and Ellis (1980) showed that the consequence of this preferential removal of the finer grain size fraction was to cause a shift towards the finer particle end. Hydroflushing was performed after the mechanical rotary sweeping and appeared to have little effect on the coarse particle sizes (> 250 μm), but removed an additional portion of the 30–250 μm particles. The resultant reduction in dust and total lead loadings were only 30 and 27 per cent respectively for the rotary sweeping, although greater efficiencies of 45 and 55 per cent were observed for hydroflushing.

Malmquist (1978) also concluded that street sweeping was relatively inefficient at removing the highly contaminated fine dust, but observed higher removal efficiencies of 57 per cent for suspended solids, 65 per cent Pb, 31 per cent Zn and 61 per cent Cu. Similar total solid removal efficiencies were reported by Bender and Terstriep (1984) ranging from 14 to 56 per cent. However, they pointed out that the < 250 μm fraction was relatively unaffected by street cleaning. The reduction in mean lead concentration for subsequent wash-off ranged from 5 to 50 per cent. Although there appears to be some debate over the removal efficiency of street cleaning after a single pass (5–65 per cent for Pb), Sartor and Gaboury (1984) have demonstrated that the governing factor for the overall efficiency is the relative frequency of cleaning and storm events. The manner in which street cleaning is practised in the UK, with a typical cleaning interval of 2 months, prevents it from being an effective technique for removing road surface pollutants. However, in some cases it may increase the efficiency of wash-off by breaking up large particles and redistributing them throughout the street (Sartor and Garboury, 1984).

Much less attention has been paid to the efficiency of indoor cleaning techniques. Cannell *et al.* (1987) conducted experiments on the decontamination of carpet tiles impregnated with a 0.6 μm fluorescent tracer, using a light industrial/commercial dry vacuum cleaner. They found that for carpets vacuumed for just 20 s m^{-2}, 61 per cent of the tracer was removed from carpets which had not been trodden on, while only 37 per cent could be removed from those which were traversed 60 times. Further more intensive decontamination (5 minutes) removed totals of 72 and 52 per cent for the two carpeted surfaces.

6.4 RETENTION BEHAVIOUR

6.4.1 INTRODUCTION

The retention behaviour of urban pollutants will naturally be a function of the input and removal rates of the processes previously described. There have been a number of studies, particularly since Chernobyl, on the retention behaviour of radionuclides adsorbed to impervious urban surfaces. In addition, the migration behaviour of pollutants through the vadose zone is generally well understood (e.g. Yaron *et al.*, 1984), but little is known about the retention of particulate bound contaminants on impervious urban surfaces.

6.4.2 PERVIOUS URBAN SURFACES

The migration of radionuclides through soil is dependent primarily on the adsorption equilibria with the components present in a soil column (e.g. clay minerals, zeolites, hydroxides of iron and humic substances) and the porewater velocity.

Molecular diffusion, plant root uptake and volatilization may also have significant influences. Migration rates are therefore quite variable, e.g. 0.6–17.9 cm y^{-1} for $^{239+240}Pu$ (Coughtrey *et al.*, 1984).

Long term migration rates, derived from weapons fallout ^{137}Cs, range from 0.07 to 1 cm y^{-1} for the various soil horizons in a forest, compared to 0.2–0.4 cm y^{-1} for grassland (Schimmack *et al.*, 1989). Initial migration rates determined for Chernobyl-derived ^{137}Cs were several orders of magnitude higher than these values (forest, 2600 cm y^{-1}; grassland, 1800 cm y^{-1}). This was attributed to a high porewater velocity and the non-attainment of sorption equilibrium. Residence times of 2.5 y cm^{-1}, 2.1 y cm^{-1} and 3.3 y cm^{-1} have been calculated for ^{137}Cs, ^{90}Sr and iodine respectively (Coughtrey and Thorne, 1982). This results in mean residence times of 75, 63 and 99 y for the top 30 cm of soil.

The reduction in dose due to weathering above a soil plot contaminated with ^{137}Cs was investigated by Gale *et al.* (1964). Migration of the ^{137}Cs resulted in an exponential decrease of the dose rate, due to increased absorption of the gamma rays by overlying soil layers. There was a high migration rate for the first four years, followed by a much longer term retention, which may even have been the radioactive decay of ^{137}Cs.

6.4.3 IMPERVIOUS URBAN SURFACES

Warming (1982), Sandalls *et al.* (1986) and Gjørup *et al.* (1986) undertook pre-Chernobyl experiments on both natural and artificial weathering of radionuclides from a variety of urban surfaces. These studies were followed up with an assessment of decontamination techniques. The dose rate above areas of asphalt and concrete contaminated with ^{86}Rb (applied in solution) exhibited an exponential decrease with time (Warming, 1982; Gjørup *et al.*, 1986). Rain was found to be the most effective agent at removing ^{86}Rb and losses were tentatively attributed to wash-off rather than migration through the surface. The age of the surface appeared to have some effect, with contaminant removal occurring more readily from young surfaces. About 30 per cent of the ^{86}Rb on asphalt underwent removal, with a half-life independent of radioactive decay, ranging from 10 to 200 d, while for concrete surfaces 50 per cent had half-lives ranging from 80 to 200 d. The remainder was assumed to be removed over the much longer timescale reported by Gale *et al.* (1964) (92 y half-life). Three decontamination techniques were attempted: fire hosing, addition of potassium fertilizer and vacuuming. Only fire hosing appeared to be reasonably successful, with a 50 per cent removal efficiency if attempted within 14 days of the initial contamination. However this was reduced to < 10 per cent if left for 40 days. Jacob *et al.* (1987, 1990) measured the reduction in dose above paved areas in Munich, Germany following the Chernobyl accident, and fitted a regression model describing the exponential decay in activity.

The percentage of weapons fallout ^{137}Cs and cosmogenic ^{7}Be retained by roofing material was investigated by Gjørup *et al.* (1986). They showed that the slope of

Table 6.2 Percentage of wet deposited Chernobyl radionuclides intercepted by various roof surfaces (Roed, 1987b)

Roofing material	Slope	Retention by roof (%)	
		^{137}Cs	^{131}I
Cement tile	45°	58	0
Red tile	45°	68	43
Corrugated eternite	45°	87	0
Silicone-treated eternite	45°	22	0
Corrugated eternite	30°	80	0
Silicone-treated eternite	30°	18	0

the roof and type of roofing material were both important parameters. The retention by different roofing material was in the order: red tile > eternite > cement tile > silicone-treated eternite.

The validity of these results was established by measurements made soon after the Chernobyl accident (Roed, 1987b). Table 6.2 shows that for 45° slopes, silicone-treated eternite again had the least capacity for retaining ^{137}Cs (22 per cent of deposition), while untreated eternite had the greatest (87 per cent deposition). The other Chernobyl radionuclides showed similar behaviour except for ^{131}I which was only retained by the red tile.

Studies on the subsequent weathering of the adsorbed radionuclides showed virtually no removal of ^{137}Cs, with < 5 per cent washed off from August to December 1986. Sandalls *et al.* (1986) and Wilkins (1987) found slightly higher removal rates from a variety of building materials. The removal of activity was found to be a function of the total rainfall and therefore the retention behaviour was expressed in terms of the integrated rainfall (Figure 6.2). Wilkins (1987) conducted laboratory experiments on caesium deposited as a dry aerosol and then subjected to simulated rainfall. There appeared to be an appreciable removal of the activity by the first irrigation, but thereafter the retention behaviour followed that of wet deposition (Figure 6.2). Further experiments showed that no difference in behaviour could be discerned between old and new building materials or inland and coastal samples. Examination of the retention of Chernobyl radiocaesium by roofing materials collected from intact buildings supported these results (Nicholson, 1988c).

The overall efficiencies with which building materials intercept and retain depositing radionuclides may be expressed as the interception/retention factor (IRF) (Sandalls, 1987a):

$$\text{IRF} = \frac{(\text{Quantity of radionuclide found on specimen (Bq m}^{-2})}{\text{Quantity of radionuclide incident on specimen (Bq m}^{-2})} \qquad (6.2)$$

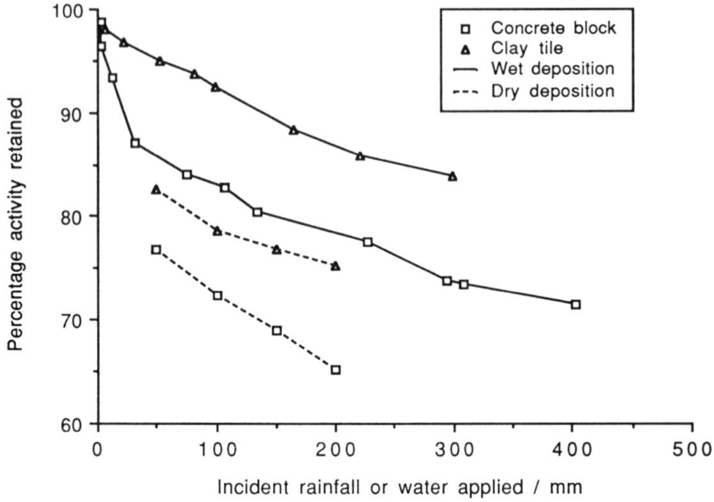

Figure 6.2 Detection behaviour of both wet and dry deposited ^{134}Cs on building materials. After Wilkins, 1987.

Sandalls (1987a) determined these IRF values for urban surfaces in Cumbria, 20 weeks after Chernobyl. He concluded that the overall retention ranged from 5 per cent for glazed roof tiles and up to 60 per cent for unglazed, while housebricks ranged from 18 to 34 per cent. Studies on decontamination of building materials by Sandalls (1987b) provide an insight into the retention behaviour of radionuclides. Non-porous materials such as glass showed little affinity for caesium and thus efficient decontamination could be achieved with pure water. Porous materials, however, exhibited a greater retention capacity for caesium. Since decontamination by ammonium nitrate appears particularly successful for concrete (> 90 per cent removal in 20 h), but less so for clay tiles (40 per cent in 20 h), it may be concluded that the availability and selectivity of ion exchange sites governs the interception and retention behaviour of radionuclides.

Allott *et al.* (1990) have derived the environmental half-lives and mean residence times of Chernobyl radiocaesium-contaminated dust in the UK town of Barrow-in-Furness. The radiocaesium levels in dust were observed to follow an exponential decrease with time. The environmental half-lives describing this behaviour ranged from 190 to 370 d for a variety of internal and external sites. Mean residence times for the dust, derived from these half-lives, ranged from 150 to 250 d. The broadly similar half-lives observed for the quite different sites within this town emphasized the pervasive nature of dust and its associated contaminants and the universal nature of the environmental processes controlling its behaviour in the urban environment.

6.5 DECONTAMINATION STRATEGIES

6.5.1 INTRODUCTION

Since urban environments may potentially become contaminated from accidental releases of radioactivity and the artificial radionuclides released may have a long-term retention behaviour (Section 6.4), various decontamination procedures have been considered. The aim of these procedures is to reduce the level of contamination and consequently shorten the period of time for which interdiction is necessary.

6.5.2 DECONTAMINATION TECHNIQUES

From a knowledge of the retention behaviour of artificial radionuclides in the urban environment, a range of decontamination procedures may be identified. Certain techniques are simply the rigorous application of normal practices, such as road sweeping, digging gardens or vacuuming indoors. Others are more comprehensive and even destructive, including removal of the road surface (road planing), ploughing of parks and roof replacement.

6.5.3 COST AND EFFECTIVENESS OF TECHNIQUES

In order to assess the overall cost of applying a decontamination technique to a city, compared to the dose reduction achievable for the population, it is necessary to consider the following factors:

1. Distribution of the deposited material with respect to different urban surfaces.
2. The contribution of the deposited artificial radionuclides on these surfaces to the dose rate.
3. The decontamination or dose reduction achievable on the individual surfaces, using appropriate methods.
4. The cost required to treat different urban surfaces.
5. The relative proportion of the urban surface types within a city.

The relative cost and decontamination achievable for various techniques has been determined for ^{137}Cs in relation to the unit area of the surface treated (Roed, 1988; Tawil *et al.*, 1985). These data have been used to calculate the overall percentage dose reductions which may be achieved by the decontamination techniques (Roed and Andersson, 1991; Robinson *et al.*, 1990). The models employed took account of the different interception and retention abilities of the urban surfaces, the shielding properties of the buildings and population habits (e.g. where they spend their time). The overall relative costs of treating unit area of city were derived from the individual surface data and the proportion of the surface within the city.

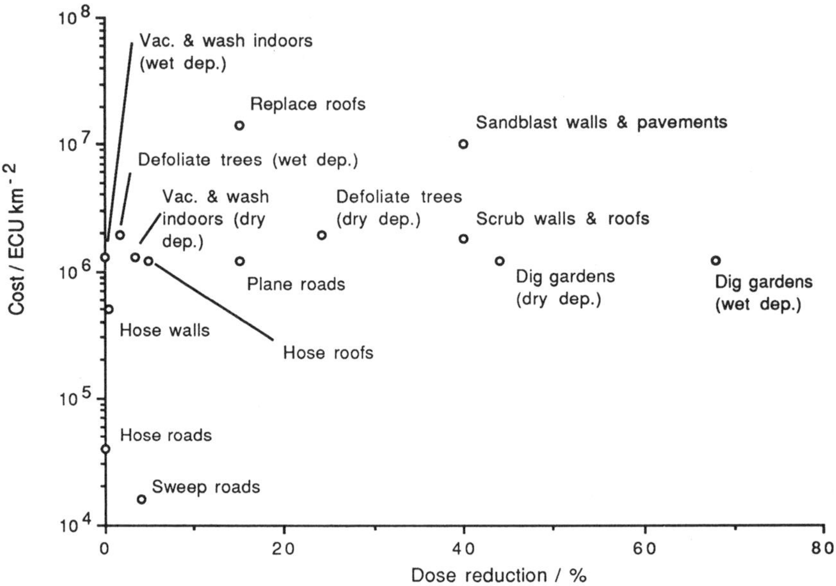

Figure 6.3 Relative cost and effectiveness of different decontamination techniques for the urban environment.

Comparing predicted levels of dose reduction with the city costs (Figure 6.3), the most cost-effective techniques appear at the bottom of the graph. The two techniques which are apparently the most cost-effective of those considered are digging gardens and road sweeping. Replacing roofs would be least cost-effective.

6.5.4 STRATEGY PLAN

In defining a strategy which optimizes decontamination in the urban environment, the cost and effectiveness of the procedures must be considered along with other factors, such as practicability and time required to complete the treatment. These have been incorporated into the schemes presented in Table 6.3 (Roed and Andersson, 1991). The relatively inexpensive procedures of digging the garden and

Table 6.3 Decontamination strategies for the urban environment

Priority	Dry deposition	Wet deposition
First	Digging gardens Defoliation of trees	Digging gardens
Second	Road cleaning	Road cleaning
Third	Hose roofs	Removal of surface debris

defoliating trees may reduce the total dose to about 30 per cent and should therefore be of first priority. Decontamination of walls and internal surfaces, however, would have little effect on dose and should be given the lowest priority.

6.6 SUMMARY

Dusts and pollutants may enter an urban environment by either wet deposition, mechanical transport or erosion of building material. The major contamination processes in the outdoor environment appear to be wet and dry deposition, with the former leading to much greater deposition per unit time. Dry deposition is highly variable due to the complex variety of surfaces and wind domains, whilst wet deposition may possibly be enhanced for cities with populations greater than one million, because of their influential increase on precipitation events. Contamination of the indoor environment appears to occur mainly via the mechanical transport of dust on the bottom of shoes.

A major difference between urban and rural areas results from the different interception abilities of pervious and impervious surfaces. In open rural areas the wet deposition of pollutants usually results in fairly uniform contamination. However, wash-off from impervious urban surfaces causes the surface contamination within cities to be highly variable.

The retention behaviour of pollutants deposited on urban surfaces will be dependent on the nature and rate of removal processes. Weathering in the outdoor environment is probably governed by wash-off, whilst cleaning constitutes the major removal process in houses. Resuspension in urban environments may be greater due to vehicular disturbances; however, this has to be balanced against the possibility of a reduction in wind-induced resuspension, resulting from lower windspeeds. Contaminants deposited on pervious surfaces will be subject to long-term weathering processes with the effective half-life of ^{137}Cs after 4 years being little different from its radioactive half-life. A faster weathering rate has been observed for the dose due to radionuclides on road surfaces, whilst half-lives of 190–370 d have been reported for dust contaminated with artificial radionuclides.

These studies form the basis of modelling artificial radionuclide behaviour in the urban environment and may therefore be used to predict the dose reduction achievable for different decontamination techniques. When the cost of the techniques is considered, it is apparent that digging gardens and defoliating trees should be of first consideration in a decontamination programme.

Artificial radionuclides have been used successfully in the study of individual urban environment pathways, with particular attention being paid to dry deposition and mechanical transport. Future studies on wet deposition, resuspension and wash-off would benefit from the use of these tracers. However, the greatest value and future scope of artificial radionuclides as tracers in the urban environment is probably in the study of the retention behaviour of pollutants, so that urban models may be improved.

ACKNOWLEDGEMENTS

We are indebted to Professor J. B. Ellis (Middlesex Polytechnic), Dr K. W. Nicholson, Dr F. J. Sandalls and Mr J. A. Garland (AEA Technology, Harwell) for their constructive criticisms and extremely helpful comments on this chapter. The financial support of the National Environment Research Council to R.W.A. is gratefully acknowledged.

7 Dosimetry and the Assessment of Environmental Effects of Radiation Exposure

Co-ordinator: **D. S. Woodhead**

7.1 INTRODUCTION

The authorized disposal of radioactive wastes into the environment, or the accidental release of radionuclides from a nuclear facility have, as an inevitable consequence, the potential for increasing the radiation exposure not only of man, but also of domestic animals and native populations of wild organisms. For authorized disposals, controls are exercised to reduce the exposure of critical groups of humans to levels as low as reasonably achievable and, in any event, to maintain them (on average) within the appropriate recommended annual dose limit (at present 1 mSv a^{-1} (ICRP, 1985)). The received wisdom is that such measures are likely to provide adequate protection for other species, although not necessarily for individual members of those species (ICRP, 1977, 1991). The ICRP have provided no evidence to support this belief, and no standards have been recommended for dose rate limits which would provide adequate protection for organisms other than man. This position clearly rests on a number of assumptions:

1. that humans occupy, or otherwise utilize the resources of, the environment contaminated by the waste to a similar extent as do populations of other organisms;
2. that there are no mechanisms or pathways for the accumulation of radionuclides in some environmental compartment which lead to much-enhanced exposure of endemic populations, but which have no relevance to man; and
3. that other organisms are not markedly more radiosensitive than man.

The first two assumptions relate to the likelihood and extent of the radiation exposure, i.e. the problem of dosimetry, and the third concerns the radiobiological response, i.e. the assessment of the potential consequential effects.

In terms of making an assessment of the possible environmental impact of a disposal practice or a potential accident it is logical to deal with the problems in this order, i.e. radiation dosimetry and then radiobiological response. It should be recognized, however, that there is feedback between these two aspects of the overall problem—the assessment of the potentially damaging impact on the environment—which dictate the nature and relative importance of the information required in each case.

Radioecology after Chernobyl
Edited by F.E. Warner and R.M. Harrison
 Published by John Wiley & Sons Ltd

For example, for controlled radioactive waste disposals it is, in general, reasonable to conclude that the concomitant exposure of wild organisms will be of a similar magnitude to that experienced by humans, i.e. low dose rate and protracted in time. With the assumption that wild organisms are not more radiosensitive than humans, this means that lethal effects are unlikely to be a problem. This conclusion does, however, expose the existence of different perceptions of the risks resulting from the exposure of humans and those for all other organisms. For humans, the practice of radiological protection focuses on the individual for whom the risk of health effects arising from radiation exposure should be as low as reasonably achievable and, in any event, kept below some limiting value which is agreed to be acceptable. For the great majority of other species, it is the population (undefined at present, but see below) which would be considered the appropriate object for protection, with the fate of individuals usually being of much lesser concern; exceptions might be individuals of rare or endangered wild species or valuable domestic animals. This different perspective immediately alters the radiation responses which need to be considered for impact assessment and the provision of adequate protection. The survival of a population requires that its ability to maintain itself, through reproduction, within the normal range of variability and in the face of all the other vicissitudes of the natural environment, should not be affected. Notwithstanding these conclusions, it must be stressed that there can be no effects due to increased radiation exposure at the population level (or, indeed, at any higher community or ecosystem level) without the appearance of clear responses in individual organisms (Haux and Forlin, 1988). Thus it is the processes of gametogenesis in individuals and embryonic development, within the population, which are important and become the targets of dosimetric attention (IAEA, 1988, 1992b; NCRP, 1992).

Accident situations are likely to be different in a number of respects. It is probable that the release would be restricted in time and the magnitude of the resulting environmental dose rates, and their variation over time, could be much greater than would be the case for authorized disposals. For example, if, as was the case at the Chernobyl power plant (UNSCEAR, 1988b), the release were to consist of large quantities of short-lived radionuclides and, relatively, much smaller quantities of long-lived radionuclides then the dose rates in the immediate vicinity would be high initially but would fall fairly rapidly to much lower values, which would decline rather slowly due to both physical decay and the influence of environmental processes causing dispersion and dilution.

7.2 DOSIMETRY

Radiation dosimetry is the process of determining the energy absorbed in a specified target from a radiation field which, for waste disposal operations or accident situations, is generated by non-uniform, time-dependent distributions of radionuclides both in the external environment and within the organism of interest. For mobile animals, the incident radiation field from the external sources is also modified by their behaviour.

As indicated above, the relevant biological responses of wild organisms to radiation exposure suggest that the appropriate targets for dosimetry are likely to be the ovaries, stamens, germinating seeds and other propagative tissues in plants, and the gonads and the developing embryos in animals. These biological entities show enormous variation throughout the plant and animal kingdoms both in their geometrical form and their capacity to accumulate radionuclides from the external environment.

The incident radiation field consists of X- and γ-rays, β-particles and α-particles which have characteristic path lengths in tissue ranging from metres to approximately 0.1 mm. Thus it is clear that the spatial distributions of the radionuclides, relative to the defined targets, also need to be determined on these scales if dose (rate) estimates are to be made.

These factors concerning the spatial and temporal distributions of the radionuclides and the geometric variability of the targets mean that it is very rarely possible to determine the dose rate by a direct (or indirect) instrumental method; also, it is clearly impossible to adopt this approach for assessments either in the pre-operational phase of a waste-disposal practice or of the potential consequences of an accident. In these circumstances reliance must be placed upon dosimetric models, that is, computational methods which use information concerning both the distributions and behaviour of radionuclides in the environment and the absorption processes acting upon radiations in tissue, to estimate the radiation dose (rate) to the specified target. In view of the enormous number of species constituting the biosphere and the wide range of physical and chemical conditions prevailing in the niches occupied, it is inevitable that simplifications and generalizations would have to be made. In addition, the required full range of information on both the internal and external distributions of the radionuclides and their environmental behaviour is very rarely available even for a single species.

The processes by which energy is transferred from the radiation field and absorbed in the target tissue depend on the nature and energy of the radiation. A theoretical analysis of the energy deposition along the paths of α- and β-particles and γ- and X-rays is possible but the resulting mathematical expressions are complex and, due to the stochastic nature of the processes involved, are not straightforward to apply for the purposes of dose estimation, particularly in an environmental context. As a consequence, simpler empirical formulae, with energy-dependent parameters, have been developed which describe the dose distributions around point sources (Berger, 1968, 1971; Loevinger *et al.*, 1956; Harley and Pasternak, 1972; Woodhead, 1979). These can then be used to estimate the dose (rate) at a given point by integration of the expressions over an extended source distribution (Loevinger *et al.*, 1956; Woodhead, 1979). The determination of the dose (rate) at different points within a target volume then allows an estimate to be made of the mean dose (rate) to the tissue or organ (Brownell *et al.*, 1968; Ellett and Humes, 1971).

These dosimetric approaches can be developed to deal with all the more or less complex situations which are to be found in the natural environment. The techniques and models introduced above provide estimates of the absorbed dose (rate), i.e. the radiation energy absorbed per unit mass of tissue. Radiobiological studies show, however, that different radiations can produce different degrees of effect for the same absorbed doses and that the variations are very largely dependent upon the ionization density produced along the particle tracks. This phenomenon is evaluated in terms of the relative biological effectiveness (RBE) of the radiation, defined as:

$$\text{RBE} = \frac{\text{Absorbed dose of 250 keV X-rays required to produce a given effect}}{\text{Absorbed dose of specified radiation required to produce the same effect}}$$

Thus the biologically effective dose is the product of the RBE and the absorbed dose. The RBE values obtained experimentally for a given radiation type vary, however, with the particular biological effect considered and the dose rate. In order to obtain a general perception of the biologically effective dose, independent of any specific effect and dose rate, quality factors (QF) have been defined in terms of ranges of ionization density (ICRP, 1977) such that

$$\text{Weighted absorbed dose} = \text{QF} \times \text{absorbed dose.}$$

Although the QF have, strictly, only been defined for use in human radiation protection applications, the values have been developed from RBEs obtained from experiments with a wide range of organisms and it is assumed that they can be applied in an environmental context. Such considerations are necessary due to the actual or potential contributions of α-radiation, for which a QF value of 20 is recommended, to the internal dose rate of many organisms.

For the purpose of dosimetry, the most difficult problem is to provide all the detailed information, concerning the distributions of the radionuclides in space and time, which is required for the application of the models. For an established waste disposal practice, relevant and specific information can be accumulated for the purpose of estimating the dose rates to particular organisms which might potentially be at risk (Woodhead, 1970, 1974, 1986; Pentreath *et al.*, 1973, 1980); in exceptional circumstances it might be possible to confirm the results of the dose calculations by direct measurements (Woodhead, 1973). For pre-operational studies and the assessment of potential accidents a different approach is necessary. In both cases the radionuclide source must be defined both in terms of the spectrum of radionuclides present, and their emission rates over time. Appropriate models must then be applied to determine the development of the dispersed radionuclide distributions in the environment, including the accumulation by organisms. Available transport models are concerned entirely with producing a basis for estimates of dose

Parts of system included in model	Model	Major processes included in model
Canister and lining Waste form	Waste package ↓ rates of release of radionuclides into the ocean, as a function of time ↓	Canister corrosion Degradation of package linings and caps Release of radionuclides from waste forms
Bottom sediments Benthic boundary layer (water and particulates) Open ocean (water and suspended particulates)	Ocean dispersion and Sedimentation ↓ Radionuclide concentrations in water and sediments, as a function of time ↓	Diffusion and advection Interactions between radionuclides and suspended particulates and bottom sediments
Exposure pathways - sea foods, beaches, atmosphere, salt, water. Marine organisms	Dose to man and organisms	Reconcentration of radionuclides in marine organisms, beach sediments, aerosols Radionuclide intake and metabolism by man and organisms

Figure 7.1 Modelling framework

rates to human populations but, interpreted with care, can also provide information of use for environmental dosimetry. The assessment made of the potential impact of the dumping of packaged low-level radioactive waste in the deep north-east Atlantic Ocean may be quoted as an example (OECD/NEA, 1985). The modelling framework is indicated in Figure 7.1, from which it can be seen that the output of the transport model was a prediction of the radionuclide concentrations in water and sediment as a function of time. The highest concentrations are predicted to occur in the deep ocean in the immediate vicinity of the dumpsite and these were used as the basis for the estimates of the dose rates to the faunal types listed, together with their dimensions, in Table 7.1. The choice of organisms was based upon the following considerations:

1. *Fish*: radiobiological studies have indicated that these are probably the aquatic organisms most sensitive to radiation exposure (IAEA, 1976, 1988). Assessments for bathypelagic and benthic types allow the contribution from γ-emitters in the sediment to be highlighted.

Table 7.1 Dimensions of the generalized geometrical models adopted to represent deepsea organisms

Organism	Mass	Length of major axes of the ellipsoid
Fish	1 kg	$45.0 \times 8.7 \times 4.9$ cm
Large crustacean	2 g	$3.1 \times 1.6 \times 0.78$ cm
Mollusc	1 g	$2.5 \times 1.2 \times 0.62$ cm
Small crustacean	16 mg	$0.62 \times 0.31 \times 0.16$ cm

2. *Large crustaceans*: these organisms generally have higher concentration factors than fish, thus increasing the importance of internal exposure relative to external sources. Again, assessments for bathypelagic and benthic types show the importance of both β- and γ-emitters in the sediment.
3. *Molluscs*: these organisms generally have higher concentration factors than the previous two groups and, with smaller size, show the effect of these factors on dose rate; and
4. *Small crustaceans*: assuming that the data available for surface-living zooplankton are applicable, these organisms would have the highest concentration factors for most elements. Also, being the smallest organisms, they show most clearly the effects of size on the relative contributions of internal and external sources.

In the absence of more detailed information, the radionuclides were assumed to be uniformly distributed throughout the bodies of the organism. The volume-averaged absorbed fractions for γ-radiation, and the absorbed fractions at the centre of the volumes for β-radiation, are shown in Figures 7.2a and 7.2b respectively, for the four organism geometries (see OECD/NEA, 1985 for more details). The results of calculations made for the natural radionuclides, past dumping and 5 years of expected future dumping (which did not take place) are given in Table 7.2. Clearly, these organisms, and their chosen geometries, cannot accurately represent all those present in the deep sea, but the dose rates estimated on the basis of these models *do* provide a realistic indication of the range of exposures likely to be experienced in the deep sea from the dumping of radioactive wastes.

The effect of a non-uniform distribution of the radionuclide body burden implied by a uniform whole-body concentration factor has been investigated for the cases where a small, centrally situated organ either preferentially accumulates, or discriminates against, β-emitting isotopes. At low β-energies the dose rate scales directly with the radionuclide concentration, but at higher energies the dose rate falls below proportionality for preferential nuclide accumulation and increases above proportionality where there is discrimination; the size of the organ relative to the body size also influences the dose rate (Pentreath and Woodhead, 1988).

An alternative approach which has been adopted (IAEA, 1992b; NCRP, 1992) is to estimate the radionuclide concentrations in the environment which would,

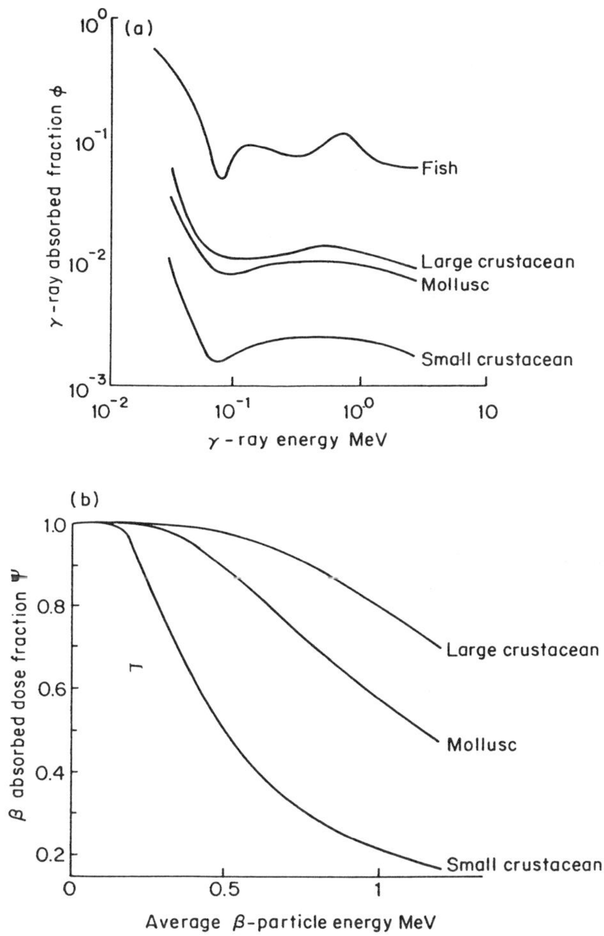

Figure 7.2 (a) Absorbed dose fraction for γ-radiation (b) Absorbed dose fraction for β-radiation.

through reasonable human use of that environment for food production, leisure etc., result in the limiting dose equivalent of 1 mSv a^{-1} being received by a critical group. These environmental concentrations may then be used as the basis for dose estimates to populations of wild organisms also inhabiting the area. The quality of the results thus obtained depends, amongst other things, upon the validity of both the assumptions made concerning the behaviour of the radionuclides in the environment and the models used to describe this behaviour and generate predictions of radionuclide distributions. Table 7.3 gives the estimates of the weighted absorbed dose rates to plants and animals which have been obtained for a controlled release to the atmosphere using this approach (IAEA, 1992b).

Table 7.2 Estimates of the weighted absorbed dose rates ($nGy\ h^{-1}$) to deepsea organisms

Organism	Natural background	Past dumping	Past dumping plus 5 years expected future dumping
Fish			
Bathypelagic	28–460	52	73
Benthic	85–1400	73	100
Large crustaceans			
Bathypelagic	930–3600	360	460
Benthic	990–4600	380	500
Molluscs	930–3400	8000	10 000
Small crustaceans			
Bathypelagic	250–2900	1200	1600
Benthic	330–4200	1300	1700

Table 7.3 Estimates of weighted absorbed dose rates to plants and animals for controlled atmospheric releases which deliver an effective dose equivalent of 1 $mSv\ a^{-1}$ to humans

Nuclide	Deposition rate ($Bq\ m^{-2}\ d^{-1}$)	Equilibrium concentration in plant tissue ($Bq\ kg^{-1}$ dry)	Upper-bound estimate of dose rate to plant tissue ($mGy\ d^{-1}$)	Equilibrium concentration in animal tissue ($Bq\ kg^{-1}$ wet)	Upper-bound estimate of dose rate to animal tissue ($mGy\ d^{-1}$)
^{3}H (as water)	6.6×10^{5}	1.8×10^{6}	0.14	1.8×10^{6}	0.14
^{14}C	—	6.0×10^{5}	0.44	3.6×10^{5}	0.27
^{32}P	5.4×10^{1}	3.2×10^{3}	0.049	6.8×10^{1}	0.001
^{95}Zr	4.9×10^{3}	1.9×10^{5}	0.90	5.2×10^{2}	0.048
^{131}I	1.0×10^{3}	5.8×10^{3}	0.029	1.7×10^{2}	0.0014
^{137}Cs	2.8×10^{1}	1.9×10^{3}	0.13	3.9×10^{2}	0.074
^{239}Pu	1.4×10^{-1}	7.2	0.011	6.9×10^{-5}	0.00026

It is clear from the foregoing that the major limiting factor in the determination of dose rates to wild organisms in contaminated environments is the availability of the necessary data on radionuclide distributions in the organisms and their environment. Where data are available, dosimetry models of sufficient sophistication have been developed to make complete use of the information provided on radionuclide distributions (Woodhead, 1979, 1986; Pentreath *et al.*, 1973; Hoppenheit *et al.*, 1980).

7.3 RADIATION EFFECTS

The effects of ionizing radiation have been studied in a wide range of plant and animal species at all scales of biological organization from biomolecules through cellular organelles, whole cells, tissues, organs, whole organisms, populations and, finally, at the community level. It is not possible to provide a comprehensive review of all the available information here (but see, for example, Bacq and Alexander, 1961, for a general introduction at the lower levels of organization; Whicker and Fraley, 1974 (plant communities) and Turner, 1975 (animal populations)). A general conclusion which can be drawn from these data, however, is that, because ionizing radiation acts by depositing energy and inducing change at the molecular level, there can be no effect at any higher level of organization without detectable change at successively lower levels (Haux and Forlin, 1988). In particular, radiation damage will not become apparent at the population level unless there is a clear response, in the individuals making up the population, in those characteristics which influence the maintenance of the population including individual morbidity, fertility (gametogenesis), fecundity (the production of viable offspring) and the gene pool. It follows that if the radiation dose (rate) either from a waste disposal practice, or arising from an accident, is sufficiently low that there can be no significant effects on these individual characteristics, then there can be no impact on the population, which, as already indicated, is generally accepted to be the object of protection.

This apparently conclusive statement does raise two questions:

1. what is the relevant population; and
2. for this population, what is the relevant dose (rate)?

In principle, the population could be defined as all the members of a particular species, but in practice this is rarely helpful. Ecologists have frequently found that the appropriate definition depends on the particular problem (and species) under study, and that it will be determined, among other factors, by the scale of interest in both space and time, thus often leading to a more limited concept (Bakker, 1971; Gross, 1986). A useful definition suggests that the population would be an essentially self-sustaining unit of a particular species, i.e. immigration into, emigration from, and interbreeding with, other populations of the species would be very minor factors in the overall maintenance of this population (Bakker, 1971). Thus it is clear, for example, that all the perch, brown trout or mussels in a freshwater lake could be considered as single populations, whereas ducks overwintering on the lake might be only a small proportion of a much larger interbreeding population. The more mobile the organism (including, for example, aerial dispersion of pollen and seeds, and water transport of gametes, eggs and larvae in the sea), the larger the space scale which needs to be considered in attempting to define the appropriate population.

Because the source of the radionuclides is likely to be restricted in space and time, it is probable that concentration gradients will exist in the environment and that the highest dose rates (all other considerations being equal) will arise in the immediate vicinity of the release. Thus it may be concluded that a range of dose rates will be experienced within a population, although of lesser extent, perhaps, for a mobile species as compared with a sedentary species, and that the increased radiation exposure could be delivered to either all the members of the population, e.g. many aquatic species inhabiting a pond or lake, or only a part of a population for many terrestrial and marine species. Given the large data requirements and obvious complexity involved in estimating average values, over the populations and through time, for the dose rates (or lifetime doses) it is usual to adopt a simpler approach. An assessment is made of the maximum dose rate likely to be experienced by individuals of a limited number of species which might be at risk in the contaminated environment, either because they belong to the most radiosensitive species or because their physiology and life-style are such that they will experience the greatest radiation exposures (see example above for the deep sea). It may then be reasonably assumed that, if the estimated dose rates are less than those which have been shown experimentally to produce no significant radiation effects, there can be no risk of damage to the population (or any higher level of organization).

7.3.1 MORBIDITY AND MORTALITY

For organisms other than man, morbidity (and mortality) are usually only considered as a consequence of acute (short term) high dose radiation exposure. Laboratory or controlled field studies of radiation-induced mortality usually determine the total dose required to kill 50 per cent of the organisms—the LD_{50}—within a specified period of time immediately post-exposure. Table 7.4 gives LD_{50} data for a variety of organisms and provides a qualitative indication of the variations in radiosensitivity which exist between groups (Woodhead, 1971; Myers, 1989). Part of the wide variation in apparent radiosensitivity is due to the use of a 30 day time period for assessing the expression of the radiation-induced mortality (giving the $LD_{50/30}$) although this is strictly applicable only to small mammals where essentially all the deaths occur within this interval. For fish and other aquatic organisms it has been shown that a 50–60 day period is necessary to encompass the acute response—a consequence of their poikilothermic metabolism (White and Angelovic, 1966)—giving an increase in apparent radiosensitivity. Adjustment of the data for this factor reduces, but does not eliminate, the general tendency for radiosensitivity to increase with the increasing biological complexity of the organism. Another factor which influences the relative radiosensitivity between groups, e.g. insects and mammals, arises from the developmental progression through the life cycle. In the former, most of the cellular proliferation and differentiation occurs in the developing embryo, whereas for the latter, it continues in certain tissues

Table 7.4 Acute lethal doses of radiation

Class of organism	Lethal dose, Gy	
	Adults	Developing embryos
Mammals	2–15	1 (mouse)
Birds	5–20	7 (chicken)
Amphibians	7–50	—
Fishes	7–600	0.2–1 (salmon, trout and plaice)
Reptiles	10–40	—
Crustaceans	15–600	6 (production of young amphipods)
Molluscs	100–1000	—
Echinoderms	390	—
Insects	20–3000	1–2 (weevil, fruit fly, wasp)
Higher plants (trees, shrubs and herbs)	7–800	—
Primitive plants (mosses, lichens and algae)	30–>12 000	—
Protozoa	100–6000	—
Bacteria	50–1000	—
Viruses	200–10 000	—

throughout life. In general, developing embryos are more radiosensitive than fully formed adults and across groups of organisms this again tends to reduce the overall range of radiosensitivity; the developing vertebrate embryo does, however, retain its position of greatest radiosensitivity.

Such acute, high dose exposures are only likely to arise as a result of a large, accidental release of radionuclides. In this circumstance, the radiation exposure may range from acute (i.e. delivered in an interval which is short compared with the time of development of the biological response) to chronic (i.e. the exposure continues over a significant fraction of the lifetime of the organism) depending on the organism under consideration and the quantities and relative proportions of short-lived (~ days) and long-lived (~ years) radionuclides present in the release.

Protraction of radiation exposure generally increases the total dose necessary to cause mortality (or any other effect). This is a consequence of two complementary factors: at the cellular level, sub-lethal damage is reparable; and at the tissue level, lethally damaged cells can be replaced through cell proliferation to maintain a viable degree of organ (and organism) function. Where the exposure extends over a significant part of the lifetime of the organism, the total accumulated dose required to kill may be two to ten times greater than the acute LD_{50} (Page, 1968; UNSCEAR, 1982a; Woodhead, 1984; IAEA, 1992b). The chronic, low-level dose rates which have been estimated to arise from controlled disposals of radionuclides to the environment (see Section 7.2) are very unlikely to lead to total accumulated doses to organisms which could result in significant mortality.

7.3.2 GAMETOGENESIS

Available reviews (Whicker and Fraley, 1974; IAEA, 1992b) do not provide any detailed discussion of the effects of radiation on gametogenesis in plants and there appear to be few data in the literature. A single reference (Woodwell and Rebuck, 1967) notes that flowering and seed set occurred in the herbaceous plant, *Carex pensylvanica*, in the dose rate range 0.17–16 Gy d^{-1}. This plant species was, however, fairly resistant to the damaging effects of radiation in somatic tissue, so this rather low radiosensitivity is probably not typical of the plant kingdom.

For animals it is necessary to consider spermatogenesis and oogenesis separately because the developmental processes for the production of the mature gametes are generally different for the two sexes.

7.3.2.1 The testes

Exposure of the testes to ionizing radiation produces sterility which, depending on the dose rate and total dose, may be temporary or permanent. Acute doses which are a small fraction of the LD_{50} may induce short periods of delayed sterility in mammals due to the damage to the most sensitive differentiating spermatogonial cell-type. Larger doses produce more immediate and longer-lasting periods of temporary sterility but recovery is possible due to the relatively radioresistant stem cell population. Permanent sterility is ultimately produced by doses greater than about half the LD_{50} depending on the species (BEIR, 1980; UNSCEAR, 1982b).

Insects may be sterilized with acute doses which are much less than those required to kill them or otherwise affect their behaviour (Bacq and Alexander, 1961).

Similar conclusions have been drawn for aquatic organisms, and fish have been shown to be the most radiosensitive of the aquatic fauna for which data are available (IAEA, 1976; Woodhead, 1984).

Of more relevance to the majority of environmental situations is the potential influence of chronic, low-level exposure. The response of the testis is somewhat unusual in that protraction of a given total dose does not always lead to a reduction in the effect (UNSCEAR, 1982b). The most sensitive mammalian species appears to be the dog for which a life-time mean dose rate greater than 1.2 mGy d^{-1} (50 μGy h^{-1}) has been determined to be the threshold for deleterious effects (Casarett, 1964; BEIR, 1980; UNSCEAR, 1982b). In fish also there appears to be considerable inter-species variability in the response of the testis to chronic irradiation with the most sensitive species examined showing minor effects at a mean dose rate of 1.1 mGy h^{-1} (Rackham *et al.*, 1992).

At dose rates above the threshold for effects, but below that at which a continuing decline into sterility occurs, there is evidence for a homeostatic response, through either mobilization of reserve proliferative capacity in the stem cell compartment or adjustment of the cell cycle kinetics, or a combination of both, leading to the maintenance of sperm production at equilibrium level below normal in both mammals (Hsu and Fabrikant, 1976; Erickson, 1978; UNSCEAR, 1982b) and fish (Hyodo-Taguchi and Egami, 1977).

The developing testis in the mammalian fetus also shows wide interspecies variations in radiosensivity, with the pig showing the greatest response at 2.5 mGy d^{-1}, the lowest dose rate employed, and being substantially more radiosensitive than the guinea pig, rat and mouse at 10 mGy d^{-1} (UNSCEAR, 1986).

7.3.2.2 The ovary

The responses of the resting oocytes in adult mammals show wide variations between species with the acute dose required for sterilization being a small fraction of the LD_{50} for the mouse and substantially greater than the LD_{50} for the monkey (UNSCEAR, 1982b). Available information indicates less variation between fish species where the sterilizing dose is of the same order as the LD_{50} (Woodhead, 1984). As might be expected, due to the cell division and differentiation in progress, the embryonic and neonatal ovaries are found to be more radiosensitive (UNSCEAR, 1982b; Woodhead, 1984).

For mammals, the available reviews of the effects of chronic irradiation on the adult ovary present the data in terms of offspring production (fecundity, see below), rather than as the numbers of oocytes at different stages of maturation present (UNSCEAR, 1982b). The developing ovary in the mammalian fetus is sensitive to the effects of continuous irradiation (although apparently less so than the developing testis) with a dose rate of 10 mGy d^{-1} throughout gestation having a substantial effect in the pig, guinea pig, rat and mouse. The pig shows the greatest sensitivity with a significant response at the lower dose rate of 2.5 mGy d^{-1} (UNSCEAR, 1986). Protraction of ovarian exposure in fish produces an apparent reduction in radiosensitivity, in contrast to the response of the testis. The most sensitive fish studied to date (*Ameca splendens*) has shown a substantial reduction in all stages of oocyte maturation at a dose rate of 1.1 mGy h^{-1} over 290 days (Rackham *et al.*, 1992); other species, however, have accumulated high total doses at higher dose rates over a greater proportion of the life with less damage becoming apparent (Woodhead, 1984). Exposure of chinook salmon embryos during the 80 day incubation period did not produce any effects on gonad development at dose rates below 100 mGy d^{-1} (Woodhead, 1984).

The potential significance of life history attributes for the response to chronic irradiation has been noted by Turner (1975). Organisms with an extended time to sexual maturity might be at greater risk if, as seems possible, the cumulative dose is a more important determinant of ovarian damage than dose rate *per se*—animals with a shorter maturation time being able to reproduce before radiation-induced sterility intervenes (see also Myers, 1989).

7.3.3 FECUNDITY

If fecundity is taken to mean the number of viable offspring produced per female, it encompasses not only the effects of radiation in reducing fertility, but also both the induction, in otherwise apparently normal gametes, of mutations incompatible with embryonic development and neonatal survival, and the direct effects of radiation on embryonic development.

The total number of resting oocytes present in the ovary at sexual maturity generally far exceeds the number likely to mature and become potentially available for reproduction. A radiation-induced reduction in the number of immature oocytes does not, therefore, necessarily lead to a like reduction in the potential fecundity although it may result in a shortening of the span of reproductive life (BEIR, 1980; UNSCEAR, 1986; Myers, 1989).

In the mouse, for which the most data are available, dominant lethal mutation (as measured by the incidence of non-viable embryos conceived in subsequent matings with unirradiated females) are produced less efficiently by chronic than by acute exposure of the male gametes; also the post-meiotic cell stages are more sensitive than the premeiotic and spermatogonial cells. The induction rate for chronic exposure at 34–60 mGy d^{-1} was estimated to be 1.4×10^{-6} mGy^{-1} per gamete for premeiotic cells and 5×10^{-5} mGy^{-1} per gamete for post-meiotic cells (UNSCEAR, 1982c). Available evidence indicates that the female mouse is much less sensitive, such that the effective rate of induction when both sexes are irradiated is half that for the males alone (BEIR, 1980). Data for other small mammals yield induction rates of similar magnitude (UNSCEAR, 1982c). It has also been concluded that aquatic organisms show a similar sensitivity for the radiation-induction of dominant lethal mutations as do the more intensively studied small mammals (Woodhead, 1984).

Experimental studies indicate that the developing embryo is more sensitive to the effects of acute irradiation, and sometime substantially so, compared with the adult organism. For insects the LD_{50} may be reduced by a factor in the region of 10^{-3} while for vertebrates the factor may be in the range 0.05 to 0.9—largely reflecting the different basic biologies of the organisms (Myers, 1989). The many studies which have been made show that there is a generally declining trend of radiosensitivity with time post-fertilization but that, for the vertebrates in particular, there are phases in embryonic and fetal development, correlated with specific organogenic processes, which show increased radiosensitivity (UNSCEAR, 1977b, 1986; Woodhead, 1984). Again, protraction of a given total exposure, as expected, generally reduces the response (Myers, 1989). Lower doses and dose rates, while not producing immediately lethal effects, might produce teratological responses which may be incompatible with long-term survival in wild animals.

A number of experiments have shown that, despite chronic irradiation, the germ line can be maintained over many generations with minimal effects being apparent in the reproductive performance. Mice of four different strains have been exposed to an average of 20 mGy d^{-1} over 10 generations (for a total accumulated dose to the germ line of 15 Gy) and have shown no effects on reproductive performance in the first litters (Stadler and Gowan, 1964). Similar results have been obtained for the rat (Brown *et al.*, 1964).

The lifetime (> 950 days) offspring production by pairs of the fish, *Poecilia reticulata*, was reduced by 57 per cent at an average dose rate of 1.7 mGy h^{-1} (the lowest used). It was concluded that this effect, which was progressive through life, resulted from both a reduction in fertility and the induction of dominant lethal mutations (Woodhead, 1977).

7.3.4 GENETIC EFFECTS

In wild populations, natural selection operates to adapt the gene pool transmitted from generation to generation to the local environment. This is not to say, however, that genetic uniformity exists because neither the environment nor the resulting selection pressure is stable over space and time. Thus there is a dynamic equilibrium between genetic variability and selection pressure. In the great majority of populations there is an enormous amount of genetic variability which is constantly being sorted into new combinations in each generation (Myers, 1989).

Irradiation of germ cells may produce both gene mutations and chromosome aberrations (changes in chromosome morphology or number). Such mutations are not different in kind from those occurring naturally but the proportions of the different types induced by irradiation may well differ (BEIR, 1980; UNSCEAR, 1977c). The induced mutations can have an impact which varies between lethality (as in the dominant lethals discussed above) and the apparently trivial, e.g. a change in eye colour (UNSCEAR, 1977c). Nevertheless, all are subject to selection pressure, and mutations are lost from the gene pool as a consequence of a reduction in fertility in, or by the pre-reproductive death of, individuals carrying the mutations.

Despite the variations in radiosensitivity, both of different gene loci within a species, and between species, the total dose required to double the spontaneous mutation rate appears to be constrained within the surprisingly narrow range of 0.3–3 Gy across a wide variety of organisms (Myers, 1989). Thus, for a given dose rate, the induction of genetic change will be much less significant for organisms with a short generation time than for long-lived, late-maturing species. Experimental studies of mice and fruit flies (*Drosophila* spp.), irradiated in each of many generations with doses equal to, or greater than, the doubling dose, have detected no effects on the health and general fitness of the populations (Myers, 1989).

7.4 CONCLUSIONS

Previous, more thorough, reviews of the available data, of which only a limited selection has been considered here, have concluded that there would be no significant effects in wild populations of:

1. terrestrial animals if the dose rate to the most exposed individuals were to be less than 1 mGy d^{-1} (0.04 mGy h^{-1}); and

2. terrestrial plants, and aquatic plants and animals if the dose rates to the most exposed individuals were to be less than 10 mGy d^{-1} (0.4 mGy h^{-1}) (IAEA, 1976, 1988, 1992b; NCRP, 1992).

Assessments of the dose rates actually (or potentially) arising from controlled waste disposal operations indicate that, with very few exceptions, these values have not been (or would not be) exceeded; often there is a large margin of safety (IAEA, 1976, 1992b; NCRP, 1992; OECD/NEA, 1985; Woodhead, 1970, 1973, 1974, 1984, 1986). The one major exception relates to dumping of packaged low-level radioactive waste into the deep ocean where, because of the remoteness of the release point from man and his food chains, and the time taken for radionuclide dispersion to take place, it is possible to envisage dumping rates, controlled on the basis of ultimate human exposure, at which very high dose rates could be delivered to benthic organisms at the dumpsite (IAEA, 1988).

The situation for a major nuclear accident is, as Chernobyl has shown, quite different. Effects were readily apparent in the pine trees close to the plant, and less dramatic responses have been detected in animals. Even here, however, it is debatable whether the long-term survival of the populations has been put at serious risk. Studies in progress should provide much useful information for the resolution of this concern.

8 Appendices

Co-ordinators: **L. J. Appleby and R. M. Harrison**

Contributors: **J. Hamilton-Taylor, R. Kirchmann, M. Kelly, P. J. Mitchell, C. S. Shapiro, E. Voice and D. S. Woodhead**

8.1 INTRODUCTION TO RADIOACTIVE PROCESSES

Variants of every element exist which have the same number of protons in the nucleus but a different number of neutrons. These variants are known as isotopes. Many are unstable, transmuting into other types of nuclei, a process which is accompanied by the emission of radiation, in the form of particles or electromagnetic energy. Such unstable nuclei may occur naturally or form as a result of human activities. Instability of atomic nuclei may arise from excess energy within the nucleus, or imbalance between the number of protons and neutrons. Disintegration of an unstable nucleus may occur spontaneously or follow interaction with a nuclear particle. Forms of radiation which may be emitted from nuclei include alpha particles (identical with ^{4}He nuclei), beta particles (electrons), gamma radiation (very short wavelength electromagnetic radiation) or neutrons. Alpha particles often carry the most energy but because of their double positive charge, interact strongly with matter and are readily stopped—for example, by a sheet of paper. Beta particles, with a single negative charge, are usually more penetrating than alpha particles and can be stopped by a thin sheet of metal.

The neutron is uncharged and only interacts with matter when passing close to the atomic nuclei, so that neutron radiation is very penetrating. Gamma radiation is also chargeless and is the most penetrating of all, and when at high energy (short wavelength) can require many centimetres of heavy metal or metres of concrete to stop. Neutrinos, emitted in beta decay, with zero charge and mass, do not interact appreciably with matter and are ignored in this study. The emission of alpha particles is common only for nuclides of mass numbers above 209 and atomic numbers above 82.

The presence of natural radiation in the environment arises from cosmic and terrestrial sources. Earth is being constantly bombarded with cosmic rays (high-energy protons and charged nuclei) of solar and galactic origin. These cosmic rays cause direct exposure but also give rise to secondary radioactivity from their interaction with stable elements in the upper atmosphere forming radionuclides. Measurable doses to humans result from ^{3}H, ^{7}Be, ^{14}C and ^{22}Na arising in this

Radioecology after Chernobyl
Edited by F.E. Warner and R.M. Harrison
 Published by John Wiley & Sons Ltd

way. Moreover, such secondary radioactivity may enter global biogeochemical cycles; for example by incorporation of ^{14}C into plants during photosynthesis and into the oceans by absorption. Tritium enters into the global hydrological cycle.

Many artificially created transuranic elements, such as plutonium, neptunium and americium, occurred on Earth at one time, but decayed to zero long ago by virtue of their short half-lives. Some transuranium elements continue to be produced in very small amounts by the absorption in uranium isotopes of naturally occurring neutrons.

Artificial radionuclides may be released into the environment as a result of deliberate, accidental or inadvertent actions such as from nuclear fuel cycle operations, reactor accidents, spillages or detonation of nuclear weapons. Whilst there has been a decline of environmental radioactivity attributable to nuclear weapons since the ban on atmospheric testing in 1963, anthropogenic radioactivity remains of great importance. This results from the growth of the nuclear power industry and the increasing use of radionuclides in fields such as medicine and biology, research, space exploration and industry (e.g. radiographic examination of structures). Radionuclides are also used in consumer products such as smoke detectors and luminous watches. The importance of radionuclide discharge to the environment from the nuclear fuel cycle, and accidents such as those which occurred at Kyshtym, Chernobyl and Windscale, are considered in detail in Chapter 2.

Radionuclides would be injected into the environment in a nuclear war. A speculative war scenario involving the direct targeting of nuclear facilities (i.e reactors, spent fuel storage, fuel reprocessing plants, high-level waste facilities) reveals that this could release quantities of fission products comparable with weapon debris from a major nuclear exchange (SCOPE, 1989).

The movement of radioactivity among the terrestrial ecosystems, the oceans and the atmosphere by interactive physico-chemical and biological processes (known as biogeochemical cycling) involves many scientific disciplines, from biology and physical chemistry to hydrology, meteorology and oceanography.

The elucidation of pathways for man-made radionuclides contributes to the protection and enhancement of the quality of the environment. Knowledge of transport mechanisms in the environment is essential for the safe disposal of nuclear wastes from civil power generation.

Artificial radionuclides are also excellent environmental tracers, far easier to study than the corresponding stable nuclides. The knowledge gained from radionuclides gives understanding of the behaviour of the stable elements. For example, Chernobyl emissions have improved understanding of wet and dry deposition processes, and Sellafield effluents have proved a useful tracer for sediment behaviour in the Irish Sea. Hence radionuclides are an excellent tool in generally advancing knowledge of biogeochemical cycling.

8.2 UNITS

The intensity of a radioactive source is determined by the number of atoms present and their decay constant. The former is proportional to the mass of radionuclide divided by the atomic weight, the latter is the fraction of the total number of atoms disintegrating in unit time. Decay rate is more commonly expressed as half-life, the time taken for half the atoms of a given radioactive isotope to decay.

The unit of activity in the SI system is the becquerel (Bq), one disintegration per second. The earlier unit, the curie (Ci), was the activity of the radon in equilibrium with one gram of radium, later defined as 3.7×10^{10} disintegrations per second, but this became unwieldy to use when very low activities assumed importance.

Units of dose quantify the absorption of ionising radiation in material or tissue, usually in terms of energy implanted. The unit is the rad (radiation absorbed dose), defined as the liberation of 100 ergs per gram of absorber, or the gray, one joule per kilogram in the SI system. Since the forms and energies of the radiation—alpha, beta, gamma, X-rays, etc.—cause different degrees of biological damage for the same absorbed dose, a relative biological effectiveness (RBE) factor is applied in practical health physics, comparing the effect of the radiation with that of 200 keV X-rays. This gives the equivalent dose in units of rem (Roentgen equivalent man, rad × RBE), or the sievert (gray × RBE) in SI units.

Table 8.1 Radiation units

Quantity	SI Unit	Symbol	Non-SI Unit	Symbol	Conversion
Activity	becquerel	Bq	curie	Ci	1 Bq = 2.7×10^{-11} Ci
Absorbed dose	gray	Gy	rad	rad	1 Gy = 100 rad
Dose equivalent	sievert	Sv	rem	rem	1 Sv = 100 rem

8.3 METHODS OF ANALYSIS

The general principle underlying most methods of detection of nuclear radiation is that whatever the form of the radiation, it gives up part or all of its energy to the detecting medium either by ionizing it directly or by causing the emission of a particle which in its turn produces ionization in the medium. The ionization thus produced is then detected by one of an increasing variety of techniques, the most widely used of which are briefly reviewed in this section.

Although it is feasible to determine some radionuclides directly or by methods requiring the minimum sample preparation, e.g., gamma-spectrometry, where the radionuclide of interest is present at a very low concentration or emits no penetrating radiation, pre-concentration and radiochemical separation from a sample matrix containing interfering radioelements is usually necessary. In such cases the aim of the analyst is to extract from the sample sufficient of the radioelement of interest with a quantified chemical recovery and in a form suitable for nuclear counting.

Normally there are four steps in the measurement of environmental radioactivity: (a) selecting the measuring system; (b) collecting a suitable sample; (c) concentrating and/or separating the desired radioelement by physical and/or radiochemical means; and (d) assaying the radioactivity in the source with a detector that has been calibrated with an appropriate radioactivity standard (NCRP, 1978).

8.3.1 SELECTION OF A LOW-LEVEL MEASUREMENT SYSTEM

The choice of system for a particular application is usually based on a number of factors (NCRP, 1978) of which the most important are: (a) the quantity of sample available; (b) the nature and intensity of the radiation to be measured; (c) the time available for counting; (d) the efficiency; (e) the background; and (f) the cost. The sensitivity of a particular system is usually expressed as a lower limit of detection (LLD). A frequently used criterion is to express three standard deviations of the background in terms of activity as the LLD. However, the IAEA (1989) have recommended the use of the following expression:

$$(\text{LLD}) = 4.66\ S/eP$$

where S is the estimated standard error of the net count-rate, e is the counting efficiency for the specific nuclide energy and P is the absolute transition probability for emission of the above energy. Since P is physical quantity and cannot be adjusted to optimize the LLD, S must be minimized and e maximized. This leads to the following expression for the figure-of-merit (FOM) of the system (Mishra, 1991):

$$(\text{FOM}) = S_a^2/B$$

where S_a is the sample count-rate with background and B is the background count-rate. To maximize the figure of merit, S_a should be made as large and B as small as possible. Since S_a occurs as the square, more benefit is gained by increasing S_a than by decreasing B. In other words, the main emphasis must be on selecting a system with the highest possible efficiency. Thus, the requirements of low-level counting systems are (a) high counting efficiencies, (b) low background count rates and (c) high stability, since long counting periods are often necessary.

Exact defining equations for the limits for qualitative detection and quantitative determination in radioactivity measurements, which are frequently quoted, have been given by Currie (1968). The statistical aspects of interpreting the results of counting experiments designed to detect low levels of radioactivity (when the detector efficiency is known), for both the case of the well-known and the poorly known background, have been discussed in detail by Sumerling and Darby (1981).

8.3.2 SAMPLING

Before reviewing the radiochemical and radiometric techniques currently in use, it is appropriate to emphasize the fundamental importance of the protocols used to obtain environmental samples. Sampling is often considered to be the weakest link in the chain of planning–sampling–analysis–reporting activities, and with some justification. Accurate and precise environmental sampling is an essential prerequisite to analysis and the development of meaningful sampling protocols necessitates careful planning of the practical procedures used in sample collection, handling and transfer. To this end, a preliminary sampling experiment can provide the experience and validation necessary to design larger-scale sampling protocols which will permit meaningful evaluations of radionuclide concentrations, their distributions in time and space, and their impact on the environment and humans. The many variables and techniques required to design and conduct reliable sampling programmes have been thoroughly reviewed in a recent publication by the American Chemical Society, entitled *Principles of Environmental Sampling* (Keith, 1988). Although aimed at sampling for the purpose of chemical rather than radionuclide analysis, many of the protocols and techniques discussed are equally applicable to the latter. Particular attention is given to sampling equipment and techniques, and sample preservation involving water, air and stacks, biota, and solids, sludges and liquid wastes, as well as quality assurance and quality control considerations, and the reader is directed to this and other sources (Kratochvil and Taylor, 1981; Green, 1979; Gy, 1983) for further reading.

The importance of assessing and controlling sample contamination has long been recognized and is yet another aspect which is worthy of the maximum consideration. Some of the problems arising in the whole process from planning of experiments and collection of samples through handling and radiochemical manipulations to the ultimate interpretation of results have been considered by Harvey and Lovett (1991).

8.3.3 DETECTION AND MEASUREMENT OF ALPHA-EMITTING RADIONUCLIDES

The extremely short range of alpha particles in matter dictates that the amount of material between the sample and the detector be minimized as much as possible. Various methods are used to achieve this condition including evacuating the air between the sample and the detector, optimizing the counting geometry, removing as much of the inactive part of the sample as possible and mounting the source inside the detector. Alpha particle measurements are normally classified into two categories, namely, (a) gross alpha counting and (b) alpha-particle spectrometry.

8.3.3.1 Gross alpha counting

Polycrystalline ZnS(Ag), the first solid material to be used as a particle detector, is still the most widely used scintillator for gross alpha counting. For high sensitivity the source thickness should never be permitted to exceed the range of the alpha particles. Discs of transparent plastic coated with silver-activated ZnS and photomultiplier tubes coupled to appropriate electronic circuitry remain in regular use. The background count-rates over the entire alpha-energy range are relatively low and of the order of one event per hour for ZnS(Ag) on a thin plastic disc of 24 mm diameter (NCRP, 1978).

Gas-filled ionization chambers and proportional counters have been used to measure alpha activity in thick or thin solid sources mounted inside the detector or outside next to a thin entrance window. The most modern systems have the capability of recording alpha and beta particles simultaneously using the technique of pulse shape discrimination.

Conventional liquid-scintillation counting has been used to assay gross alpha activities in certain environmental samples. However, due to the high background count-rates, the technique is seldom indicated for the measurement of low-level sources.

Etchable plastic nuclear track detectors such as lexan and CR-39 have been used for many years to detect very low alpha activities in environmental samples, provided the exposure periods are sufficiently long, typically weeks or longer. The method is a form of autoradiography. The tracks produced in these dielectric materials are usually less than 0.1 μm in length and special techniques must be employed to treat the exposed material. Chemical etching, electrochemical etching and grafting are the three methods most widely used for track enlargement (Mishra, 1991). Various automated techniques for counting track etch detectors have been reported (Mishra, 1991) and include optical scanning, light scattering, electrical conductivity measurement, spark counting and etched-pit counting using an alpha particle flux.

8.3.3.2 Alpha-particle spectrometry

In the case of alpha-particle spectrometry, maximum energy resolution is achieved with a source which approximates as closely as possible to that of an infinitely thin, weightless source on a perfectly flat substrate. If absorption corrections are to be less than 1 per cent, the source thickness should be less than 1 per cent of the average range of the alpha particles. Methods for the preparation of alpha-particle-emitting sources are summarized below.

A wide range of different detector types are employed in alpha-particle spectrometry today including semiconductor detectors, liquid scintillation counters, ionization chambers, proportional counters and magnetic spectrometers. The most commonly used are silicon surface barrier (SSB) detectors, though passivated ion-implanted silicon (PIPS) detectors are beginning to supersede them, having the advantages of higher resolution and more rugged construction and, hence, the exposed surface can be wiped to decontaminate it. In a typical system, the source (e.g. electroplated stainless steel disc or polypropylene membrane filter) is counted, under conditions of 'close geometry', with a PIPS detector (active area 300–600 mm^2) coupled in series to a pre-amplifier, spectroscopy amplifier and multi-channel analyser (MCA), either free-standing or PC-based. Although typical counting times are long (days to weeks), spectrum stability is seldom a problem and, thus, the use of digital spectrum stabilization techniques is unnecessary.

Low-level liquid scintillation counting, using pulse-shape discrimination to separate alpha- from beta-particle-induced events and anti-coincidence shielding to reduce background, is also employed (Yu *et al.*, 1991). The disadvantages are much inferior energy resolution and sensitivity compared to PIPS- or SSB-based systems and problems associated with chemiluminescence in the scintillator cocktail.

8.3.3.3 Mass spectrometry

Conventional thermal ionization mass spectrometry (TIMS) or inductively coupled plasma (ion-source) spectrometry (ICP-MS) is sometimes the method of choice for very low-level measurements of long-lived radionuclides, as the sensitivity of these techniques for nuclides such as ^{239}Pu, ^{240}Pu, ^{237}Np and ^{99}Tc is some orders of magnitude greater than that of the standard radiometric techniques usually used (Halverson, 1984; Buesseler and Halverson, 1987). Because of isobar and chemical interferences it is necessary, particularly when using TIMS, to first carry out a high-quality chemical separation of the element of interest. The instrumentation itself is also very expensive. ICP-MS, on the other hand, offers an attractive alternative, being considerably less expensive and requiring less complex sample preparation—solution sources can be used. Its sensitivity, however, is inferior to that of TIMS and its use more effective at $A > 80$ due to the presence of background ions below this mass range. Sensitivities of < 1 pg/g are readily achievable for heavy elements. The technique has been used to determine with accuracy the $^{239}Pu/^{240}Pu$ ratio in environmental samples (Kim *et al.*, 1989) (^{239}Pu and ^{240}Pu cannot be resolved by conventional high-resolution alpha spectrometry).

Resonance ionization mass spectrometry (RIMS) has recently been shown to be an ultra-sensitive technique for the detection of trace elements. The method is based on the stepwise excitation of atoms from a defined state to highly excited states by resonant absorption of photons followed by ionization with subsequent mass analysis of the ions. A detection limit for plutonium of 10^7 atoms has been reported.

8.3.3.4 Source preparation techniques in alpha spectrometry

Techniques for the preparation of sources suitable for alpha spectrometric measurements have been reviewed by Lally and Glover (1984) and include vacuum sublimation, direct evaporation using a spreading agent such as tetraethylene glycol, electrodeposition, electrospraying and electrostatic precipitation. For metrology measurements with pure solutions either vacuum sublimation, direct evaporation or electrodeposition is usually used. However, for low-level radiochemical analysis the favoured techniques are either electrodeposition of actinides onto a polished stainless steel or platinum disc from an aqueous solution containing one or more ammonium salts to which a complexing agent such as EDTA is sometimes added, or coprecipitation with rare earth metals in various media.

8.3.4 RADIOCHEMICAL SEPARATION TECHNIQUES

Radiochemical analysis differs from classical analytical chemistry in that the mass of the substance being determined is extremely small. It is usually in the sub-microgram or, indeed, the sub-femtogram range when dealing with environmental samples contaminated with artificial radioactivity. The selection of chemical techniques for the analysis of environmental matrices should be based on their ability to eliminate potential gravimetric and radiometric interferences. Other factors, such as their flexibility and suitability within a given laboratory, should also be taken into account. Two main approaches exist in the analysis of actinides in the environment, namely liquid/liquid solvent extraction and ion exchange separation. A third possibility, that of column reversed-phase partition chromatography, also known as extraction chromatography, is less widely used.

Since most chemical separations are not quantitative, a known quantity of an isotopic, non-interfering, radiometric or stable yield monitor must be added to the sample at the outset. The least desirable technique is to determine the recovery in a series of spiked samples and to use the resultant, averaged value to assess the activities of future samples. Where the recovery is to be measured gravimetrically, such factors as constant composition and hygroscopicity must be taken into account. Moreover, chemical equilibrium must be established with the determinant at the earliest possible stage and certainly before any loss of either determinant or tracer occurs (Harvey and Lovett, 1984). Details of the radiochemical separation techniques currently employed are beyond the scope of this limited review and may be found elsewhere (De Regge and Boden, 1984; Singh, 1988).

8.3.5 DETECTION AND MEASUREMENT OF BETA-EMITTING NUCLIDES

8.3.5.1 Actively shielded gas flow counting

For the low-level measurement of solid (planchet) sources anti-coincidence shielded gas flow proportional counters (or Geiger–Mueller counters) have been in use for many years. Typically, a large-area gas flow counter acts as a common guard detector located above a set of ultra-flat gas flow proportional counter tubes. The whole assembly is encased in a 10-cm thick shielding of interlocking lead bricks. A number of planchets can be measured simultaneously and detection limits for beta radiation (3 sigma accuracy) of 15 mBq for a 100-minute counting time using

60 mm diameter planchets are quite usual. These systems are also suitable for gross alpha measurement, in which case detection limits are significantly lower.

8.3.5.2 Liquid scintillation counting

Liquid scintillation counting systems equipped with active shielding are now available, whose performance characteristics greatly outstrip those of conventional liquid scintillation counters. Not only does the anti-coincidence guard counter suppress cosmic radiation and much of the residual environmental gamma radiation, but it also rejects the cross-talk from Cerenkov light-induced pulses which are caused primarily by cosmic radiation in the photomultiplier tubes themselves. Careful design and the selection of low-activity components ensure that the intrinsic level of radioactivity is minimized. In one system the use of specially designed high-purity Teflon/copper counting vials is recommended as a further aid in reducing background. For example, in the case of tritium counting using a window which allows 25 per cent efficiency, backgrounds as low as 0.4 cpm (figure-of-merit, E^2/B = 1850) have been reported (Schonhofer and Henrich, 1987). These systems are now used to measure a wide variety of beta- (and alpha-)emitting nuclides, including ^{14}C, $^{90}Sr/^{90}Y$, ^{137}Cs, ^{226}Ra, ^{241}Pu and ^{241}Am, and are also ideally suited to Cerenkov counting.

8.3.5.3 Cerenkov counting

Cerenkov radiation may be observed when a charged particle, such as an electron, passes through a dielectric medium at a velocity, v, greater than the velocity of light in the medium. The resulting pulse of light is called 'Cerenkov emission'. Cerenkov counting has a number of advantages in the assay of radionuclides, not least in the ease of sample preparation. In the case of water, it is found that the threshold energy for Cerenkov production is 263 keV for electrons.

The intensity of Cerenkov emission, and hence the detectability, is governed by the margin by which the electron's energy exceeds the threshold value. A starting point for the evaluation of the intensity may be found in the work of Frank and Tamm (1937). Ross (1969) calculated a series of Cerenkov photon yields for electrons up to 4 MeV. He found, for example, that 89 per cent of beta emissions from ^{90}Y are available for Cerenkov production.

8.3.6 DETECTION AND MEASUREMENT OF GAMMA-EMITTING NUCLIDES

Traditionally, scintillation detectors, in particular thallium-activated sodium iodide (NaI(Tl)) crystals, have been used in gamma-ray spectrometry. The latter's high light yield, excellent linearity and high atomic number of its iodine constituent makes sodium iodide a very efficient and versatile tool in many applications. However, due to its limited energy resolution (~7–8 per cent FWHM at 662 keV), the deconvolution of complex spectra is by no means a simple matter. In the past two decades, semiconductor detectors, based initially on lithium-drifted germanium but now almost exclusively on intrinsic germanium, have taken the place of the scintillation detector in high-resolution analysis applications. Greatly improved energy resolution (1.7–2.3 keV FWHM at 1.332 MeV) is one of the principal characteristics of these detectors. Moreover, intrinsic (p-type) germanium detectors with efficiencies (relative to a 3×3 inch NaI detector) in excess of 100 per cent have now become available commercially.

Although, in general, the analysis of a sample by gamma spectrometry is considered to be non-destructive, certain sample/source preparation steps are essential for precise measurements. For example, it is necessary that the sample be completely homogenized and the measurement carried out in the same geometry as used in the efficiency calibration. The full energy peak efficiency as a function of energy must be determined experimentally using sources whose activities are traceable to national or international activity standards. Ideally, the calibration source and the samples to be measured should also have the same chemical composition and density. If this is not the case, corrections must be made for differences in the degree of self-attenuation. Corrections may also have to be made for coincidence summing, which occurs with radionuclides which emit gamma rays in cascade and which is particularly important for low source–detector distances. Chance coincidence summing (pulse pile-up) corrections may also be necessary at high count rates.

To suppress background radiation and thus improve sensitivity, all gamma detectors must be surrounded by a passive shield made from 'aged' lead or pre-World War II battleship steel. Actively shielded gamma-ray spectrometers, in which the intrinsic germanium detector is surrounded by a large NaI(Tl) coincidence shield, have found application in ultra-low-level environmental studies (Hotzl and Winkler, 1981; Wogman, 1981). Greatly enhanced background reduction and Compton continua suppression are achieved when operating such systems in anti-coincidence mode. For example, Hotzl and Winkler (1981) have demonstrated that the identification and measurement of ^{85}Sr (514 keV) in soil samples is greatly helped by operating in AC mode as the strongly interfering annihilation peak at 511 keV is drastically reduced.

8.3.7 CONCLUSIONS

The relatively brief discussion given above is not intended to be a comprehensive review of all of the methods of analysis presently in use. It will, however, have succeeded in its purpose if it has provided an insight into the diversity and complexity of the methods employed to assay trace concentrations of radioactivity in the environment. New and improved techniques continue to be developed and the reader's attention is directed to those periodicals and proceedings in which the latest scientific and technical advances are traditionally published. Of particular interest are the *Journal of Radioanalytical and Nuclear Chemistry* (articles), the *Journal of Nuclear Instruments and Methods in Physics Research*, *Health Physics* and proceedings of conferences organized by international (and national) bodies such as the IAEA, ICRM, IRPA, NCRP (US), etc., in the general fields of low-level measurements and their application to environmental radioactivity studies.

8.4 GLOSSARY OF TERMS

Absorbed dose: Quantity of energy deposited from incident radiation per unit mass of absorber. SI unit: gray, symbol Gy = 1 joule per kg.

Actinaria: An order of coelenterates, the sea anemones.

Actinides: Series of 15 elements with atomic numbers 89 (actinium) to 103, and including uranium and the transuranics.

Activation: Induction of radioactivity in a stable element by irradiation, usually with neutrons.

Activation product: Radionuclide generated by irradiation of a stable nuclide.

Activity: The intensity or strength of a radioactive source; the number of atoms disintegrating per unit time. SI unit is the becquerel, symbol Bq, = 1 disintegration per second. The earlier unit is the curie, symbol Ci, = 3.7×10^{10} Bq.

Adsorption: Uptake of a substance by physical or chemical reaction on the accessible surface of a solid, or at a liquid interface.

Aerosol: Solid or liquid particles suspended in a gas.

Alpha particle: Doubly positively charged particle, an ^4He nucleus, comprising two protons plus two neutrons, emitted during the decay of some radionuclides.

Anion: Ion with a negative charge.

Atom: The smallest unit of an element, comprising a positive nucleus with orbiting electron(s).

Atomic number: Number of protons in the nucleus of an atom. Symbol Z.

Becquerel: SI unit of radioactivity. Symbol Bq = 1 disintegration per second.

Benthic: Pertaining to, or with the characteristics of, the benthos.

Benthos: The animal and plant life on the sea bottom.

Beta particle: An elementary particle emitted during the decay of some radionuclides. Usually it is the electron, the fundamental unit of negative electric charge. More rarely its charge is positive, when it is known as the positron.

Biological cycling: General term meaning biologically mediated transformations and pathways.

Biota: The flora and fauna of a given region.

Bioturbation: The mechanical disturbance of sediment by animal movement.

Boundary layer: The layer of a moving fluid with modified dynamics immediately adjacent to a solid surface.

BWR: Boiling water reactor. Nuclear reactor in which water is boiled within the core to raise steam.

Can: Pin or tube, usually of metal, containing the fuel in a nuclear reactor.

Cation: Ion with a positive charge.

Cephalopoda: A class of marine molluscs with the foot modified into tentacles, with or without shell, e.g. squid, octopod, nautilus.

Chlorite: (1) A salt of chlorous acid; (2) a geological mineral, a complex hydrated silicate of Al, Fe and Mg.

Cladding: The material, usually a metal, forming the fuel can in a nuclear fission reactor, resistant to chemical and physical attack and preventing fuel corrosion and fission-product release.

Collective dose: Term often used for collective effective dose.

Collective effective dose: Product of the mean effective dose to a group or population from a given source and the number of individuals in the group. SI unit; person-sievert.

Collective equivalent dose: Product of the average equivalent dose to an exposed group and the number of individuals in the group.

Colloid: A state of subdivision of matter, with particle sizes ranging from molecular dimensions up to one micron (10^{-6} m).

Committed effective dose: Summation of weighted committed organ or tissue equivalent doses resulting from an intake of radioactive material.

Committed equivalent dose: Integral of equivalent dose-rate in a given tissue or organ over time following an intake of radioactive material.

Concentration factor: Ratio of element or nuclide in the consumer (or a specific tissue, organ, etc.) to that in what is consumed, or to that in the environmental medium.

Control rods: Rods of neutron-absorber for controlling the power of a nuclear fission reactor.

Coolant: Liquid or gas for removing heat from a nuclear reactor.

Core: That part of a nuclear reactor in which the fission process takes place.

Coriolis: An effect or 'force' deflecting the path of a moving body as a result of the Earth's rotation.

Coryphaenidae: A genus of pelagic fishes.

Critical group: Sub-group of the public most affected by a given release of radioactivity.

Curie: Non-SI unit of radioactivity, symbol Ci, defined as 3.7×10^{10} distintegrations per second.

Daughter: Stable or radioactive nuclide resulting from the decay of a parent radionuclide, sometimes down a chain of radioactive daughters.

Decapoda: (1) An order of Eucarida, including shrimps, prawns, lobsters, crayfish and crabs; (2) an order of Dibranchia, including squid and cuttlefish.

Decay: The transformation of a radionuclide, usually by the emission of radiation, into a more-stable nuclide (although this may be via less-stable intermediate daughters).

Decay-corrected fallout: Measured contamination corrected for decay to give activities at time of initial fallout.
Decay product: Nuclide generated by decay of radionuclide.
Decontamination: The removal or 'clean-up' of radioactivity.
Deposition velocity: The ratio of the activity deposited per unit surface area per second to the activity per unit volume of air in contact.
Diagenesis: Chemical modifications taking place over time in buried sediments by interaction with pore waters.
Diagenetic: Consequent on diagenesis.
Discharges: Releases of effluents (gaseous, aerosol or liquid) from industrial operations.
Disintegration: See Decay.
Dispersion coefficient: A measure (m^2s^{-1}) of hydrodynamic dispersion resulting from molecular diffusion and processes associated with the flow of water.
Distribution coefficient: The ratio of the concentrations of a given solute dissolved in two immiscible liquids at equilibrium.
Dose: See under Absorbed dose.
Dose rate: Absorbed dose per unit time.
Ecosystem: An environmental unit, including all relevant physical features and all living organisms which function within it.
Eddy diffusion: Rapid diffusion in turbulent flow by the mixing of eddies.
Effective dose: Sum of weighted equivalent doses in all tissues and organs in the body.
Electron: A stable elementary particle, the basic unit of negative electric charge.
Element: Substance comprising only atoms of the same atomic number.
Enrichment: Process of increasing the proportion of a chosen isotope of an element.
Equivalent dose: Product of the absorbed dose averaged over a tissue or organ and the quality factor of the radiation. SI unit is the sievert, symbol Sv.
Fallout: Airborne or terrestrial radioactive debris from nuclear detonations or accidents.
Fast reactor: Type of nuclear reactor utilizing high-energy (i.e. fast-moving) neutrons to cause fission.
FBR: Fast breeder reactor. Nuclear reactor in which the fast neutrons from fission convert non-fissile nuclides (e.g. ^{238}U) to fissile nuclides (e.g. ^{239}Pu), thus generating more fuel than is consumed.
Fermentation: The decomposition of organic substances by micro-organisms or enzymes, usually under anaerobic conditions and evolving heat and gas.
Fission: The exoenergetic break-up of the nucleus of an atom into two or more smaller nuclei. It may be spontaneous or follow the absorption of an energetic particle, usually a neutron in a fission reactor, e.g. $^{235}U + {}^{1}n = {}^{95}Mo + {}^{139}La + 2{}^{1}n + 200$ MeV.
Fission product: Nuclides (radioactive or stable) resulting from fission, directly or by the subsequent decay of primary fission fragments.

Free energy: Thermodynamic potential; the capacity of a system to perform work.
Fucus: A genus of branched flat-fronded seaweeds.
Fuel cycle: The cyclic process of fabricating fuel elements, using them in a reactor and reprocessing to extract the unused portion of the fuel (plus any plutonium) for re-fabrication and re-use.
Fuel element: Rod or pin of clad fuel, or an integral assembly of such rods.
Fuel, nuclear: Fissionable material, such as ^{235}U, processed into suitable form for reactor use.
Fusion: The forming of a single atomic nucleus from two light nuclei, with the release of energy, e.g. $^{2}H + ^{3}H = ^{4}He + ^{1}n + 17.6$ MeV.
Gamma ray: Photon of high-energy (= short-wavelength, < 0.01 nm). electromagnetic radiation emitted in some radioactive decay processes.
Gray: SI Unit of absorbed dose, symbol Gy, equal to 1 joule of energy per kilogram.
Half-life: Time in which half the nuclei of a given radioisotope disintegrate.
Halocline: Zone where salinity changes markedly with ocean depth.
High-level waste: Highly radioactive heat-generating fission and activation products from the reprocessing of spent nuclear fuel.
Holocene: 'Wholly-recent', a geological epoch extending from 10 000 BP to the present day.
Holothurioidea: A class of the phylum Echinodermata, the sea cucumbers.
Hot particles: Radioactive fallout particles of a few microns or less in dimensions but with activities as much as several orders of magnitude higher than the generality of fallout.
Humic: Pertaining to the humus, the biologically generated organic soil fraction.
Hypogene: Geological processes within or below Earth's crust.
Illite: A group of mica-like sedimentary clay minerals.
Intermediate-level waste: Radioactive waste from nuclear plant, etc., having an activity too high for direct discharge or simple burial, i.e. greater than 4×10^{9}Bq of alpha per tonne, or 1.2×10^{10}Bq beta–gamma, but not generating sufficient heat to be classified as high-level waste.
Interstitial water: Water between the particles of a sediment.
Ion: Electrically charged atom or group of atoms.
Ion-exchange: The reversible interchange of ions between a solution and a solid.
Ionization: The adding or removing of electrons(s) from a neutral atom or molecule to make an ion.
Irradiation: Exposure of material to ionizing radiation.
Isohalines: Profiles joining points of equal salinity.
Isotopes: Nuclides of a given chemical element which, while having the same number of protons in their nuclei, have different numbers of neutrons and hence different atomic masses.
Kaolinite: A geological mineral, a finely crystalline hydrated aluminium silicate.

Langmuir circulation: A type of wind-driven cellular circulation pattern, observed in the surface layers of lakes and oceans. Cell convergence often marked by visible lines of buoyant particulate matter.

Lanthanides: Series of 15 elements with atomic number 57 (lanthanum) to 71 (lutetium), otherwise known as the 'rare earths'.

Leaching: Removal of the soluble component of a porous solid by the permeation of water.

Ligand: Molecule, ion or group bound (chelated) to the central atom in a co-ordination compound.

Limnetic: Pertaining to the pelagic zone (upper and middle depths) of a freshwater body.

Lithology: The characteristics of a rock (colour, structure, grain size, etc.) as evident to the eye.

Lithostatic: Pertaining to ground pressure, arising from superimposed strata or rock movement.

Low-level waste: Radioactive waste from nuclear plant, etc., having an activity sufficiently low for direct discharge or simple burial, i.e. less than 4×10^9 Bq of alpha or 1.2×10^{10} Bq of beta-gamma per tonne.

LWR: Light water reactor. Nuclear reactor using ordinary ('light') water as moderator and core coolant.

Macrobenthos: Animals of size > 0.5 mm living on and in bottom sediments.

Marl lakes: Lakes in which calcium carbonate is precipitated, generally through biological mediation.

Mass number: Total number of protons plus neutrons in the nucleus of a given atom. Symbol A.

Maxmuelleria: '*Maxmuelleria lankesteri*': A large echiurid worm living in burrows in marine sediments in the Irish Sea and responsible for the bioturbation of the sediment.

Meromictic: Pertaining to permanently stratified lakes.

Mixohaline: Brackish water in the salinity range of 0.5 to 30 parts per thousand.

Moderator: Material used in the core of a fission reactor to reduce the energy (velocity) of the neutrons generated by fission.

Molecule: The smallest particle of a compound capable of independent existence while still retaining its chemical properties.

Monazite: A complex mineral phosphate, primarily a thorium ore, but also the principal source of lanthanides.

Natural radiation: Radiation in the environment from natural radioactive elements and cosmic rays.

Neustonic: Pertaining to micro-organisms floating or swimming on the water surface.

Neutron: Uncharged elementary particle of approximately unit atomic mass.

Nucleus: The positively charged core of an atom, composed of protons and neutrons.

Nuclide: An isotope of an element, characterized by its unique total and ratio of protons and neutrons. (Sometimes, in addition, by the energy state of the nucleus.)
Parent radionuclide: A radionuclide which yields a daughter nuclide on disintegration.
Person-sievert: Unit of collective dose obtained by multiplying the average effective eqivalent dose by the number of people exposed.
Photic: Responding to light.
Photic zone: The uppermost layer of a body of water through which enough sunlight penetrates for photosynthesis to occur.
Phytoplankton: Passive or weakly motile floating plant life.
Plankton bloom: An occurrence (usually seasonal) of very large numbers of planktonic algae.
Pleistocene: A recent geological epoch extending over the last 2–3 million years up to the Holocene, 10 000 years BP.
Plutonium: A transuranic chemical element of the actinide series, atomic number 94.
Primary production: Biomass per unit area produced by photosynthesis (or less commonly chemosynthesis).
Proton: Positively charged elementary particle of approximately unit atomic mass.
Purex: Chemical process for the separation of uranium and plutonium from spent nuclear reactor fuel.
PWR: Pressurized water reactor. Nuclear reactor in which the primary core coolant is water maintained under high pressure to prevent boiling.
Quality factor: Factor to take account of the different degrees of damage to tissue by different types and energies of radiation.
Rad: Non-SI unit of absorbed dose. 1 rad = 100 erg g^{-1} = 0.01 J kg^{-1} = 0.01 gray.
Radioactivity: Emission of particles or electromagnetic radiation from the nuclei of unstable atoms.
Radiobiology: The study of the use, involvement or consequences of radioactivity in biological systems.
Rainout: The removal of aerosols from the atmosphere by rain.
RBMK: Russian design of nuclear reactor with graphite moderator and ordinary ('light') water boiling directly in the core to raise steam.
Rem: Non-SI unit of dose. The product of the absorbed dose and the quality factor. 1 rem = 0.01 sievert.
Reprocessing: Separation of spent reactor fuel into unused uranium, plutonium, and waste fission products, usually by chemical methods.
Resuspension factor: The ratio of the concentration of a radionuclide in air to the ground surface contamination from which the airborne activity arises.
Salps: Small pelagic filter-feeding gelatinous invertebrates of the families Salpidae and Doliolidae.
Seiche: A resonance oscillation in a fluid body; a standing wave so formed in an enclosed or semi-enclosed water body.

Shear stress: (1) The force deforming a solid body by displacing a plane relative to parallel planes; (2) the drag exerted on a boundary by contiguous fluid flow.

Shear velocity: A parameter ('fictitious' velocity) whose square equals the quotient of shear stress and fluid density.

Sievert: SI unit of equivalent radiation dose. The product of the absorbed dose in grays and the quality factor of the radiation. Symbol Sv.

SI units: Système Internationale, the agreed system of scientific units based primarily on the kilogram, metre, second, ampere, kelvin, mole and candela.

Smectite: A series of clay minerals, such as montmorillonite and saponite, with high absorptive and ion-exchange properties.

Somatic change: Change occurring in biological tissue other than the germ cells.

Sorption: A general term for the processes of adsorption, absorption, and persorption.

Spent fuel: Nuclear fuel in which neutron-absorbing fission products have accumulated sufficiently to inhibit the chain reaction.

Stochastic: Pertaining to random variables.

Stratigraphic: Of the form and relationships of sedimentary rock strata.

Telluric: (1) Pertaining to Earth; (2) derived from or containing the chemical element tellurium.

Thermocline: A layer in a body of water with vertical temperature gradient greater than in overlying and underlying layers.

TR: (Tritium ratio) represents one ^{3}H atom in 10^{18} atoms of ^{1}H or an activity of 0.118 Bq kg^{-1} in water.

Transuranic: Chemical elements with atomic numbers greater than that of ^{92}U.

Tritium: Hydrogen isotope, ^{3}H, with nucleus of one proton and two neutrons. It is radioactive with a half-life of 12.4 years.

Vitrify: (1) To fuse to a glass; (2) a process of incorporating highly radioactive fission products into a corrosion-resistant glassy matrix for disposal.

Void coefficient: Quantifies the effect of voids (vapour bubbles in liquid moderator or coolant) on the power level of a nuclear reactor.

Zeolites: A group of aluminosilicate minerals with pronounced ion-exchange and reversible dehydration properties.

Zooplankton: Microscopic animals moving in aqueous ecosystems.

References and Bibliography

Aarkrog, A. (1983). Translocation of radionuclides in cereal crops. In Coughtrey, P. J., Bell, J. N. B. and Roberts, T. M. (Eds) *Ecological Aspects of Radionuclide Releases*: 81–90. Blackwell Scientific Publications, Oxford.

Aarkrog, A., Boelskifte, S., Dahlgaard, H., Dunniec, S., Hallstadius, L., Holm, E. and Smith, J. N. (1987). Technetium-99 and caesium-134 as long distance tracers in Arctic waters. *Est., Coastal Shelf Sci.*, **24**, 637–647.

Ad Hoc Interagency Dose Assessment Group (1979). Population dose and health impact of the accident at the Three Mile Island Nuclear Station. US Nuclear Regulatory Commission Report NUREG-0558.

Aitken, M. J. (1978). Archaeological involvements of physics. *Phys. Lett.*, **40C**, 277–351.

Alberts, J. J., Pinder, J. E., Bowling, J. W., Nelson, D. M. and Orlandini, K. A. (1986). $^{239,240}Pu$, ^{241}Am and ^{232}Th in lakes: the effects of seasonal anoxia. *J. Environ. Radioact*, **4**, 167–176.

Alberts J. J., Wahlgren M. A., Orlandini K. A. and Durbahn C. A. (1989). The distributions of $^{239,240}Pu$, ^{238}Pu, ^{241}Am and ^{137}Cs among chemically defined components of sediments, settling particulates and net plankton of Lake Michigan. *J. Environ. Radioact.*, **9**, 89–103.

Alexakhin, R. M. (1992). Countermeasures of the Chernobyl accident consequences. *Proc. EC Workshop on Relative Effectiveness of Agricultural Countermeasure Techniques*, Brussels.

Alexakhin, R. M. (in press). Countermeasures in agricultural production as a powerful factor in liquidation of the Chernobyl accident consequences. *Proc. EC Workshop on Relative Effectiveness of Agricultural Countermeasures Techniques*, Brussels.

Al-Khayat, T. A. H. (1989). Dispersion of radioactive emission from a nuclear fuel fabrication plant. PhD thesis, Lancaster University.

Aller, R. C. (1983). The effects of macrobenthos on chemical properties of marine sediments and overlaying water. In McCall, P. and Tevesz, M. J. S. (Eds) *Animal–sediment Relations*: 53–102. Plenum Press, New York.

Allott, R. W. (1990). Behaviour of radionuclide contaminated dust in the urban environment of Barrow-in-Furness. PhD thesis, Lancaster University.

Allott, R. W., Hewitt, C. N. and Kelly, M. R. (1990). The environmental half-lives and mean residence times of contaminants in dust for an urban environment: Barrow-in-Furness. *Sci. Tot. Environ.*, **93**, 403–410.

Al Rayes, A. H., Ronneau, C., Apers, D. and Myttenaere, C. (1988). Ecological behaviour of thermo-generated caesium aerosols. *Dokl. Akad. Nauk, USSR*, **8**, 75–79.

Andersson, T., Forsgren, G., Hakanson, L., Malmgren, L. and Nilsson, A. (1990). Radioactive cesium in fish in Swedish lakes after Chernobyl. SSI-Report 90-04. National Institute of Radiation Protection. Stockholm, Sweden. (In Swedish, English summary.)

Andolina, J. and Guillitte, O. A. (1990). Methodological approach of soils sampling and analysis in the study of radionuclides transfers forest ecosystems. In Desmet, G., Nassimebeni, P. and Belli M. (Eds) *Transfer of Radionuclides in Natural and Seminatural Environments*: 161–168. Elsevier, London and New York.

ApSimon, H. M. and Mahadeva, N. (1989). The behaviour of ruthenium isotopes released from Chernobyl. Paper RE.20.89 presented at SCOPE-RADPATH First Workshop, 29–2 June 1989.

ApSimon, H. M. and Simms, K. L. (1986). Estimating the effects of rain and snow on potential reactor accident consequences. *Nucl. Energy*, **25**(4), 235–242.

ApSimon, H. M. and Wilson, J. J. N. (1987). Modelling atmospheric dispersal of the Chernobyl release over Europe. *Boundary Layer Met.*, **41**, 123–133.

ApSimon, H. M., Goddard, A. J. H. and Wrigley, J. (1985). Long-range atmospheric dispersion of radioisotopes: The Mesos Model. *Atmos. Environ.*, **19**, 99–111.

ApSimon, H. M., Gudiksen, P., Khitrov, L., Rodhe, H. and Yoshikawa, T. (1988). Modelling the dispersal and deposition of radionuclides. *Environment*, **30**(5), 17–20.

ApSimon, H. M., Wilson, J. J. N. and Simms, K. L. (1989). Analysis of the dispersal and deposition of radionuclides from Chernobyl across Europe. *Proc. R. Soc. Lond., A* **425**, 365–405.

ApSimon, H. M., Barker, B. M., Kayin, S., and Stott, P. A. (1991a). Wet deposition of radionuclides in precipitation systems. *Atmos. Environ.* (submitted).

ApSimon, H. M., Kruse-Plass, M. and Whitcombe, G. (1991b). Studies of deposition in hill cloud. *Il Nuovo Cimento*, **13C**, 955–971.

Arkhipov, N. P., Meshalkin, G. S., Arkhipov, A. N., Burov, N. I., Fedotov, I. S., Mishenkov, N. N. (1991). Measures (and their effectiveness) to improve the radioecological situation given the particular features of the contamination caused by the Kyshtym and Chernobyl accidents. Comparative Assessment of the Environmental Impact of Radionuclides Released During Three Major Nuclear Accidents: Kyshtym, Windscale, Chernobyl. CEC, Luxembourg. EUR 13574.

Armi, L. and D'Asaro, E. (1980). Flow structures of the benthic ocean. *J. Geophys. Res.*, **85**, 469–484.

Arnold, L. (1990). Volatilisation of radionuclides from soil and plants: Sn and I. *Nirex Safety Studies*, NSS R219, UK Nirex Ltd, Harwell.

Arthur, W. J. III and Alldredge, A. W. (1979). Soil ingestion by mule deer in northcentral Colorado. *J. Range Mgmt.*, **32**(1), 67–71.

Arthur, W. J. III and Alldredge, A.W. (1982). Importance of plutonium contamination on vegetation surfaces at Rocky Flats Colorado. *Environ. Exp. Bot.*, **22**(1), 33–38.

Arthur, J. W. III and Markham, O. D. (1983). Small mammal soil burrowing as a radionuclide transport vector at a radioactive waste disposal area in southeastern Idaho. *J. Environ. Qual.*, **12**(1), 117–122.

Assinder, D. J., Kelly, M. and Aston, S. R. (1984). Conservative and non-conservative behaviour of radionuclides in an estuarine environment, with particular respect to the behaviour of plutonium isotopes. *Environ. Technol. Lett.*, **5**, 23–30.

Assinder, D. J., Kelly, M. and Aston, S. R. (1985). Tidal variations in dissolved and particulate phase radionuclide activities in the Esk Estuary, England, and their distribution coefficients and particulate activity fractions. *J. Environ. Radioact.*, **2**, 1–22.

ATOM (1981). Fires at La Hague. *Atom*, **300**, 278.

ATOM (1983). France can be 'rightfully proud'—Minister. *Atom*, **317**, 58.

ATOM (1989). Radioactive waste the French way. *Atom*, **389**, 28–33.

Auerbach, S. I. (1987). Comparative forest ecosystem behaviour of three long-lived radionuclides. In *The Cycling of Long-lived Radionuclides in the Biosphere: Observations and Models*. CEC, Luxembourg.

Auerbach, S. I. (1990). Background information relating to the Soviet 1957 accident in the Southern Urals. Presentation at SCOPE–RADPATH First Case-study Meeting, 26–30 March 1990, University of Lancaster.

Auerbach, S. I., Olson J. S., and Waller, H. D. (1964). Landscape investigations using caesium-137. *Nature*, **201**, 761–764.

Auxier, J. A., Berger, C. D., Eisenhauser, C. M., Gesell, T. F., Jones, A. R. and Masterson, M. E. (1979). Report of the Task Group of Health Physics and Dosimetry. In Report of the Public Health and Safety Task Force to the President's Commission on the Accident at Three Mile Island.

Avoine, J. (1986). Sediment exchange between the Seine Estuary and its adjacent shelf. *J. Geol. Soc. Lond.*, **144**, 135–148.

Aycoberry, C. and Rougeau, J. P., (1988). Closing the nuclear fuel cycle: The industrial demonstration. *Trans. Am. Nucl. Soc.*, **56**, (Suppl. 1.1), 531–535.

Bacq, Z. M. and Alexander, P. (1961). *Fundamentals of Radiobiology*, 2nd edn. Pergamon Press, Oxford.

Baehr, W. (1989). Industrial vitrification processes for high-level liquid waste solutions. *IAEA Bull.*, **31**(4), 43–46.

Bair, W. J. (1960). Radioisotope toxicity from pulmonary absorption. In Caldecott, R. S. and Snyder, L. A. (Eds) *Radioisotopes in the Biosphere*. University of Minneapolis, Minnesota. Comstock Publishing Associates, Ithaca and London.

Bakker, K. (1971). Some general remarks on the concepts 'population' and 'regulation'. In den Boer, P. J. and Gradwell, G. R. (Eds) *Dynamics of Populations*: 565–567. Centre for Agricultural Publishing and Documentation, Wageningen.

Beasley, T. M. and Jennings, C. D. (1984). Inventories of $^{239,240}Pu$, ^{241}Am, ^{137}Cs and ^{60}Co in Columbia River sediments from Hanford to the Columbia River Estuary. *Environ. Sci. Technol.*, **18**, 207–212.

Beasley, T. M., Ball, L. A., Andrews, J. E. and Halverson, J. E. (1981). Hanford derived plutonium in Columbia River sediments. *Science*, **24**, 913–915.

Beasley, T. M., Carpenter, R. and Jennings, C. D. (1982). Plutonium, ^{241}Am and ^{137}Cs ratios, inventories and profiles in Washington and oregon continental shelf sediments. *Geochim. Cosmochim. Acta*, **46**, 1931–1946.

BEIR (1980). *The Effects on Populations of Exposure to Low Levels of Ionizing Radiation: 1980*. National Academic Press, Washington, DC.

Belli, M., Blasi, M., Borgia, A., Deluisa, A., Menegon, S., Michelutti, G., Nazzi, P., Pividori, G. and Sansone, U. (1990). The behaviour of caesium in mountain soils. In Desmet, G., Nassimebeni, P. and Belli, M. (Eds) *Transfer of Radionuclides in Natural and Semi-natural Environments*: 1443–1151, Elsevier, London and New York.

Bender, G. M. and Terstriep, M. L. (1984). Effectiveness of street sweeping in urban runoff pollution control. *Sci. Tot. Environ.*, **33**, 185–192.

Beniston, G. and Prouteau, L. (1990). Recycling impacts. Fuel Cycle 90 Conference, US Council for Energy Awareness, Nashville (Ten.) 25–28 March.

Beniston, G., Blanchon, T. and Lenail, B. (1987). The French programme concerning the final part of the nuclear fuel cycle. International Symposium on the Final Part of the Fuel Cycle—Strategies and Options, Vienna, Austria, 11–15 May 1987, IAEA, Vienna. (In French.)

Bentley, H. W., Phillips, F. M., Davis, S. N., Habermehl, M. A., Airey, P. L., Calf, G. E., Elmore, D., Gove, H. E. and Torgersen, T. (1986). Chlorine-36 dating of very old groundwater 1. The Great Artesian Basin, Australia. *Water Resources Res.*, **22**, 1991–2001.

Beresford, N. A. (1989). The transfer of Ag-110m to sheep tissues. *Sci. Tot. Environ.*, **85**, 81–90.

Beresford, N. A., Livens, F. R. and Howard, B. J. (1987). Radioecology of Cs-134 and Cs-137 in upland sheep pasture systems. Report to MAFF 1113. ITE Grange-over-Sands.

Berg, M. T. (1990). A system approach to the assessment of risk from radionuclides deposited in forest ecosystems from nuclear power accidents. Dissertation proposal submitted to the Department of Industrial Engineering at the University of Pittsburgh.

Berger, M. J. (1968). Energy deposition in water by photons from point isotropic sources. *J. Nucl. Med.*, **9** (Suppl. 1), 15–25.

Berger, M. J. (1971). Distribution of absorbed dose around point sources of electrons and β-particles in water and other media. *J. Nucl. Med.*, **12** (Suppl. 5), 5–23.

Berner, R. A. (1980). *Early Diagenesis*. Princeton University Press, New Jersey.

Berry, R. J. and Coulston, D. J. (1989). How good is BNFL's health and safety. *Atom*, **390**, 26–30.

BIOMOVS, (1991). BIOMOVS Scenario A4. Multiple model testing using Chernobyl fall-out data of ^{131}I in forage and milk and ^{137}Cs in forage, milk, beef and grain. Technical Report No. 13. Swedish Institute of Radiation Protection, Stockholm, Sweden.

Birkeland, P. W. (1974). *Pedology, Weathering and Geomorphological Research*. Oxford University Press, New York.

Bishop, G. P. (1989). Review of biosphere information: biotic transport of radionuclides as a result of mass movement of soil by burrowing animals. NSS/R194. UK Nirex Ltd, Harwell.

Bobovnicova, T. I., Virchenko, Y. P., Konoplyov, A. V., Siverina, A. A. and Shkuratova, I. G. (1990). Chemical forms of long-lived radionuclides and their transformation in soils of the 30 km zone of the Chernobyl APS. *Pochvovedeniye*, **10**, 20–25.

Bokori, E. K., Szerbin, P. and Fülöp, N. (1989). Behaviour of Cs-134, Cs-137 and Sr-90 in soils. Proceedings of the Conference on Radioecology. Institute of Radioecology and Applied Nuclear Techniques, Kosice, Czechoslovakia.

Bondietti, E. A., Hoffman, F. O. and Larsen, I. L. (1984). Air-to-vegetation transfer rates of natural submicron aerosols. *J. Environ. Radiation*, **1**, 5–27.

Booker, D. V., (1958). *Physical Measurements of Activity in Samples from Windscale*. AERE, HP/R 2607. HMSO, London.

Borole, D. V., Krishnaswami, S. and Somayajulu, B. L. K. (1982). Uranium isotopes in rivers, estuaries and adjacent coastal sediments of western India: their weathering, transport and oceanic budget. *Geochim. Cosmochim. Acta*, **46**, 125–137.

Bourdeau, P., Cavalloro, R., Myttenaere, C. and Verfaillie, G. (1965). Movement of fallout radionuclides in ecosystems of the Po valley, Italy. *Health Phys.*, **11**, 1429–1444.

Bowden, K. F. (1980). Physical factors: salinity, temperature, circulation, and mixing processes. In Olausson, E. and Cato, I. (Eds) *Chemistry and Biogeochemistry of Estuaries*: 37–70. John Wiley & Sons, Chichester.

Bowen, V. T., Noshkin, V. E., Livingstone, H. D. and Volchok, H. L. (1980). Fallout radionuclides in the Pacific Ocean: vertical and horizontal distributions, largely from GEOSECS stations. *Earth Planet. Sci. Lett.*, **49**(2), 411–434.

Brisbin, L., Beyers, R. J., Dapson, R. W., Geiger, R. A., Gentry, J. B., Gibbons, J. W., Smith, M. H. and Woods, S. K. (1974). Patterns of radiocesium in the sediments of a stream channel contaminated by production reactor effluents. *Health Phys.*, **27**, 19–27.

British Nuclear Fuels plc (1989a). *Health and Safety Annual Report 1988*. British Nuclear Fuels plc, Risley.

British Nuclear Fuels plc (1989b). *Nuclear Energy*. British Nuclear Fuels plc, Risley.

British Nuclear Fuels plc (1989c). *Nuclear Fuel Reprocessing Technology*. British Nuclear Fuels plc, Risley.

Brittain, J. E. (1991). Long-term development of radiocesium in trout from the Upper Heimdals Lake, Norway. To be published. University of Oslo, Norway. (In Norwegian.)

Brown, S. O., Krise, G. M., Pace, H. B. and de Boer, J. (1964). Effect of continuous irradiation on reproductive capacity and fertility of the albino rat and mouse. In Carlson, W. D. and Gassner, F. X. (Eds) *Effects of Ionizing Radiation on the Reproductive System*: 103–110. Macmillan, New York.

Brownell, G. L., Ellett, W. H. and Reddy, A. R. (1968). Absorbed fractions for photon dosimetry. *J. Nucl. Med.*, **9** (Suppl. 1), 27–39.

Brucher, H., Niepaus, D. and Nommensev, O. (1981). Feasibility and consequences of sea disposal of krypton-85. Proc. Impacts of radionuclides releases into the marine environment: 673–684. IAEA-SM-248/136, Vienna.

Brunskill, R. T. (1967). The relationship between surface and airborne contamination. In Fish, B. R. (Ed.) *Surface Contamination*: Pergamon Press, Oxford.

Buesseler, K. O. and Halverson, J. E. (1987). *J. Environ. Radioact.*, **5**, 425–444.

Buesseler, K. O. and Sholkovitz, E. R. (1987). The geochemistry of fallout plutonium in the North Atlantic: 1. A pore water study in shelf, slope and deep-sea sediments. *Geochim. Cosmochim. Acta*, **51**, 2605–2622.

Buesseler, K. O., Livingston, H. D., Honjo, S., Hay, B. J., Konuc, T. and Kempe, S. (1990). Scavenging and particle deposition in the Southwestern Black sea. Evidence from Chernobyl radiotracers. *Deep-sea Res.*, **37**, 413–430.

Buldakov, L. A., Demin, S. N., Kostyuchenko, V. A., Koshurnikova, N. A., Krestinina, L. Yu., Saurov, M. M., Ternovskij, I. A., Tokarskaya, Z. B. and Shvedov, V. L. (1989). Medical consequences of the radiation accident in the Southern Urals in 1957. Paper presented to IAEA, Vienna, November.

Bunzl, K. and Kracke, W. (1984). Distribution of Pb-210, Po-210, stable lead and fall-out Cs-137 in soil, plants and moorland sheep of a heath. *Sci. Tot. Environ.*, **39**, 143–159.

Bunzl, K. and Kracke, W. (1986). Accumulation of fallout Cs-137 in some plants and berries of the family Ericaceae. *Health Phys.*, **50**, 540–542.

Bunzl, K., Schimmack, K., Kreutzer, K. and Schierl, R. (1988). The migration of fallout Cs-134, Cs-137, and Ru-106 from Chernobyl and of Cs-137 from weapons testing in a forest soil. *Pflanzenenähr. Bodenk.*, **152**, 39–44.

Burton, P. J. (1988). Remobilisation of actinides in the estuary at Ravenglass. In Wade, B. (Ed.) *Radiological Protection Annual Report 1988*: 115–117. United Kingdom Atomic Energy Authority, Harwell.

Burton, P. J. and Yarnold, L. P. (1988). Remobilisation of actinides from intertidal sediments of the Ravenglass Estuary. *Sci. Tot. Environ.*, **69**, 239–260.

Bustad, L. K., Wood, D. H., Elefson, E. E., Ragan, H. A. and Mc. Clellan, R. D. (1964). I-131 in milk and thyroid of dairy cattle following a single contamination event and prolonged daily administration. In Bustad, L. K. (Ed.) *Biology of Radioiodine*. Proceedings of Hanford Symposium on the Biology of Radioiodine: Pergamon Press, Oxford.

Calmet, D. and Gueguenia t, P. (1985). The release of radioactive liquid effluents from the La Hague irradiated fuel reprocessing centre (France) and the radiological evolution in the marine environment. In IAEA (Ed.) *Behaviour of Radionuclides Released in Coastal Waters*, IAEA Techdoc 329: 111–144. IAEA, Vienna. (In French.)

Calmet, D., Charmasson, S., Gontier, G. and Daburon, M. L. (1988). The impact of Chernobyl fallout on Mytilus sp. collected from the French coast. In IV Symposium International de Radioecologie: Impacts des Accidents d'Origine Nucleaire sur l'Environnement, 14–18 March 1988.

Cannell, R. J., Goddard, A. J. H. and ApSimon, H. M. (1987). Contamination of dwellings by particulate matter: ingresss and distribution within the dwelling. *Radiat. Prot. Dos.*, **21**, 111–116.

Caput, C., Camus, H., Gauther, D. and Belot, Y. (1989). Lavage de l'iodide par les precipitations. Experiences recentes in situ. In FAO/IAEA/UNEP/WHO (Eds) Symposium on Environmental Contamination Following a Major Nuclear Accident, Vienna, Austria, 16–20 October. IAEA, Vienna.

Carlsson, S. (1976). Cesium-137 in a dysoligotrophic lake—a radiological field study. PhD thesis, Radiation Physics Department, University of Lund, Sweden.

Carpenter, R. and Peterson, P. L. (1989). Chemical cycling in Washington's coastal zone.

In Landry, M. R. and Hickey, B. M. (Eds) *Coastal Oceanography of Washington and Oregon*: 367–509. Elsevier, Amsterdam.

Carpenter R., Beasley, T. M., Zahnle, D. and Somayajulu, B. L. K. (1987). Cycling of fall-out (Pu, ^{241}Am, ^{137}Cs) and natural (U, Th, ^{210}Pb) radionuclides in Washington continental slope sediments. *Geochim. Cosmochim. Acta*, **51**, 1897–1921.

Carr, A. P. and Blackley, M. W. L. (1985). Implications of sedimentological hydrological processes on the distribution of radionuclides in a saltmarsh near Sellafield, UK. 197. Inst. Oceanogr. Sci. UK, Taunton.

Casarett, G. W. (1964). Long-term effects of irradiation on sperm production of dogs. In Carlson, W. D. and Gassner, F. X. (Eds) *Effects of Ionizing Radiation on the Reproductive System*: 127–146. Macmillan, New York.

Cataldo, D. A. and Vaughan, B. E. (1980). Interaction of airborne plutonium with plant foliage. In Hanson, W. C. (Ed.) *Transuranic Elements in the Environment*: 288–289. DOE/TIC-22800. US Dept. Energy, NTIS, Springfield, VA.

Cawse, P. A. and Horrill, A. D. (1986) A survey of caesium-137 and plutonium in British soils in 1977. AERE R 10155, Harwell.

CEC (1987). Radioactive transfer factors for animal feeding stuffs and animal products. Report of the Group of Experts Appointed under the Terms of Article 31 of the Euratom Treaty. CEC Doc. No. 7682/87, 20-5-87.

CEC (1990). *The Radiological Exposure of the Population of the European Community from Radioactivity in North European Marine Waters—Project Marina*. CEC, Luxembourg, EUR-12483.

CEC (1991). *Evaluation of Data on the Transfer of Radionuclides in the Foodchain*. Post-Chernobyl Action. EUR-12550. Commission of the European Community, Brussels.

Cerling, T. E. and Turner, R. R. (1982). Formation of freshwater Fe–Mn coatings on gravel and the behavior of ^{60}Co, ^{90}Sr and ^{137}Cs in a small watershed. *Geochim. Cosmochim. Acta*, **46**, 1333–1343.

Chadwick, R. C. and Chamberlain, A. C. (1970). Field loss of radionuclides from grass. *Atmos. Environ.*, **4**, 51–56.

Chamberlain, A. C. (1953). Aspects of travel and deposition of aerosol and vapour clouds. AERE HP/R 1261.

Chamberlain, A. C. (1967). Transport of *Lycopodium* spores and other small particles to rough surfaces. *Proc. Roy. Soc., A* **296**, 45–70.

Chamberlain, A. C. (1970). Interception and retention of radioactive aerosols by vegetation. *Atmos. Environ.*, **4**, 57–78.

Chamberlain, A. C. (1975). The movement of particles in plant communities. In Monteith, J. L. (Ed.) *Vegetation and the Atmosphere*, 1. *Principles*: 155–203. Academic Press, London.

Chamberlain, A. C. (1981). Emission of fission products and other activities during the accident to Windscale pile No. 1 in October 1957. AERE Report M-3194. AERE, Harwell.

Chamberlain, A. C. (1991). *Radioactive Aerosols*. Cambridge University Press, Cambridge.

Chamberlain, A. C. and Dunster, H. J. (1958). Deposition of radioactivity in north-west England from the accident at Windscale. *Nature*, **182**(4636), 629–630.

Changnon, S. A. (1980). Is your weather being modified? *Proc. Am. Soc. Civil Eng. J. Irrig. Drain Div.*, **106**, 37–48.

Chester, C. V. and Chester, R. O. (1976). Civil defense implications of the US Nuclear power industry during a large nuclear war in the year 2000. *Nucl. Technol.*, **31**, 326–338.

Christensen, G. (1989). Personal communication. Institutt for energiteknikk. Kjeller, Norway.

Cigna Rossi, L. (1971). Radioactivity levels in some lichens. *Giorn. Fis. Sanitar. Protez. Contro Radiaz.*, **15**, 124–129.

Clark, M. J. and Smith, F. B. (1988). Wet and dry deposition of Chernobyl releases. *Nature*, **332**, 245–249.

Clarke, R. H. (1974). An analysis of the 1957 Windscale accident using the WEERIE Code. *Ann. Nucl. Sci. Eng.*, **1**, 73–88.

Clarke, R. H. (1989). Current radiation risk estimates and implications for the health consequences of Windscale, TMI and Chernobyl accidents. In Crosbie, W. A. and Gittus, J. H. (Eds) *Medical Response to Effects of Ionising Radiation*: 102–118. Elsevier, London.

Clifton, R. J. and Hamilton, E. I. (1982). The application of radioisotopes in the study of estuarine sedimentary processes. *Est. Coastal Shelf Sci.*, **14**, 433–446.

Clint, G. M., Dighton, J. and Rees, S. (1991). Influx 137-Cs into hyphae of basidiomycete fungi. *Mycol. Res.*, **95**, 1047–1051.

Cloke, J., McKay, W. A., and Liss, P. S. (1991). Laboratory investigations into the effect of marine organic material on the sea salt aerosol generated by bubble bursting. *Marine Chem.*, **34**, 77–95.

Clough, W. S. (1973). Transport of particles to surfaces. *J Aerosol Sci.*, **4**, 227–234.

Clough, P. N. (1990). Severe accidents. In Dunderdale J. (Ed.) *Energy and the Environment*: pp. 103–127, Royal Society of Chemistry, Cambridge.

Cochran, J. K., Carey, A. E., Sholkowitz, E. R. and Surprenant, L. D. (1986). The geochemistry of uranium and thorium in coastal marine sediments and sediment pore waters. *Geochim. Cosmochim. Acta*, **50**, 633–680.

Comans, R. N. J., Middleburg, J. J., Zonderhuis, J., Woittiez, R. W., De Lange, G. J., Das, H. A. and Van Der Weijden, C. H. (1989). Mobilisation of radiocaesium in porewater of lake sediments. *Nature*, **339**, 367–369.

Comar, C. L., Wasserman, R. H. and Twardock, A. R. (1961). Secretion of calcium and strontium into milk. *Health Phys.*, **7**, 69–80.

Command 302 (1958). The accident at Windscale no. 1 pile on 10 October 1957. HMSO, London.

Conkic, L. J., Skrbic, Z., Slivka, J., Veskovic, M. and Bikit, I. (1990). Elimination of long lived fission products from river sediment. *Water Res.*, **24**, 333–335.

Corbett, J. O. (1980). The enhancement of natural radiation dosage by coal-fired power generation in the United Kingdom. UK Central Electricity Generating Board, Internal Report RD/B/N4760.

Cornett, R. J. and Chant, L. (1988). Residence times in freshwaters and accumulation in Shield lake sediments. *Canad. J. Fish. Aquat. Sci.*, **45**, 407–415.

Coughtrey, P. J. (1988). Models for radionuclide transport in soils. *Soil Use Mgmt.*, **4**, 84–90.

Coughtrey, P. J. (1990). Radioactivity transfer to animal products. EUR 12608. Commission of the European Communities, Luxembourg.

Coughtrey, P. J. and Thorne, M. C. (1982). *Radionuclide Distribution and Transport in Terrestrial and Aquatic Ecosystems*, Vol. 1. A. A. Balkema, Rotterdam.

Coughtrey, P. J., Thorne, M. C., Jackson, D. C., Jones, C. H. and Kane, P. (1983–1985). *Radionuclide Distribution and Transport in Terrestrial and Aquatic Ecosystems*, Vols. 1 to 6. A. A. Balkema, Rotterdam.

Coughtrey, P. J., Jackson, D., Jones, C. H., Kane, P. and Thorne, M. C. (1984) *Radionuclide Distribution and Transport in Terrestrial and Aquatic Ecosystems*, Vol. 4. A. A. Balkema, Rotterdam.

Coughtrey, P. J., Kirton, J. A. and Mitchell, N. G. (1989). Transfer of radioactive caesium from soil to vegetation and comparison with potassium in upland grasslands. *Environ. Pollut.* **62**, 4, 281–315.

Coughtrey, P. J. Jackson, A. and Beetham, C. J. (1990a). *Radionuclide Distribution and*

Transport in Terrestrial and Aquatic Ecosystems after the Chernobyl Disaster. Commission of the European Communities, Luxembourg. EUR 13436.

Coughtrey, P. J., Kirton, J. A. and Mitchell, N. G. (1990b). Evaluation of food-chain transfer data for use in accident consequence assessment. In *Methods and Codes for Assessing the Offsite Consequences of Nuclear Accidents*. EUR 13013. Commission of the European Communities, Brussels.

Coughtrey, P. J., Kirton, J. A. and Mitchell, N. G. (1990c). Caesium distribution and cycling in upland pastures of N. Wales and Cumbria. In Desmet, G., Nassimebeni, P. and Belli, M. (Eds) *Transfer of Radionuclides in Natural and Semi-natural Environments*: 259–266. Elsevier, London.

Crabtree, J. (1959). The travel and diffusion of the radioactive material emitted during the Windscale accident. *Quart J. Roy. Met. Soc.*, **85**, 362–370.

Cremers, A., Elsen, P., De Preter, P. and Maes, A. (1988). Quantitative analysis of radiocaesium retention in soils. *Nature*, **335**, 247–336.

Cremers, A., Elsen, A., Valcke, E., Wauters, J., Sandalls, F. J. and Gaudern, S. L. (1990). The sensitivity of upland soils to radiocaesium contamination. In Desmet, G., Nassimbeni, P. and Belli, M. (Eds) *Transfer of Radionuclides in Natural and Semi-natural Environments*: 238–248. Elsevier, London.

Crick, M. J. and Brown, J. (1990). Expert: a model for evaluating exposure from radionuclide material deposited in the urban environment. NRPB-R235.

Crick, M. J. and Linsley, G. S. (1982). An assessement of the radiological impact of the Windscale reactor fire, October 1957. NRPB-R135, NRPB, Chilton.

Crick, M. J. and Linsley, G. S. (1983). Addendum to report NRPB-R135, NRPB, Chilton.

Crout, N. M. J., Galer A., Howard, B. and Beresford, N. A. (1990). A comparative assessment of the impact of Windscale and Chernobyl accidents on radiocaesium levels in lamb in West Cumbria using the RUINS model. In *Comparative Assessment of the Environmental Impact of Radionuclides Released During Three Major Nuclear Accidents: Kyshtym, Windscale and Chernobyl*. Commission of the European Communities, Luxembourg, EUR 13574.

Cuff, Y. S. (1989). The SPADE2 suite of codes: development of time-dependent models for specific animal products. ANS Report No. 2017-R1. ANS Consultants Ltd, Epsom.

Currie, L. A. (1968). Limits for qualitative detection and quantitative determination. *Anal. Chem.*, **40**, 586–593.

Cushing, C. E. (1976). *Radioecology and Energy Resources*. Dowden, Hutchinson & Ross, Stroudsburg, USA.

Cutshall, N., Renfro W. C., Evans, D. W. and Johnson, V. G. (1973). ^{65}Zn in Oregon-Washington continental shelf sediments. In Nelson, P. J. (Ed) Radionuclides in Ecosystems. *Proceedings of Third National Symposium on Radioecology*: 294–702. CONF 710501-P2, Oak Ridge, Tennessee.

Davison, W., Hilton, J., Hamilton-Taylor, J., Kelly, M., Livans, F., Rigg, E., Cassick, T. R. and Singleton, D. L. (1992). The transport of Chernobyl-derived ^{137}Cs through two freshwaters lakes in Cumbria, UK. *J. Environ. Radioact.* (In press).

Delaune, R. D., Patrick, W. H. Jr and Buresh, R. J. (1978). Sedimentation rates determined by ^{137}Cs dating in a rapidly accreting salt marsh. *Nature*, **275**, 532–533.

Delmas, J., Grauby, A., Paulin, R., Rinaldi, C., Ndoye, R. and Seckgassama, S. (1987). Etude experimentale du transfert du ^{90}Sr et du ^{137}Cs d'un sol de la region du Cap Vert a l'arachide. *Radioprotection*, **22**, 333–40.

Department of the Environment (1987). Household Particulate Survey. (Radioactivity in housedust in West Cumbria and its significance). Report no. DOE/RW/87033.

Derbetin, K. and Helmer, R. (1988). *Gamma- and X-ray Spectrometry with Semiconductor Detectors*. North-Holland, Amsterdam.

De Regge, P. and Boden, R. (1984). Review of chemical separation techniques applicable to alpha spectrometric measurements. *Nucl. Instr. Meth. Phys. Res.*, **223**, 181–187.

Desmet, G. and Myttenaere, C. (1988). Considerations on the role of natural ecosystems in the eventual contamination of man and his environment. *J. Environ. Radioact.*, **6**, 197–202.

Desmet, G., Nassimbeni, P. and Belli, M. (Eds) (1990). *Transfer of Radionuclides in Natural and Semi-Natural Environments*. Elsevier, London.

Devell, L. (1991). Composition and properties of the plume and fallout materials. In Moberg, L. (Ed) *The Chernobyl Fallout in Sweden—Results from a Research Programme on Environmental Radiology*: 29–46. Swedish Radiation Protection Institute.

Devell, L., Aarkrog, A., Blomqvist, L., Magnussen, S. and Tveten, V. (1986). How the fallout from Chernobyl was detected and measured in the Nordic Countries. *Northern Europe*, **11**, 16–17.

Dickson, D. (1989). Kyshtym 'almost as bad as Chernobyl'. *New Sci.*, **124** (1696/1697), 3–4.

Dickson, R. R., Gurbutt, P. A. and Kershaw, P. J. (1986). *Interim Oceanographic Description of the North-East Atlantic Site for the Disposal of Low-level Radioactive Waste*, 2. OECD Nuclear Energy Agency, Paris.

Dighton, J., Clint, G. M. and Poskitt, J. (1991). Uptake and accumulation of radiocaesium by grassland soil fungi: a potential pool of Cs immobilisation. *Mycol. Res.*, **95**, 1051–1956.

DOE (1984). An incident leading to contamination of the beaches near to the British Nuclear Fuels Limited Windscale and Calder Hall Works, Sellafield November 1983. Department of Environment, UK, London.

Donn, S. G. (1990). The environmental impact of radioactive waste disposal facilities. Paper RL.32.90 presented at SCOPE–RADPATH First Case-study Meeting, 26–30 March 1990, University of Lancaster.

Donoghue, J. F., Bricker, O. P. and Olsen, C. R. (1989). Particle-borne radionuclides as tracers for sediment in the Susquehanna River and Chesapeake Bay. *Est., Coastal Shelf Sci.*, **29**, 341–360.

Dreicer, M., Hakonson, T. E. White, G. C. and Whicker, F. W. (1984). Rainsplash as a mechanism for soil contamination of plant surfaces. *Health Phys.*, **46**, 177–188.

D'Souza, T. J. and Mistry, K. B. (1980). Behaviour of gamma-emitting fission products ^{106}Ru, ^{125}Sb, ^{134}Cs and ^{144}Ce deposited on established pastures in tropical environs. *J. Nucl. Agric. Biol.*, **9**, 50–3.

D'Souza, T. J., Fagniart, E. and Kirchmann, R. (1980). Effects of clay mineral type and organic matter on the uptake of radiocaesium by pasture plants. BLG 538 (October). Centre d'Etude de L'Energie Nucleaire, Belgium.

D'Souza, T. J., Kirchmann, R. and Lehr, J. J. (1972). Distribution of radiostrontium and radiocaesium in the organic and mineral fractions of pasture soils and their subsequent transfer to grasses. Isotopes and radiation in soil–plant relationships including forestry. IAEA-SM-151/4. IAEA, Vienna.

Duinker, J. C. (1980). Suspended matter in estuaries. In Olausson, E. and Cato, I. (Eds) *Chemistry and Biogeochemistry of Estuaries* 121–152. John Wiley & Sons, Chichester.

Duinker, J. C., Nolting, R. F. and Michel, D. (1982). Effects of salinity, pH and redox conditions on the behaviour of Cd, Zn, Ni and Mn in the Scheldt Estuary. *Thalassia Jugoslav.*, **18**, 191–202.

Dunster, H. J., Howells, H. and Templeton, W. L. (1958). District surveys following the Windscale accident, October 1957. In *Proc. Second U.N. Conference on Peaceful Uses of Atomic Energy*, Vol. 15: 296–308. UNO, Geneva.

Duursma, E. K., Frissel, M. J., Guary, J. C., Martin, J. M., Nieuwenhuize, J., Pennders, R. M. J. and Thomas, A. J. (1985). Plutonium in sediments and mussels of the Rhine–

Meuse–Scheldt Estuary. In: *The Behaviour of Radionuclides in Estuaries*: 71–106. XII/380/85-EN. Commission of the European Community, Luxembourg.

Dyer, K. R. (1986). *Coastal and Estuarine Sediment Dynamics*. John Wiley & Sons, Chichester.

Eakins, J. D. and Lally, A. E. (1984). The transfer to land of actinide-bearing sediments from the Irish sea by spray. *Sci. Tot. Environ.*, **35**, 23–32.

Eakins, J. D., Pattenden, N. J., Cambray, R. S., Lally, A. E. and Playford, K. (1981). Studies of environmental radioactivity in Cumbria Part 2–Radionuclide deposition in soil in the coastal region of Cumbria. Report AERE-R9873, Harwell.

Eakins, J. D., Burton, P. J., Humphreys, D. G. and Lally, A. E. (1985). The remobilisation of actinides from contaminated intertidal sediments in the Ravenglass Estuary. In *The Behaviour of Radionuclides in Estuaries*: 107–122. XII/380/85-EN. Commission of the European Community, Luxembourg.

Eakins, J. D., Morgan, A., Baston, G. M. N., Pratley, F. W., Strange, L. P. and Burton, P. J. (1990). Measurements of alpha emitting plutonium and americium in the intertidal sands of west Cumbria. *J. Environ. Radioact.*, **11**, 37–54.

Edgington, D. N. and Robbins, J. A. (1975). The behaviour of plutonium and other long-lived radionuclides in Lake Michigan: II. Patterns of deposition in the sediments. In *Impacts of Nuclear Releases into the Aquatic Environment*. Proc. Int. Symp., Ontaniemi, Finland. IAEA, Vienna.

Efremenkov, V. M. (1989). Radioactive waste management at nuclear plants. *IAEA Bull.*, **31**(4), 37–42.

van Egmond, N. D. and Kesseboom, H. (1983). Mesoscale air pollution dispersion models—Eulerian GRID Model. *Atmos. Environ.*, **17**, 257–266.

Eisenbud, M. (1987). *Environmental Radioactivity—From Natural, Industrial and Military Sources*, (3rd edn.), Academic Press, London.

Elbaz-Poulichet, F., Huang, W. W., Jednacek-Biscan, J., Martin, J-M. and Thomas, A. J. (1982). Trace metal behaviour in the Gironde estuary: the problem revisited. *Thalassia Jugoslav.*, **18**, 61–95.

Elderfield, H. and Hepworth, A. (1975). Diagenesis, metals and pollution in estuaries. *Mar. Pollut. Bull.*, **6**, 85–87.

Ellett, W. H. and Humes, R. M. (1971). Absorbed fractions for small volumes containing photon-emitting radioactivity. *J. Nucl. Med.*, **12** (Suppl. 5), 25–32.

Ellis, J. B. (1986). Pollution aspects of urban runoff. In Torno, H. C., Marsalek, J. and Desbordes, M. (Eds) *Urban Runoff Pollution*, Springer-Verlag, Berlin.

Ellis, J. B., Revitt, D. M., Harrop, D. O. and Beckwith, P. R. (1987). The contribution of highway surfaces to urban stormwater sediments and metal loadings. *Sci. Tot. Environ.*, **59**, 339–349.

Erickson, B. H. (1978). Effect of continuous gamma-radiation on the stem and differentiating spermatogonia of the adult rat. *Mutation Res.*, **52**, 117–128.

Eriksson, E. (1977). Direct uptake by vegetation of deposited materials. 1. Retention of nuclides and simulated fallout particles in pasture grass. SLU-IRB-42. Agricultural College, Uppsala, Sweden.

Evans, D. W. and Cutshall, N. H. (1973). Effects of ocean water on the soluble-suspended distribution of Columbia River radionuclides. In *Radioactive Contamination of the Marine Environment*: 125–140. IAEA, Vienna.

Evans, D. W., Alberts, J. J. and Clark, A. (1983). Reversible ion-exchange fixation of cesium-137 leading to mobilisation from reservoir sediments. *Geochim. Cosmochim. Acta*, **47**, 1041–1049.

Evans, J. C., Bryce, R. W., Bates, D. J. and Kemner, M. L. (1990). Hanford Site Ground-Water Surveillance for 1989. PNL-7396, Pacific Northwest Laboratory, Richland, Washington.

Fish, B. R., Walker, R. L., Royster, G. W. and Thompson, J. L. (1967). Redispersion of settled particulates. In Fish, B. R. (Ed.) *Surface Contamination*: Pergamon Press, Oxford.

Fontes, J-Ch. and Garnier, J-M. (1979). Determination of the initial ^{14}C activity of the total dissolved carbon: a review of the existing models and a new approach. *Water Resources Res.*, **15**, 399–413.

Förstner, U. and Schoer, J. (1984). Diagenesis of chemical associations of ^{137}Cs and other artificial radionuclides in river sediments. *Environ. Technol. Lett.*, **5**, 295–306.

Förstner, U. and Wittmann, G. T. W. (1979). *Metal Pollution in the Aquatic Environment*. Springer-Verlag, Berlin.

Fowler, S. W., Buat-Menard P., Yokoyama Y., Ballestra, S., Holm, E. and Nguyen, H.V. (1987). Rapid removal of Chernobyl fallout from Mediterranean surface waste by biological activity. *Nature*, **329**, 56–58.

Fowler, S. W., Ballestra, S., La Rosa, J., Holm, E. and Lopez, J. J. (1990a). Flux of transuranium nuclides in the northwestern Mediterranean following the Chernobyl accident. *Rapp. Comm. Int. Mer. Medit.*, **32**(1), 317.

Fowler, S. W., Ballestra, S. and Villeneuve, J. P. (1990b). Flux of transuranium nuclides and chlorinated hydrocarbons in the northwestern Mediterranean. *Cont. Shelf Res.* (In press).

Frank, I. and Tamm, I. (1937). Coherent visible radiation of fast electrons passing through matter. *Dokl. Akad. Nauk. SSSR*, **14**, 109–114.

Frissel, M. J. (1990). The uptake of Pu, Am and Np by crops from soils, including a comparison of data obtained from Sellafield sediments and other sources. VII Report, IUR Working Group on Soil-to-Plant Transfer. Soil Adhesion. RIVM-LSO, Bilthoven, Netherlands.

Frissel, M. J. and Bergeijik van, K. E. (1989). Mean transfer values derived by simple regression analyses. VI Report, IUR Working Group on Soil-to-Plant Transfer. RIVM-LSO, Bilthoven, Netherlands.

Frissel, M. J. and Pennders, R. M. J. (1983). Models for the accumulation and migration of Sr-90, Cs-137, Pu-239,240 and Am-241 in the upper layers of soil. In Coughrey, P. J., Bell, J. N. B. and Roberts, T. M. (Eds.), *Ecological Aspects of Radionuclide Release*. Blackwell Scientific Publications, Oxford.

Frissel, M. J. and Pennders, R. M. J. (1991). The role of various conditions on the uptake of radiocaesium and strontium by food crops, with special reference to Chernobyl caesium. *Geochemical Pathways of Artificial Radionuclides Migration in the Biosphere*. USSR Acad. Sci. (In preparation).

Frissel, M. J., Blaauboer, R. O., Köster, H. W., Leenhouts, H. P., Stoutjesdijk, J. F. and Vaas, L. H. (1989). Radioactive contamination by food and the intake by man. *Radiat. Phys. Chem.*, **34**(2), 327–336.

Fukai, R., Ballestra, S., Vas, D. (1983). Characteristics of the vertical transport of transuranic elements through the Mediterranean water column. In *VI Journes d'Etudes sur les Pollutions Marines en Mediterranne*: 95–101. Monaco: CIESM.

Gale, H. J., Humphreys, D. L. O. and Fisher, E. M. R. (1964). Weathering of caesium-137 in soil. *Nature*, **201**, 257–261.

Garland, J. A. (1979). Resuspension of particulate matter from grass and soil. AERE-R9452, HMSO, London.

Garland, J. A. (1980). Surface deposition from radioactive plumes. Seminar on Radioactive Releases and their Dispersion in the Atmosphere Following a Hypothetical Reactor Accident. CEC, Luxembourg.

Garland, J. A. and Cambray, R. S. (1988). Deposition, resuspension and the long term variation of airborne radioactivity from Chernobyl. 4th International Symposium on Radioecology, Centre d'etudes nucleaires de Cadarache, Cadarache.

Garland, J. A. and Pattenden, N. J. (1990). Resuspension following Chernobyl. In *Methods and Codes for Assessing the Offsite Consequences of Nuclear Accidents*. EUR 13013. Commission of the European Communities, Brussels.

Garland, J. A., Nicholson, K. W and Derwent, R. G. (1987). The deposition of trace substances from the atmosphere. In Grefen, K. and Lobel, J. (Eds) *Environmental Meteorology*: pp. 141–160. Kluwer, Dordrecht.

Garten, C. T. (1980). Technetium-99 cycling in deciduous forests. Review and ecosystem model development. *Environ. Int.*, **13**, 311–321.

Gatz, D. F. (1977). Scavenging ratio measurements in METROMEX. In Semonin, R. G. and Beadle, R. W. (Eds) *Precipitation Scavenging (1974)*. Proceedings of a Symposium held at Champaign, Illinois. Conf-741003.

Gerusky, T. M. (1988). The accident at Three Mile Island—1979. In Carter, M. W. (Ed.-in-Chief), *Radionuclides in the Food Chain*: 157–171. International Life Sciences Institute Monographs, Springer-Verlag, New York.

Gesell, T. F. and Prichard, H. M. (1975). The technologically enhanced radiation environment. *Health Phys.*, **28**, 361–366.

Gillette, D. A., Blifford, I. H. and Fryrear, D. W. (1974). The influence of wind velocity on the size distributions of aerosols generated by the wind erosion of soils. *J. Geophys. Res.*, **79**, 4068–4075.

Girardi, F., Graziani, G., Pagliari, V., Myttenaere, C. and Sinnaeve, J. (1987). Development, application and perspectives of a data bank of environmental radioactivity levels in the European community. IAEA Seminar on Applications of Computer Technology to Radiation Protection, Bled Yugoslavia, 22–26 June. IAEA Sr-136/62, 1979.

Gjørup, H. L., Jensen, P. H., Lauridsen, B., Roed, J. and Warming, L. (1986). Deposition, retention and decontamination of radioactive material on urban surfaces. The Danish program. In Luykx, F. and Sinnaeve, J. (Eds) *CEC/LUX Workshop on Methods for Assessing the Off-site Radiological Consequences of Nuclear Accidents*.

Graham, M. J., Hall, M. D., Strait, S. R. and Brown, W. R. (1981). *Hydrology of the Separations Area*. RHO-ST-42, Rockwell Hanford Operations, Richland, Washington.

Grebenshchikova, N. V., Firsakova, S. K., Timofeev S. F., Novik, A. A. and Palekshanova, G. I. (1991). Transfer patterns of radionuclides into crops in the condition of radioactive contamination after the Chernobyl accident. *Comparative Assessment of the Environmental Impact of Radionuclides Released During Three Major Accidents: Kyshtym, Windscale and Chernobyl*. CEC, Luxembourg. EUR 13574.

Green, R. H. (1979). *Sampling Design and Statistical Methods for Environmental Biologists*. John Wiley, New York.

Groom, J. M. and Rasdale, H. L. (1980). A model of radiocaesium cycling in a sand hills–Turkey oak (*Quercus laevis*) ecosystem. *Ecol. Mod.*, **11**, 55–65.

Gross, L. G. (1986). Ecology: an idiosyncratic overview. In Hallam, T. G. and Levin, S. A. (Eds) *Mathematical Ecology*: 3–15. Springer-Verlag, Berlin.

Guary, J. C., Gueguenіat, P., Pentreath, R. J. (Eds) (1988) *Radionuclides: A Tool for Oceanography*. Elsevier, London.

Gudiksen, P. H., Harvey, T. F. and Lange, R. (1989). Chernobyl source term, atmospheric dispersion and dose estimation. *Health Phys.*, **57**(5), 697–705.

Gueguenіat, P., Gandon, R., Baron, Y., Salomon, R., Pentreath, J., Brylinski, J. M. and Cabioch, L. (1988). Use of artificial radionuclides (125-Sb, 137-Cs, 134-Cs) for the *in situ* observation of the movement of water masses in the Channel. In Guary, J. C., Gueguenіat, P. and Pentreath, R. J. (Eds) *Radionuclides: A Tool for Oceanography*: 260–270. Elsevier, London. (In French.)

Guillitte, O., Casia, M. C., Lambinon, J., Fraiture, A., Collard, J. and Kirchmann, R. (1987). La contamination des champignons sauvages en Belgique et au Grand Duché de Luxembourg après l'accident nucléaire de Tchernobyl. *Mém. Soc. Roy. Bot. Belg.*, **9**, 79–93.

Gurbutt, P. A. and Dickson, R. R. (1983). *Interim Oceanic Description of the North-East Atlantic site for the Disposal of Low-level Radioactive Waste*. Nuclear Energy Authority (NEA-OECD) Paris.

Gy, P. M. (1983). Sampling errors. They may deprive the analytical results of any meaning. *Analusis*, **11**(9), 413–440.

Hakonson, T. E. and Nyhan, J. W. (1980). Ecological relationships of plutonium in Southwest ecosystems. In Hanson, W. C. (Ed.) *Transuranic Elements in the Environment*: 403–419. DOE/TIC-22800. US Dept. Energy, NTIS, Springfield, VA.

Hakonson, T. E., Martinez, J. L. and White, G. C. (1982). Disturbance of a low-level waste burial site cover by pocket gophers. *Health Phys.*, **42**(6), 868–871.

Hall, M. J. (1984). *Urban Hydrology*. Elsevier, London.

Halverson, J. E. (1984). A review of applications of mass spectrometry to low level radionuclide metrology. *Nucl. Instr. Meth. Phys. Res.*, **223**, 349–355.

Hamilton, E. I. (1985). The disposal of radioactive wastes into the marine environment: the presence of hot particles containing Pu and Am in the source term. *Mineral. Mag.*, **49**, 117–194.

Hamilton, E. I. and Clarke, K.R. (1984). The recent sedimentation history of the Esk Estuary, Cumbria, UK. The application of radiochronology. *Sci. Tot. Environ.*, **35**, 325–386.

Hamilton, R. S., Revitt, D. M., Warren, R. S. and Duggan, M. J. (1985). Resuspension of Heavy Metals from road Surface Dust. In *Proc. 5th Int. Conf. Heavy Metals in the Environment, Athens*. CEP Consultants, Edinburgh.

Hamilton-Taylor, J., Kelly, M., Mudge, S. and Bradshaw, K. (1987). Rapid remobilisation of plutonium from estuarine sediments. *J. Environ. Radioact.*, **5**, 409–423.

Hammar, J., Notter, M. and Neumann, G. (1989). *Caesium in Trout Lakes*. Environmental Protection Agency. Stockholm, Sweden. (In Swedish.)

Handl, J. and Pfau, A. (1987). Feed milk transfer of fission products following the Chernobyl accident. *Atomkernenergie-Kerntechnik*, **49**, 171–173.

Handley, R. and Overstreet, R. (1961). Effects of various cations upon absorption of carrier-free caesium. *Plant Physiol.*, **36**(1), 66–99.

Hansen, W. G., Campbell, J. E., Fooks, J. H., Mitchell, H. C. and Eller, C. H. (1964). Farming practices and concentrations of fission products in milk. U.S. Public Health Service Publication No. 999-R-6.

Hanson, W. C. (1966). Radioecological concentration processes characterizing arctic ecosystems. In Aberg, B. and Hungate, F. P. (Eds) *Radioecological Concentration Processes*: 183–191. Pergamon Press, Oxford and New York.

Hanson, W. C. (1973). Fallout ^{90}Sr and ^{137}Cs in Northern Alaskan Ecosystems During 1959–1970. Dissertation, Colorado State University, Fort Collins.

Hanson, W. C. (1980). *Transuranic Elements in the Environment*. DOE/TIC-22800. NTIS, Springfield, VA.

Hanson, W. C. and Thomas, J. M. (1982). Prediction of ^{90}Sr body burdens and radiation dose in Anaktuvuk Pass Alaska Eskimos due to fallout. *Health Phys.*, **43**, 323–333.

Harley, N. H. and Pasternak, B. S. (1972). Alpha absorption measurements applied to lung dose from radon daughters. *Health Phys.*, **23**, 771.

Harrison, R. M. (1979). Toxic metals in street and household dusts. *Sci. Tot. Environ.*, **11**, 89–97.

Harrison, R. M. (1990). Some important air pollutants and their chemical analysis. In Harrison, R. M. (Ed.) *Pollution: Causes, Effects and Control*: Royal Society of Chemistry, Cambridge.

Harvey, B. R. and Lovett, M. B. (1984). The use of yield tracers for the determination of alpha-emitting actinides in the marine environment. *Nucl. Instr. Meth. Phys. Res.*, **223**, 224–234.

Harvey, B. R. and Lovett, M. B. (1991). Identification and control of contamination, Interference and other errors in low-level radionuclide analysis. In Garcia-Leon, M. and Madurga G. (Eds) *Low-level Measurements of Man-made Radionuclides in the Environment*: 239–262. World Scientific, Singapore.

Haugen, L. E., Bjornstad, H. E. and Brittain, J. E. (1991). Lateral transport of radiocaesium on mountain pasture, Upper Heimdalen, Norway. To be published. Agricultural University of Norway, As, Norway. (In Norwegian.)

Haux, C. and Forlin, L. (1988). Biochemical methods for detecting effects of contaminants on fish. *Ambio*, **17**, 376–380.

Hayes, D. W. and Sackett, W. M. (1987). Plutonium and cesium radionuclides in sediments of the Savannah River Estuary. *Estuar. Coastal Shelf Sci.*, **25**, 169–174.

Haywood, S. M., Simmonds, J. R. and Linsley, G. S. (1980). The development of models for the transfer of ^{137}Cs and ^{90}Sr in the pasture-cow-milk pathway using fallout data. NRPB-R110. National Radiological Protection Board, Didcot.

Heafield, W., Barlow, P. and Bryan, G. D. (1989). BNFL's policy and strategy for radioactive waste management. *Radioact. Waste Mgmt.* **2**: 29–33. BNES, London.

Hermanson, M. H. (1990). ^{210}Pb and ^{137}Cs chronology of sediments from small, shallow Arctic lakes. *Geochim. Cosmochim. Acta*, **54**, 1443–1451.

Hickey, B. M. (1989). Patterns and processes of circulation over the shelf and slope. In Landry, M. R. and Hickey, B. M. (Eds) *Coastal Oceanography of Washington and Oregon*: 41–116. Elsevier, Amsterdam.

Hill, T. A., Choularton, T. W. and Penkett, S. A. (1986). A model of sulphate protection in a cap cloud and subsequent turbulent deposition onto the hill surface. *Atmos. Environ.*, **20**, 1763–1771.

Hoek Van den, J. (1989). European research on the transfer of radionuclides to animals—a historical perspective. *Sci. Tot. Environ.*, **85**, 17–27.

Hoek Van den, J. and Binnerts, W. T. (1967). The influence of liming of grass on the secretion of Sr-90 in cow's milk. *Atompraxis*, **13**, 1–4.

Hoek Van den, J., Kirchmann, R. J., Colard, J. and Sprietsma, J. E. (1969). Importance of some methods of pasture feeding, of pasture type and of seasonal factors on Sr-85 and Cs-134 transfer from grass to milk. *Health Phys.*, **17**, 691–700.

Hoffman, F. O., Frank, M. L., Blaylock, B. G., von Bernuth, R. D., Deming, E. J., Graham, R. V., Mohrbacher, D. A. and Waters, A. E. (1989). Pasture grass interception and retention of I-131, Be-7, and insoluble microspheres deposited in rain. ORNL-6542, Oak Ridge National Laboratory, Tennessee.

Holleman, D. F., Luick, J. R. and Whicker, F. W. (1971). Transfer of radiocaesium from lichen to reindeer. *Health Phys.*, **21**, 657–666.

Holm, E. and Persson, R. B. R. (1975). Fallout plutonium in Swedish reindeer lichens. *Health Phys.*, **29**, 43–51.

Hoppenheit, M., Murray, C. N. and Woodhead, D. S. (1980). Uptake and effects of americium-241 on a brackish water amphipod. *Helgoldnder Meeresunters*, **33**, 138–152.

Horrill, A. D. (1983). The levels and point-to-point variation of the radionuclide burden in a Cumbrian saltmarsh. In Coughtrey, P. J., Bell, J. N. B. and Roberts, T. M. (Eds) *Ecological Aspects of Radionuclide Release*: 199–215. Blackwell Scientific Publications, Oxford.

Horrill, A. D., Kennedy, V. H. and Harwood, T. R. (1990). The concentrations of Chernobyl derived radionuclides in species characteristic of natural and semi-natural ecosystems. In Desmet, G., Nassimbeni, P. and Belli, M. (Eds) *Transfer of Radionuclides in Natural and Semi-natural Environments*: 27–39. Elsevier, London.

Horsley, D. M. C. and Howden, M. (1990). The reduction of radioactive discharges from

Sellafield, Process Safety and Environmental Protection. *Trans. I. Chem. E., Part B*, **68**(B2), 140–146.

Hotzl, H. and Winkler, R. (1981). The GSF anticoincidence-shielded Ge(Li) gamma-ray spectrometer and its application to the analysis of environmental samples. In *Methods of Low-level Counting and Spectrometry*: 77–91. IAEA, Vienna, STI/PUB/592.

Hove, K. and Ekern, A. (1988). Combating radiocaesium in farm animals. In Lag, J. (Ed.) *Int. Symp. on Health Problems in Connection with Radioactive Matter in Fertilizers, Soils and Rocks*. Norwegian Acad. Sci. Lett., Oslo.

Hove, K., Pedersen, O., Garmo, T. H., Hansen, H. S. and Staaland, H. (1990). Fungi: a major source of radiocaesium contamination of grazing ruminants in Norway. *Health Phys.*, **59**(2), 189–192.

Howard, B. J. (1989). A comparison of radiocaesium transfer for sheep milk and muscle derived from both field and laboratory studies. *Sci. Tot. Environ.*, **85**, 189–198.

Howard, B. J. and Lindley, D. K. (1985). Aspects of the uptake of radionuclides by sheep grazing on an estuarine saltmarsh. 2. Radionuclides in sheep tissue. *J. Environ. Radioact.*, **2**, 199–213.

Howard, B. J. and Livens, F. R. (1991). Radioactive Contamination of Tide-washed Pastures. Contract Report No. 2069, Ministry of Agriculture, Fisheries and Food.

Howard, B. J., Beresford, N. A., Burrow, L., Shaw, P. V. and Curtis, E. J. C. (1987). A comparison of caesium-137 and 134 activity in sheep remaining on upland areas contaminated by Chernobyl fallout with those removed to less active lowland pasture. *J. Soc. Radiol. Prot.*, **7**, 71–73.

Howorth, J. M. and Eggleton, A. E. J. (1988). Studies of environmental radioactivity in Cumbria, Part 12: Modelling of the sea-to-land transfer of radionuclides and an assessment of the radiological consequences. AERE, Harwell, R-11733.

Howorth, J. M. and Kirby, C. R. (1988). Studies of environmental radioactivity in Cumbria. Part II: Modelling the Dispersion of Radionuclides in the Irish sea. AEA Technology, Harwell, R-11734.

Hsu, T. S. H. and Fabrikant, J. I. (1976). Spermatogonial cell renewal under continuous irradiation at 1.8 and 45 rads per day. In *Biological and Environmental Effects of Low-Level Radiation*, Vol. 1: 157–168. IAEA, Vienna.

Hull, A. P. (1980). A critique of source term and environmental measurement at Three Mile Island. *IEEE Trans. Nucl. Sci.*, NS-27.

Hunt, G. J. (1977–1992). Radioactivity in surface and coastal waters of the British Isles. *Aquat. Environ. Monit. Rep. Ser.* MAFF Direct. Fish. Res. Lowestoft, no. 3, 4, 6, 8, 9, 11, 12, 13, 14, 18, 19, 21, 23, 29.

Hunt, G. J. (1985). Timescales of dilution and dispersion of transuranics in the Irish sea near Sellafield. *Sci. Tot. Environ.*, **46**, 261–278.

Hunt, G. J. (1988). *Radioactivity in Surface and Coastal Waters of the British Isles, 1987.* Ministry of Agriculture, Fisheries and Food, Lowestoft.

Hutchinson-Benson, E., Svoboda, J. and Taylor, W. (1985). The latitudinal inventory of caesium-137 in vegetation and topsoil in northern Canada. *Canad. J. Bot.*, **63**, 784–791.

Hvinden, T., Lillegraven, A. and Lillesaeter, O. (1964). Passage of a radioactive cloud over Norway, November 1962. *Nature*, **202**, 950–952.

Hyodo-Taguchi, Y. and Egami, N. (1977). Damage to spermatogenic cells in fish kept in tritiated water. *Radiat. Res.*, **71**, 641–652.

IAEA (1974). *Evaluation of Radiation Emergencies and Accidents.* Technical Report Series No. 152. IAEA, Vienna.

IAEA. (1976). *Effects of Ionizing Radiation on Aquatic Organisms and Ecosystems.* Technical Report Series No. 172: 1–131. IAEA, Vienna.

IAEA. (1983). Tritium. In *Isotope Techniques in the Hydrological Assessment of Potential Sites for the Disposal of High Level Radioactive Wastes*: 57–61. Technical Report Series No. 228. IAEA, Vienna.

IAEA. (1984). *The Oceanographic and Radiological Basis for the Definitions of High-level Wastes Unsuitable for Dumping at Sea.* Safety Series 66. IAEA, Vienna.

IAEA. (1985a). *The Behaviour of Radionuclides Released into Coastal Waters.* Technical Document Series No. 329. IAEA, Vienna.

IAEA. (1985b). *Sediment Kds and Concentration Factors for Radionuclides in the Marine Environment.* Technical Report Series No. 247. IAEA, Vienna.

IAEA. (1986). *International Nuclear Safety Advisory Group. Summary report on the post-accident review meeting on the Chernobyl accident.* IAEA Safety Series INSAG-1, Vienna.

IAEA. (1988). *Assessing the Impact of Deep Sea Disposal of Low-level Radioactive Waste on Living Marine Resources.* Technical Report Series No. 288. IAEA, Vienna.

IAEA. (1989). *Measurement of Radionuclides in Food and the Environment.* IAEA Technical Report Series No. 295. IAEA, Vienna.

IAEA. (1990). GESAMP Working group 25, GESAMP Reports and Studies, Number 43 on Coastal Modelling. IAEA Vienna.

IAEA (1991a). *A Practical Methodology for the Assessment of Individual and Collective Radiation Doses from Radionuclides in the Environment.* IAEA, Vienna.

IAEA (1991b). *Handbook of Parameter Values for the Prediction of Radionuclide Transfer in the Terrestrial and Freshwater Environments.* IAEA, Vienna.

IAEA. (1992a). *Inventory of Radioactive Material Entering the Marine Environment: Accident at Sea Involving Radioactive Materials.* IAEA Technical Document Series. IAEA, Vienna.

IAEA (1992b). *Effects of Ionizing Radiation on Plants and Animals at Levels Implied by Current Radiation Protection Standards.* Technical Report Series, No. 332. IAEA, Vienna.

Ibrahim, M., Barrie, L. A. and Fanaki, F. (1983). An experimental and theoretical investigation of the dry deposition of particles to snow, pine trees and artificial collectors. *Atmos. Environ.*, **17**, 781–788.

ICRP. (1977). Recommendations of the International Commission on Radiological Protection, ICRP Publication 26, *Ann. ICRP*, **1**(3), 1–53.

ICRP. (1985). Statement from the 1985 Paris meeting of the International Commission on Radiological Protection. *J. Soc. Radiol. Prot.*, **5**, 87–88.

ICRP. (1991). 1990 Recommendations of the International Commission on Radiological Protection, ICRP Publication 60, *Ann. ICRP*, **21**(1-3), 1–201.

Imboden, D. M., Weiss, R. F., Craig, H., Michel, R. L. and Goldman, C. R. (1977). Lake Tahoe geochemical study. 1. Lake chemistry and tritium mixing study. *Limnol. Oceanogr.*, **22**, 1039–1051.

Israel, Y. A. and Petrov, V. N. (1988). Procedures on assessment and measures of safety for the population after the Chernobyl accident. In Harley, J. H., Schmidt, G. D. and Silini, G. (Eds) *Radionuclides in the Food Chain*: 285–290. Springer-Verlag, New York.

Israel, Y. A., Petrov, V. N. and Severov, D. A. (1987). Radioactive fallout in the close-in area of the Chernobyl NPP. *Sov. J. Meteorol. Hydrol.*, **7**.

IUR (1982). Report on a Workshop on the Measurement of Soils to Plant Transfer Factors for Radionuclides. EURATOM-Ital Report, Wageningen, Netherlands.

IUR (1990). VII Report of the IUR Working Group on Soil-to-Plant Transfer. Soil Adhesion. RIVM-LSO, Bilthoven, Netherlands.

Izdar, E. and Murray, J. W. (Eds) (1991). *Black Sea Oceanography.* NATO AS1 Series C: Mathematical and Physical Sciences: 351. Kluwer, Dordrecht.

Jacob, P., Meckbach, R. and Müller, H. M. (1987). Reduction of external exposure from deposited Chernobyl activity by runoff, weathering, street cleaning and migration in the soil. *Radiat. Prot. Dos.*, **21**, 51–57.

Jacob, P., Meckbach, R., Müller, H. M. and Meimberg, K. (1990). Abnahme von Künstlicher Radioaktivität nach Deposition in Städtischer Umgebung. GSF Neuherberg.

Jaquish, R. E. and Bryce, R. W. (1990). Hanford Site Environmental Report for Calendar Year 1989. In Jaquish, R. E. and Bryce, R. W. (Eds) PNL-7346. Pacific Northwest Laboratory, Richland, Washington.

Jaworowski, Z. (1982). Natural and man-made radionuclides and heavy metals in the atmosphere—a global perspective. Regional Congress of the International Radiological Protection Association, Avignon.

Jeandel, C., Martin, J.-M. and Thomas, A. J. (1980). Plutonium and other artificial radionuclides in the Seine Estuary and adjacent areas. In *Techniques for Identifying Transuranic Speciation in Aquatic Environments*: 89–102. IAEA-STI/TUB-613. IAEA, Vienna.

Jeandel, C., Martin, J. M. and Thomas, A. J. (1981). Les radionucleides artificiels dans les estuaires français. In *Impacts of Radionuclide Releases into the Marine Environment*. 15–33. IAE-SM-248/123. IAEA, Vienna.

Jenkin, M. E., Cox, R. A. and Candeland, D. E. (1985). Photochemical aspects of tropospheric iodine behaviour. *J. Atmos. Chem.*, **2**, 359–375.

Johanson, K. J., Bergstrom, R., von Bothmer, S. and Karlen, G. (1990). Radiocesium in wildlife of a forest ecosystem in Central Sweden. In Desmet, G., Nassimbeni, P. and Belli, M. (Eds) *Transfer of Radionuclides in Natural and Semi-natural Environments*: Elsevier, London and New York.

Johnson, L. F., and Larkins, W. F. (1990). The generation, storage, treatment and disposal of nuclear waste at BNFL Sellafield, In Dunderdale, J. (Ed.) *Energy and the Environment*: pp. 90–102. Royal Society of Chemistry, Cambridge.

Jones, D. G., Miller, J. M. and Roberts, P. D. (1984). The distribution of caesium-137 in surface and inter-tidal sediments from the Solway Firth. *Mar. Pollut. Bull.*, **15**, 187–194.

Jones, I. S. and Pond, S. F. (1967). Some experiments to determine the resuspension factor of plutonium from various surfaces. In Fish, B. R. (Ed.) *Surface Contamination*: Pergamon Press, Oxford.

Jordan, C. F., Kline, J. R. and Sasscer, D. R. (1973). A simple model of strontium and manganese dynamics in a tropical rain forest. *Health Phys.*, **24**, 477–489.

Joshi, S. R. (1988). West Valley plutonium and americium-241 in Lake Ontario sediments off the mouth of Niagara River. *Water Air Soil Pollut.*, **42**, 159–168.

Jost, D. J., Gaggeler, H. W., Baltensperger, U., Zinder, B. and Haller, P. (1986). Chernobyl fallout in size-fractional aerosol. *Nature*, **324**, 22–33.

Jouve, A. (1990). Essai de modelisation du facteur de transfert sol plante du strontium 85. Note Ressac 14/90. CEN, Cadarache.

Jumars, P. A., Novell A. R. M. and Self R. F. L. (1981). A simple model of flow-sediment-organism interaction. *Mar. Geol.*, **42**, 155–172.

Kautsky, H. (1988). Determination of distribution processes, transport routes and transport times in the North Sea and the northern North Atlantic using artificial radionuclides as tracers. In Guary, J. C., Gueguenat, P. and Pentreath, R. J. (Eds) *Radionuclides: A Tool for Oceanography*: 271–280. Elsevier, London.

Keith, L. H. (Ed.). (1988). *Principles of Environmental Sampling*. American Chemical Society Professional Reference Book, Washington.

Kelly, M. and Emptage, M. (1991). *Radiological Assessment of the Estuary at Ravenglass: Distribution of Radioactivity and its Relationship to Sedimentary Processes*. Department of Environment, UK.

Kelly, M., Emptage, M., Mudge, S., Bradshaw, K. and Hamilton-Taylor, J. (1991). The relationship between sediment and plutonium budgets in a small macrotidal estuary: Esk Estuary, Cumbria, UK. *J. Environ. Radioact.*, **13**, 55–74.

Kemeny, J. G. (1979). Report of the President's Commission on the Accident at Three Mile Island. US Government Printing Office, Washington DC.

Kershaw, P. J., Swift, D. J., Pentreath, R. J. and Lovett, M. B. (1984). The incorporation of plutonium, americium and caesium into the Irish Sea seabed by biological activity. *Sci. Tot. Environ.*, **40**, 61–81.

Kershaw, P. J., Swift, D. J. and Denoon, D. C. (1988). Evidence of recent sedimentation in the eastern Irish Sea. *Mar. Geol.*, **85**, 1–14.

Kershaw, P. J., Pentreath, R. J., Woodhead, D. S. and Hunt, G. J. (1992). Radioactivity in the Irish Sea—A Review. Aquatic Environ. Monit. Rep. Ser., MAFF Direct. Fish. Res. Lowestoft, No. 31.

Key, R. M., Stallard, R. F., Moore, W. S. and Sarmiento, J. L. (1985). Distribution and flux of ^{226}Ra and ^{228}Ra in the Amazon River estuary. *J. Geophys. Res.*, **90**, 6995–7004.

Khitrov, L. M. (1991). Special lecture on hot particles. *Goechemical Pathways of Artificial Radionuclides Migration in the Biosphere*. USSR Academy of Sciences (In preparation).

Kim, C. K., Oura, Y., Takaku, Y., Nitta, H., Igarashi, Y. and Ikeda, N. (1989). Measurement of 240Pu/239Pu ratio by fission track method and inductively coupled plasma mass spectrometry. *J. Radioanal. Nucl. Chem., (Lett.)*, **136**(5), 353–362.

King, G. (1983). Sulfate reduction in Georgia saltmarsh soils: an evaluation of pyrite formation using ^{35}S and ^{55}Fe tracers. *Limnol. Oceanogr.*, **28**, 987–995.

Kirchmann, R., Van Gelder-Bonnijns, E., Gillard, J. and Declerq-Versele, H. (1985). Evolution of distribution in the Scheldt Estuary of radium-226 released from industrial effluents. In *The Behaviour of Radionuclides in Estuaries*: 123–142. XII/380/85-EN. Commission of the European Community, Luxembourg.

Kirk, G. J. D. and Staunton, S. (1989). On predicting the fate of radioactive caesium in soil beneath grassland. *J. Soil Sci.*, **40**, 71–84.

Kirton, J. A., Coughtrey, P. J. and Mitchell, N. G. (1990). Derivation of soil to plant concentration ratios for upland plant species in the glasshouse and comparison with field and experimental measurements. In Desmet, G., Nassimbeni, P. and Belli, M. (Eds) *Transfer of Radionuclides in Natural and Semi-natural Environments*: 267–274. Elsevier, London.

Klechkovskii, V. M., Polikarpov, G. G. and Aleksakhin, R. M. (1971). *Radioecology*. John Wiley & Sons, New York.

Kline, J. R., Colon, J. A. and Brar, S. S. (1973). Distribution of ^{137}Cs in soils and vegetation of the island of Puerto Rico. *Health Phys.*, **24**, 469–475.

Knoll, G. F. (1989). *Radiation Detection and Measurement*, 2nd edn. John Wiley, New York.

Knox, J. B., Dickerson, M. H., Greenly, G. D., Gudiksen, P. H. and Sullivan, T. J. (1981). *The Atmospheric Release Advisor Capability (ARAC): Its use during and after the Three Mile Island Accident*. Lawrence Livermore National Laboratory Report UCRL-58194.

Konoplyov, A. V. and Bobovnikova, T. S. (1991). Comparative analysis of chemical forms of long-lived radionuclides and their migration and transformation in the environment following the Kyshtym and Chernobyl accidents. *Environmental Impact of Radionuclides Released During Three Major Accidents: Kyshtym, Windscale and Chernobyl*. CEC, Luxembourg.

Konoplyov, A. V. and Bulgakov, A. A. (1991). Behaviour of radionuclides in wetland. RE.24.91. Paper presented at SCOPE–RADPATH Review and Synthesis Meeting, 13–19 April. University of Essex, UK.

Konoplyov, A. V., Borzilov, V. A., Bobovnicova, T. I., Virchenki, Y. P., Popov, V. E., Kutnyakov, I. V. and Chumichev, U. B. (1988). Distribution of radionuclides in the soil–water system fallen out in consequence of the Chernobyl failure. *Meteorol. Gidrol.*, **12**, 63–74.

Koranda, J. J., Robison, W. L., Thompson, S. E. and Stuart, M. L. (1978). *Enewetak Radioecology Research Program: I. Ecological Studies on Engebi Island 1975–76*. Lawrence Livermore National Laboratory, CA. UCRL-52409-1.

Kratochvil, B. and Taylor, J. K. (1981) Sampling for chemical analysis (Report). *Anal. Chem.*, **53**, 18, 924A–938A.

Krause, W. J., and Mundschenk, H. (1989). Determination of flow times and longitudinal dispersion coefficients in the Main River using ^{3}HHO as tracer. In Feldt, W. (Ed.) *The Radioecology of Natural and Artificial Radionuclides*: 61–66. Fachverb. Strahllenschutz e.V., 22.

Krishnaswami, S., Baskaran, M., Fowler, S. W. and Heyraud M. (1985). Comparative role of salps and other zooplankton in the cycling and transport of selected elements and natural radionuclides in the Mediterranean waters. *Biogeochem.*, **1**, 353–360.

Kulikov, N. V. and Molchanova, I. V. (1982). *Continental Radioecology*. Plenum Press, Moscow.

Lafaille, C., (1980). Radioprotection in the La Hague establishement. *Radioprotection*, **15**(4), 215–230. (In French.)

Lal, D., Malhotra, P. K. and Peters, B. (1958). On the production of radioisotopes in the atmosphere by cosmic radiations and their application to meteorology. *J. Atmos. Terrestr. Phys.*, **12**, 306–328.

Lally, A. E. and Glover, K. M. (1984). Source preparation in alpha spectrometry. *Nucl. Instr. Meths. Phys. Res.*, **223**, 259–265.

Lamarsh, J. R. (1966). *Introduction to Nuclear Reactor Theory*: 93. Addison-Wesley, Reading, Mass.

Lambert, C. E., Linden, P., Lowenthal, P. and Stern M. (1983). Benthic fronts and global excess radon distributions. *Geophys. Astrophys. Fluid Dyn.*, **25**, 309–315.

Lambert, C. E., Bischop, J. K. B., Biscaye, P. E. and Chesselet, R. (1984). Particulate aluminium, iron and manganese at the deep Atlantic boundary layer. *Earth Planet. Sci. Lett.*, **70**, 237–348.

Lange, R. (1978). A 3-dimensional particle-in-cell model for the dispersal of atmospheric pollutants and its comparison to regional tracer studies. *J. Appl. Meteorol.*, **17**, 320.

Lange, R., Dickerson, M. H., Gudiksen, P. H. (1992). *Dose Estimates from the Chernobyl Accident*. Lawrence Livermore National Laboratory Report UCRL-96934, Nuclear Technology. (In press).

Lassey, K. R. (1983). The interception and retention of aerosols by vegetation. II. A filtration model description for various aerosols on grassland. *Atmos. Environ.*, **17**, 2303–2310.

Lembrechts, J. F. (1991). Migration and cycling of radionuclides in the soil environments. IUR Summer Course Book.

Lerman, A. and Lietzke, T. A. (1975). Uptake and migration of tracers in lake sediments. *Limnol. Oceanogr.*, **20**, 497–510.

Li, Y. H. and Chan, L-H. (1979). Desorption of Ba and Ra-226 from river-borne sediments in the Hudson Estuary. *Earth Planet. Sci. Lett.*, **43**, 343–350.

Li, Y. H., Santschi, P. H., O'Hara, P., Amdurer, M. and Doering, P. (1989). The importance of a benthic ecosystem to the removal of radioactive trace elements from coastal waters. *Environ. Technol. Lett.*, **10**, 57–70.

Liden, K. (1961). Caesium-137 burdens in Swedish laplanders and reindeer. *Acta Radiol.*, **56**, 237–240.

Lindner, G., Pfeiffer, W., Robbins, J. A. and Recknagel, E. (1989). Long-lived Chernobyl radionuclides in Lake Constance: speciation, sedimentation and biological transfer. In Feldt, W. (Ed.) *Radioecology of Natural and Artificial Radionuclides*: 295–300. Fachverb. Strahllenschutz e.V., 22.

Linsalata, P., Simpson, H. J., Olsen, C. R., Cohen, N. and Trier, R. M. (1985). Plutonium and radiocaesium in the water column of the Hudson Estuary. *Environ. Geol. Sci.*, **7**, 193–204.

Linsley, G. S. (1978). *Resuspension of the Transuranium Elements—A Review of Existing Data*. NRPB-R75. HMSO, London.

Little, P. (1979). Particle capture by natural surfaces. *Agric. Aviat.*, **20**, 129–144.

Little, P. and Wiffen, R. D. (1977). Emission and deposition of petrol engine exhaust Pb.I. Deposition of exhaust Pb to plant and soil surfaces. *Atmos. Environ.*, **11**, 437–447.

Livens, F. R. and Loveland, P. J. (1988). The influence of soil properties on the environmental mobility of caesium in Cumbria. *Soil Use Mgmt.*, **4**, 69–75.

Livens, F. R., Baxter, M. S. and Allen, S. E. (1986). Physico-chemical associations of plutonium in Cumbrian soils. In Bulman, R. A. and Cooper, J. R. (Eds) *Speciation of Fission and Activation Products in the Environment*: 143–50. Elsevier, Amsterdam.

Livens, F. R., Horrill, A. D. and Singleton, D. L. (1991). Distribution of radiocaesium in the soil-plant systems of upland areas of Europe. *Health Phys.*, **60**(4), 539–545.

Lloyd, A. (1981). Fire at French nuclear plant leaks radiation. *New Sci.*, **89**(1236), 125.

Loevinger, R., Japha, E. M. and Brownell, G. L. (1956). Discrete radioisotope sources. In Hine, G. J. and Brownell, G. L. (Eds): *Radiation Dosimetry*: 693–799. Academic Press, New York.

Lowe, B. G. (1978). Levels of ^{137}Cs in soils and vegetation of West Malaysia. *Health Phys.*, **34**, 439.

Lowe, V. P. W. and Horrill, A. D. (1986). Transfer of radionuclides to man from Greylag Geese *Anser anser* and Wigeon *Anas penelope* grazing the saltmarshes at Ravenglass. *J. Environ. Radioact.*, **4**, 101–121.

Lowe, V. P. W. and Horrill, A. D. (1988). Ecological half-life of caesium in Roe deer (*Capreolus capreolus*). *Environ. Pollut.*, **54**, 81–87.

Lowe, V. P. W. and Horrill, A. D. (1991). Caesium concentrations in wild herbivores and the fox (*Vulpes vulpes*). *Environ. Pollut.*, **70**, 93–107.

Lund, L., Michelsen, O. B., Ottar, B. and Wik, T. (1962). A study of Sr-90 and Cs-137 in Norway 1957–1959. Norwegian Defence Research Establishment. Internal Report No. K-253. Kjeller, Norway.

Mackenzie, A. B. and Scott, R. D. (1982). Radiocaesium and plutonium intertidal sediments from southern Scotland. *Nature. Lond.*, **299**, 613–616.

Maeda, M. and Windom, H. L. (1982). Behaviour of uranium in two estuaries of the south-eastern United States. *Mar. Chem.*, **11**, 427–436.

Malcolm, S. J., Kershaw, P. J., Lovett, M. B. and Harvey, B. R. (1990). The interstitial water chemistry of $^{239,240}Pu$ and ^{241}Am in the sediments of the north-east Irish Sea. *Geochim. Cosmochim. Acta*, **54**, 29–35.

Malloch, A. J. C. (1972). Salt-spray deposition on the maritime cliffs of the Lizard Peninsula. *J. Ecol.*, **60**, 103–112.

Malmquist, P-A. (1978). Atmospheric fallout and street cleaning—effects on urban storm-water and snow. *Prog. Wat. Tech.*, **10**, 495–505.

Mann, W. B., Rytz, A. and Spernol, A. (1988). *Radioactivity Measurements: Principles and Practice*. Pergamon Press, Oxford.

Maringer, F. J., Rank, D. and Terlunen, J. (1989). Radioactivity in the River Danube—special investigations. In Feldt, W. (Ed.) *The Radioecology of Natural and Artificial Radionuclides*: 332–337. Fachverb. Strahllenschutz e.V., 22.

Martin, J-M., Nijampurkar, V. and Salvadori, F. (1978). Uranium and thorium isotope behaviour in estuarine systems. In Goldberg, E. D. (Ed.) *Biogeochemistry of Estuarine Sediments*: 111–127. Proc. UNESCO/SCOR Workshop, Melreux, Belgium. UNESCO, Paris.

Martin, J-M., Thomas, A. J. and Jeandel, C. (1981). Transferts atmospheriques des radionucleides artificiels de la mer vers le continent. *Oceanol. Acta*, **4**, 263–6.

Martin, M. H. and Coughtrey, P. J. (1982). *Biological Monitoring of Heavy Metal Pollution*. Elsevier, London.

Mascanzoni, D., (1990). Uptake of ^{90}Sr and ^{137}Cs by mushrooms following the Chernobyl accident. In Desmet, G. Nassimbeni, P and Belli, M. (Eds) *Transfer of Radionuclides in Natural and Semi-natural Environments*: pp. 459–467. Elsevier, London.

McCartney, M., Kershaw, P. J., Woodhead, D. S. and Denoon, D. S. (1992). Artificial radionuclides in the surface sediment of the Irish Sea, 1968–1988. *Sci. Total Environ.*

McKay, W. A. and Pattenden, N. J. (1990). The transfer of radionuclides from sea to land via the air: A review. *J. Environ. Radioact.*, **12**, 49–77.

Medvedev, Z. A. (1976). Two decades of dissidence. *New Sci.*, **72**(1025), 264–267.

Medvedev, Z. A. (1979). *Nuclear Disaster in the Urals*. Angus & Robertson, London.

Medvedev, Z. (1990). Report on Kyshtym visit and Moscow seminar. Presentation at SCOPE-RADPATH First Case-study Meeting, 26–30 March 1990, University of Lancaster.

Middleton, L. J., Handley, R. and Overstreet, R. (1960). Relative uptake and translocation of potassium and caesium in barley. *Plant Physiol.*, **35**(6), 913–918.

Milbourn, G. M. and Taylor, R. (1965). The contamination of grassland with radioactive strontium 1. Initial retention and loss. *Radiat. Bot.*, **5**, 337–347.

Miller, C. W. (1980). An analysis of measured values for the fraction of a radioactive aerosol intercepted by vegetation. *Health Phys.*, **38**, 705–719.

Mishra, U. C. (1991). Solid state nuclear track detection technique and its dosimetric applications. In Garcia-Leon, M. and Madurga, G. (Eds) *Low-level Measurements of Man-Made Radionuclides in the Environment*: 72–92. World Scientific, Singapore.

Mishra, U. C. and Sadavisan, S. (1972) Fallout radioactivity in Indian soils. *Health Phys.*, **23**, 55–62.

Mitchell, N. G. and Davis, A. (1990). Specification of models for upland environments. ANS Report No. 2199-RI. ANS Consultants Ltd, Epsom.

Mitchell, N. T. (1967–77). Radioactivity in surface and coastal waters of the British Isles, Aquat. Environ. Monit. Rep. Ser. MAFF Direct. Fish. Res. Lowestoft, no. 1, 2, 5, 7, 8, 9, 10, 12, 13.

Mitchell, P. I., Vives-Batlle J., Ryan, T. P., Schell, W. R., Sanchez-Cabeza J. A. and Vidal-Quadras, A. (1991). In Kershaw, P. J. and Woodhead, D. S. (Eds) *Radionuclides in the Study of Marine Processes*: 37–51. Elsevier, London.

Mittelstaedt, E. (1986). NOAMP Final Report. Deutschen Hydro. Inst., Hamburg.

Mollah, A. S., Rahman, M. M. and Husain, S. R. (1986). Distribution of γ-emitting radionuclides in soils at the Atomic Energy Research Establishment, Savor, Bangladesh. *Health Phys.*, **50**, 85–8.

Monte, L., Quaggia, S., Pompei, F. and Fratarcangeli, S. (1990) The behaviour of ^{137}Cs in some edible fruits. *J. Environ. Radioact.*, **11**, 207–214.

Moore, D. G. and Scott, M. R. (1986). Behaviour of 226Ra in the Mississippi mixing zone. *J. Geophys. Res.*, **90**, 14317–29.

Moore, K. S., Naegle, S. R. and Bradley, W. G. (1977). Plutonium-239 and Americum-241 contamination of small vertebrates in NAEG study areas at NTS and TTR. In White, M. G., Dunaway, P. B. and Howard, W. A. (Eds) *Environmental Plutonium on the Nevada Test Site and Environs*: 193–217. US DOE report, NVO-171, UC-2. US Department of Energy, NTIS, Springfield.

Morgan, J. E. (1990a). Review of biosphere information: retention and mobility of artificial radionuclides in soil. UK Nirex Ltd, Harwell. NSS R217.

Morgan, J. E. (1990b). Retention and mobility of artificial radionuclides in soil. UK Nirex Ltd, Harwell. NSS/R217.

Mudge, S., Hamilton-Taylor, J., Kelly, M. and Bradshaw, K. (1988). Laboratory studies of the chemical behaviour of plutonium associated with contaminated estuarine sediments. *J. Environ. Radioact.*, **8**, 217–237.

Muller, R.N., Sprugel, D.G., Wayman, C.W., Bartelt, G.E. and Bobula, C.M. (1977). Behavior and transport of industrially derived plutonium in the Great Miami River, Ohio. *Health Phys.*, **33**, 411–416.

Munson, R. D. (Ed). (1985). Potassium in agriculture. 1223. American Society of Agronomy, Madison, Wisconsin, USA.

Myers, D. K. (1989). The general principles and consequences of environmental radiation exposure in relation to Canada's nuclear fuel waste management concept. Report no. AECL-9917: 1-96, Atomic Energy of Canada Ltd, Chalk River.

Myttenaere, C. (1972). Absorption of radiocesium by flooded rice: relative importance of roots and shoot base in the transfer of radioactivity. *Plant Soils*, **36**, 215–18.

Myttenaere, C., Bittel, R., Mousny, J. M. and Dabin, P. (1975) Transfer of conventional and nuclear pollutants to irrigated rice. *Radioprotection*, **10**, 235–252.

NCRP (1976). Report No. 50. Environmental radiation measurements. National Council on Radiation Protection and Methods, Washington, DC.

NCRP (1978). Report No. 58. A Handbook of Radioactivity Measurements Procedures. National Council on Radiation Protection and Measurements. Washington, DC.

NCRP (1992). Effects of ionizing radiation on aquatic organisms. NCRP Report No. 109, NCRP, Bethesda, Maryland.

NEA (1980). *Review of the Continued Suitability of One Dumping Site for Radioactive Waste in the Northeast Atlantic*. NEA/OECD.

NEA/OECD (1983). *Sorption: Modelling and Measurement for Nuclear Waste Disposal Studies*. Summary of a Nuclear Energy Agency workshop, held 6–7 June 1983 in Paris. NEA/OECD Report RWM-4.

NEA (1985). *Review of the Continued Suitability of the Dumping Site for Radioactive Waste in the Northeast Atlantic*. NEA/OECD.

NEA (1986). Nuclear Reactor Accident Source Term Report by an NEA Group of Experts, March.

NEA (1988). *Feasibility of Disposal of High-level Radioactive Waste into the Seabed*. 8 volumes. NEA, Paris.

NEA (1991) *Influence of Seasonal and Meteorological Factors on Nuclear Emergency Planning*. NEA, Paris.

Nellis, P. M. (1990). The transfer of radionuclides from sea to land in sea spray. PhD thesis, University of Edingburgh.

Nelson, D. M. and Lovett, M. B. (1978). Oxidation state of plutonium in the Irish sea. *Nature*, **276**, 599–601.

Nelson, D. M. and Orlandini, K. A. (1986). The role of natural dissolved organic compounds in determining the concentrations of americium in natural waters. In Bulman, R.A. and Cooper, J.R. (Eds) *Speciation of Fission and Activation Products in the Environment*: 262–268. Elsevier, London.

Nelson, D. M., Larsen R. P. and Penrose W. R. (1987). Chemical speciation of plutonium in natural waters. In *Environmental Research on Actinide Elements*. Proc. Conf. Hilton Head, S.C., 7–11 November 1983: 27–48. Department of Energy CONF-841142.

Nevessi, A. and Schell, W. R., (1975). Distribution of plutonium and americium in Bikini atoll lagoon. *Health Phys*, **28**, 539–547.

Nevstrueva, M. A., Ramzeav, P. V., Moiseev, A. A., Ibatullin, M. S. and Teplykh, L. A. (1966). The nature of ^{137}Cs and ^{90}Sr transport over the lichen–reindeer–man foodchain. In Aberg, B. and Hungate, F. P. (Eds) *Radioecological Concentration Processes*: 740. Pergamon Press, Oxford.

Ng, Y. C. (1982). A review of transfer factors for assessing the dose from radionuclides in agricultural products. *Nucl. Safety*, **23**, 57–71.

Ng, Y. C., Colsher, C. S. and Thompson, S. E. (1982). *Soil-to-Plant Concentration Factors for Radiological Assessments*. Lawrence Livermore National Laboratory, Livermore, CA, UCID-19436. Prepared for US Nuclear Regulatory Commission, Final Report, NUREG/CR-2975.

Nicholson, K. W. (1987). Deposition of caesium to surfaces of buildings. *Radiat. Prot. Dos.*, **21**, 37–42.

Nicholson, K. W. (1988a). A review of particle resuspension. *Atmos. Environ.*, **22**, 2639–2651.

Nicholson, K. W. (1988b). The dry deposition of small particles: a review of experimental measurements. *Atmos. Environ.*, **22**, 2653–2666.

Nicholson, K. W. (1988c). The retention of ^{134}Cs and ^{137}Cs by roof tile. In Olast, M. and Sinnaeve, J. (Eds). *Joint CEC/OECD(NEA) Workshop on Recent Advances in Reactor Consequence Assessment*. Proceedings of Second Part of Workshop, Rome.

Nicholson, K. W. and Branson, J. R. (1990). Factors affecting resuspension by road traffic. *Sci. Tot. Environ.*, **93**, 349–358.

Nies, H. (1989). Temporal development of the radioactive contamination of the North Sea and Baltic Sea. In *Proceedings of the Commission of the European Communities Seminar on the Radiological Exposure of the Population of the European Community from Radioactivity in North European Marine Waters (Project MARINA)*: 155–170. Bruges, Belgium, 14–16 June, 1989.

Nies, H. (1990). The contamination of the North Sea by artificial radionuclides during the year 1987. *J. Environ. Radioact.*, **11**, 55–70.

Nikipelov, B. V. (1989). Experience in managing the radiological and radioecological consequences of the accidental release of radioactivity which occurred in the Southern Urals in 1957. Paper presented to IAEA, Vienna, November.

Nisbet, A. F. and Lembrechts, J. F. (1990). The dynamics of radionuclide behaviour in soil solution with special reference to the application of countermeasures. In Desmet, G., Nassimebeni, P. and Belli, M. (Eds) *Transfer of Radionuclides in Natural and Semi-natural Environments*: 371–381. Elsevier, London.

Nishita, H., Dixon, D. and Larson, K. H. (1962). Accumulation of Cs and K and growth of bean plants in nutrient solution and soils. *Plant Soil*, **17**(2), 221–242.

Nuclear Engineering (1957). Windscale—The Committee's report. *Nucl. Eng.*, **2**,(21), 495–496, 510–512.

Nye, P. H. and Tinker, P. B. (1977). *Solute Movement in the Soil Root System*. Studies in Ecology 4. Blackwell Scientific Publications, Oxford.

Nyffeler, F. and Simmons, W. (Eds) (1989). *Interim Oceanographic Description of the North-east Atlantic Site for the Disposal of Low-level Radioactive Waste*. NEA/OECD, 3.

OECD. (1983). Decision of the OECD Council of the 22nd July 1977 Establishing a Multilateral Consultation and Surveillance Mechanism for Sea Dumping of Radioactive Waste. OECD.

OECD/NEA (1985). Review of the continued suitability of the dumping site for radioactive waste in the northeast Atlantic. OECD/NEA, Paris.

O'Farrell, T.P. and Gilbert, R.O. (1975). Transport of radioactive materials by Jack rabbits on the Hanford Reservation. *Health Phys.*, **29**, 9–15.

Ogard, A. E., Thompson, J. L., Rundberg, R. S., Wolfsberg, K., Kubik, P. W., Elmore, D.

and Bentley, H. W. (1988). Migration of Chlorine-36 and tritium from an underground nuclear test. *Radiochem. Acta*, **44/45**, 213–217.

Olsen, C. R., Simpson, H. J. and Trier, R. M. (1981). Plutonium, radiocesium and cobalt in sediments of the Hudson River estuary. *Earth Planet. Sci. Lett.*, **55**, 377–392.

Olsen, C. R., Cutshall, N. H., Larsen, I. L., Simpson, H. J., Trier, R. M. and Bopp, R. F. (1984–85). An estuarine fine-particle budget determined from radionuclide tracers. *Geo-Marine Lett.*, **4**, 157–160.

Olsen C. R., Thein, M., Larsen, I. L., Lowry, P. D., Mulholland, P. J., Cutshall, N. H., Byrd, J. T. and Windom, H. L. (1989). Plutonium, lead-210, and carbon isotopes in the Savannah Estuary: riverborne versus marine sources. *Environ. Sci. Technol.*, **23**, 1475–1481.

Olson, J. S. (1965). Equations for caesium transfer in a *Liriodendron* forest, *Health Phys.*, **11**, 1385–1392.

Orlandini, K. A., Penrose, W. R. and Nelson, D. M. (1986). Pu(V) as the stable form of oxidized plutonium in natural waters. *Mar. Chem.*, **18**, 49–57.

Orlandini, K. A., Penrose, W. R., Harvey, B. R., Lovett, M. B. and Findlay, M. W. (1990). Colloidal behavior of actinides in an oligotrophic lake. *Environ. Sci. Technol.*, **24**, 706–712.

Osterberg, C. A., Carey, A. G. and Curl, H. (1963). Acceleration of sinking rates of radionuclides in the ocean. *Nature*, **200**, 1276–1277.

Packham, J. R. and Harding D. J. L. (1982). *Ecology of Woodland Processes*. Edward Arnold, London.

Page, N. P. (1968). The effect of dose-protraction on radiation lethality of large animals. In Brown D. G., Cragle, R. G. and Noona, T. R. (Eds) *Dose Rate in Radiation Biology*. UT-AEC Agricultural Research Laboratory, Oak Ridge, Tennessee, 12.1–12.19.

Palmer, H. E., Hanson, W. C., Griffin, B. I. and Roesch, W.C. (1963). Caesium-137 in Alaskan Eskimos. *Science*, **142**, 64–66.

Pasciak, W., Branagan, E. F., Congel, F. J. and Fairobent, J. E. (1981). A method for calculating doses to the population from Xe-133 releases during the Three Mile Island accident. *Health Phys.*, **40**, 457–465.

Patel, B., Mulay, C. D. and Ganguly, A. K. (1975). Radioecology of Bombay Harbour—a tidal estuary. *Est. Coastal Mar. Sci.*, **3**, 13–42.

Patel, B., Patel, S. and Pawar, S. (1978). Desorption of radioactivity from nearshore sediment. *Est. Coastal Mar. Sci.*, **7**, 49–58.

Pattenden, N. J., Cambray, R. S., Playford, K., Eakins, J. D. and Fisher, E. M. R. (1980). Studies of environmental radioactivity in Cumbria, Part 3: Measurements of radionuclides in airborne and deposited materials. AERE-R9857. HMSO, London.

Pattenden, N. J., Cambray, R. S. and Playford, K. (1989a). Environmental radioactivity in Caithness and Sutherland, Part 4: Radionuclide deposits in coastal soils 1978–1983. *Nucl. Energy*, **28**(2), 111–127.

Pattenden, N. J., McKay, W. A. and Branson, J. R. (1989b). Environmental radioactivity in Caithness and Sutherland, Part 5: Radionuclide soil deposits associated with coastal inlets. *Nucl. Energy*, **28**(4), 235–248.

Pavlotskaya, F. I., Goryachenkova, T. A., Myasoedov, B. F. and Kazinskaya, I. E. (1991). Plutonium behaviour in soils after the South Urals and Chernobyl accidents. *Comparative Assessment of the Environmental Impact of Radionuclides Released During Three Major Accidents: Kyshtym, Windscale and Chernobyl*. CEC, Luxembourg. EUR 13574.

Peirson, D. H. (1988). Artificial radioactivity in Cumbria, summary of an assessment by measurement and modelling. *J. Environ. Radioact.*, **6**, 61–75.

Penrose, W. R., Metta, D. N., Hylko, J. M. and Rinckel, L. A. (1987). The reduction of plutonium (V) by aquatic sediments. *J. Environ. Radioact.*, **5**, 169–184.

Pentreath, R. J. (1988), Radionuclides in the aquatic environment. In Harley, J. H., Schmidt, G. D and Silini, G. (Eds) *Radionuclides in the Food Chain:* 99–119. Springer Verlag, New York.

Pentreath, R. J. and Woodhead, D. S. (1988). Towards the development of criteria for the protection of marine fauna in relation to the disposal of radioactive wastes into the sea. In *Radiation Protection in Nuclear Energy*, Vol. 2: 213–243. IAEA, Vienna.

Pentreath, R. J., Woodhead, D. S. and Jefferies, D. F. (1973). Radioecology of the plaice (*Pleuronectes platessa*, L.) in the northeast Irish Sea. In Nelson, D. J. (Ed.) *Radionuclides in Ecosystems*, Vol. 2: 731–737. USAEC, Oak Ridge.

Pentreath, R. J., Woodhead, D. S., Harvey, B. R. and Ibbett, R. D. (1980). A preliminary assessment of some naturally-occurring radionuclides in marine organisms (including deep sea fish) and the absorbed dose resulting from them. In *Marine Radioecology*: 291–302. OECD/NEA, Paris.

Pentreath, R. J., Harvey, B. R. and Lovett, M. B. (1986). Chemical speciation of transuranium nuclides discharged into the marine environment. In Bulman, R. A. and Copper, J. R. (Eds): *Speciation of Fission Products in the Environment*: 312–325. Elsevier, London.

Persson, C., Rodhe, H. and de Geer, L. E. (1987). The Chernobyl accident—a meteorological analysis of how radionuclides reached and were deposited in Sweden. *Ambio*, **16**, 20–31.

Peute, A., de Meijer, R. J. and Put L. W. (1988). Radiological impact of releases by the Dutch non-nuclear industrial processes. *Radiat. Prot. Dosim.*, **24**(1/4), 425–429.

Phillips, F. M., Davis, S. N. and Kubik, P. (1990). A proposal to use chlorine-36 for monitoring the movement of radionuclides from nuclear explosions. *Groundwater Monit. Rev.*, **10**, 106–113.

Pinder, J. E., McLeod, K. W. and Adriano, D. C. (1989). The accuracy of some simple models for predicting particulate interception and retention in agricultural systems. *Health Phys.*, **56**, 441–450.

Pitty, A. F. (1979). *Geography and Soil Properties*. Methuen & Co., London.

Polikarpov, G. G., Kulebakina, L. G., Timoshchuk, V. I. and Stokozov, N. A. (1991). ^{90}Sr and ^{137}Sr in surface waters of the Dniepr River, the Black Sea and the Aegean Sea in 1987 and 1988. *J. Environ. Radioact.*, **13**, 25–38.

Postma, H. (1980). Sediment transport and sedimentation. In Olausson, E. and Cato, I. (Eds) *Chemistry and Biogeochemistry of Estuaries*: 153–186. John Wiley & Sons, Chichester.

President's Commission on the Accident at Three Mile Island (1979). Report of the Task Force on Public Health and Safety. US Government Printing Office, Washington, DC.

Presley, B. J. and Trefry, J. H. (1980). Sediment–water interactions and the geochemistry of interstitial waters. In Olausson, E. and Cato, I. (Eds) *Chemistry and Biogeochemistry of Estuaries*: 187–232. John Wiley & Sons, Chichester.

Prister, B. S. (1991). Agricultural aspects of the radiological situation in contaminated territories following the South Ural and Chernobyl accidents. *Comparative Assessment of the Environmental Impact of Radionuclides Released During Three Major Accidents: Kyshtym, Windscale and Chernobyl*. CEC, Luxembourg. EUR 13574.

Prokhorov, V. M. (1973). Ion diffusion in soils and its role in radionuclide migration. In Klechkovskii, V. M., Polikarpov, G. G. and Alexakhin, R. M. (Eds) *Radioecology*: 103–125. John Wiley & Sons, Chichester.

Prokhorov, V. M. and Ginzberg L. R. (1971). Modelling of the process of migration of radionuclides in forest ecosystems and description of the model. *Soviet J. Ecol.*, **2**(5), 396–402.

Pye, K. (1987). *Aeolian Dust and Dust Deposits*. Academic Press, New York.

Quay, P. D., Broecker, W. S., Hesslein, R. H. and Schindler, D. W. (1980). Vertical diffusion rates determined by tritium tracer experiments in the thermocline and hypolimnion of two lakes. *Limnol. Oceanogr.*, **25**, 201–218.

Rackham, B. D., Knowles, J. F. and Woodhead, D. S. (1992). A comparison of the effects of chronic gamma irradiation on the gonads of *Ameca splendens*, *Xiphophorus maculatus* and *Brachydanio rerio* (Osteichthyes: Teleostei). Fisheries Laboratory, Lowestoft.

Randa, Z., Beneda, J., Horyna J. and Klan J. (1990). Mushrooms significant source of internal contamination by radiocaesium in transfer of radionuclides in natural and semi natural environments. In Desmet, G. Nassimebeni, P. and Belli M. (Eds) *Transfer of Radionuclides in Natural and Semi-Natural Ecosystems*: 169–178. Elsevier, London.

Ratcliffe, D. A. (Ed.). (1977). *A Nature Conservation Review*. Cambridge University Press, Cambridge.

Revitt, D. M. and Ellis, J. B. (1980). Rain water leachates of heavy metals in road surface sediments. *Water Res.*, **14**, 1403–1407.

Rey, L. (1982). *The Arctic Ocean*. Comité Arctique, Monaco.

Robbins, J. A. and Edgington, D. N. (1975). Determination of recent sedimentation rates in Lake Michigan using Pb-210 and Cs-137. *Geochim. Cosmochim. Acta*, **39**, 285–304.

Roberts, L. E. J. (1990). The environmental impact of the nuclear fuel cycle. Presentation at Energy and the Environment Symposium, Leeds.

Robinson, C. A., Haywood, S. M. and Brown, J. (1990). The costs and effectiveness of various decontamination procedures. In Proceedings of CEC Seminar on Methods and Codes for Assessing the Off-site Consequences of Nuclear Accidents, Athens.

Robison, W. L., Conrado, C. L. and Stuart, M. L. (1988). *Radiological Conditions at Bikini Atoll: Radionuclide Concentrations in Vegetation, Soil, Animals, Cistern Water, and Ground Water*. Lawrence Livermore National Laboratory, Livermore, CA. UCRL-53840.

Robson, A. (1983). A comparison of the radiological impact of coal and nuclear fuel cycles in the UK. UK Central Electricity Generating Board, Internal Report TPRD/GS/E/1/83.

Roed, J. (1987a). Dry deposition in urban and rural areas in Denmark. *Radiat. Prot. Dos.*, **21**, 33–36.

Roed, J. (1987b). Runoff from and weathering of roof material following the Chernobyl reactor accident. *Radiat. Prot. Dos.*, **21**, 59–63.

Roed, J. (1988). *Using Decontamination from the Chernobyl Debris to Guide Decontamination Planning*. American Chemical Society, San Diego.

Roed, J. and Andersson, K. (1991). Strategy on reclamation of nuclear contaminated urban areas. Paper RE.42.91 presented at SCOPE–RADPATH Review and Synthesis Meeting, University of Essex, UK, 13–19 April.

Roed, J. and Cannell, R. J. (1987). Relationships between indoor and outdoor aerosol concentration following the Chernobyl reactor accident. *Radiat. Prot. Dos.*, **21**, 107–110.

Roed, J. and Goddard, A. J. H. (1990). Ingress of radioactive material into dwellings. In Proceedings of CEC Seminar on Methods and Codes for Assessing the Off-site Consequences of Nuclear Accidents, Athens.

Roed, J. and Jacob, P. (1990). Deposition on urban surfaces and subsequent weathering. In Proceedings of CEC Seminar on Methods and Codes for Assessing the Off-site Consequences of Nuclear Accidents, Athens.

Rommelt, R., Hiersche L., Schaller G. and Wirth E. (1990). Influence of soil fungi (basidiomyctes) on the migration of Cs-134+Cs-137 and Sr-90 in coniferous forest soils. In Desmet, G., Nassimebeni, P. and Belli M. (Eds) *Transfer of Radionuclides in Natural and Semi-Natural Environments*: 152–160. Elsevier, London.

Romney, E. M. and Wallace, A. (1976). Plutonium contamination of vegetation in dusty environments. In White, M. G. and Dunaway, P. B. (Eds) *Transuranics in Natural Environments*: 287–302. NVO-178. USERDA, NTIS, Springfield.

Ronneau, C. (1990). Transfer of radioactivity in forest ecosystems. Report of Contract No. 90-PR-008, CEC DGXI.

Ronneau, C., Cara, J. and Apers, D. (1987). The deposition of radionuclides from Chernobyl to a forest in Belgium. *Atmos. Environ.*, **21**, 1467–1468.

Ross, H. H. (1969). Measurement of a-emitting nuclides using Cerenkov radiation. *Anal. Chem.*, **41**, 1260–1265.

Rotblat, J. (1981). *Nuclear Radiation in Warfare*. Stockholm International Peace Research Institute, Taylor & Francis, London.

Rougeau, J. P. and Lallement, R. (1990). *The Nuclear Fuel Cycle: A Mature Advancing Industry*. 7th Pacific Basin Nuclear Conference, 4–8 March 1990, San Diego, California, USA.

Russell, E. W. (1973). *Soil Conditions and Plant Growth*. 10th edn. Longman, London.

Rutgers van der Loeff M. M. and Lavaleye M. S. S. (1986). *Sediments Fauna and the Dispersal of Radionuclides at the Northeast Atlantic Dumpsite for Low-level Radioactive Waste*. Report of the Dutch DORA program. Netherlands Institute for Sea Research.

Salisbury, E. (1952). *Downs and Dunes. Their plant life and its environment*. Bell, London.

Salo, A. (1983). Personal communication. Finnish Centre for Radiation and Nuclear Safety. Helsinki, Finland.

Sandalls, F. J. (1987a). Removal of radiocaesium from urban surfaces. *Radiat. Prot. Dos.*, **21**, 137–140.

Sandalls, F. J. (1987b). Radiocaesium on urban surfaces after Chernobyl. *Radiat. Prot. Dos.*, **21**, 65–68.

Sandalls, F. J., Stewart, S. P. and Wilkins B. T. (1986). Natural and forced decontamination of urban surfaces contaminated with radiocaesium. In Luykx, F. and Sinnaeve J. (Eds). *CEC/LUX Workshop on Methods for Assessing the Off-site Radiological Consequences of Nuclear Accidents.*

Sandalls, F. J., Eggleton, A. E. J. and Gaudern, S. L. (1988). Uptake of radiocaesium by upland herbaceous vegetation in relation to soil type. UKAEA Report AERE R 13389.

Santschi, P. H., Nyffeler U. P., Anderson R. F., Schiff S. L., O'Hara P. and Hesslein R. H. (1986). Response of radioactive trace metals to acid-base titrations in controlled experimental ecosystems: evaluation of transport parameters for application to whole-lake radiotracer experiments. *Canad. J. Fish. Aquat. Sci.*, **43**, 60–77.

Santschi, P. H., Hoehn, E., Lueck, A. and Farrenkothen, K. (1987). Tritium as a tracer for the movement of surface water and groundwater in the Glatt Valley, Switzerland. *Environ. Sci. Technol.*, **21**, 909–916.

Santschi, P. H., Bollhalder, S., Zingg, S., Luck, A. and Farrenkothen, K. (1990). The self-cleaning capacity of surface waters after radioactive fallout. Evidence from European waters after Chernobyl, 1986–1988. *Environ. Sci. Technol.*, **24**, 519–527.

Sartor, J. D. and Gaboury, D. R. (1984). Street sweeping as a water pollution control measure: lessons learned over the past 10 years. *Sci. Tot. Environ.*, **33**, 171–183.

Saxen, R. (1990). Radioactivity of surface water and freshwater fish in Finland in 1987. Supplement 3 to Annual Report STUK-A74. STUK-A77. Finnish Centre for Radiation and Nuclear Safety, Helsinki, Finland.

Saxen, R. and Aaltonen, H. (1987). Radioactivity of Surface Water in Finland after the Chernobyl Accident in 1986. Supplement 5 to annual report STUK-A55. STUK-A60. Finnish Centre for Radiation and Nuclear Safety, Helsinki, Finland.

Saxen, R. and Rantavaara, A. (1987). Supplement 6 to Annual Report STUK-A55. STUK-A61. Centre for Radiation and Nuclear Safety, Helsinki, Finland.

Schell, W. R. and Myttenaere C. (1989). Radionuclide contamination of natural ecosystem with emphasis on forest ecosystems. *A Review of Radionuclide Deposition, Transport and Removal Processes in Forest Ecosystems: Validation of Model Predictions (VAMP).* IAEA, Vienna.

Schell, W. R. and Tobin, M. J. (1990). Deposition and mobility of chemical elements in forest and wetland environments. In Desmet, G., Nassimebeni, P. and Belli, M. (Eds) *Transfer of Radionuclides in Natural and Semi-Natural Environments*: 118–128. Elsevier, London.

Schell, W. R., Tobin, M. J., Mitchell, P. I., Sanchez-Cabeza, J. A. and Vidal-Quadras, A. (1989). The determination of plutonium-241 in samples of *Fucus* from the Irish coast. In Proc. 34th Annual Conf. on Bioassay, Analytical and Environmental Radiochemistry, Las Vegas, 17–21 October.

Schimmack, W., Bunzl, K. and Zelles, L. (1989). Initial rates of migration of radionuclides from the Chernobyl fallout in undisturbed soils. *Geoderma*, **44**, 211–218.

Schlosser, P., Stute, M., Dorr, H., Sonntag, C. and Munich, K. 0. (1988). Tritium/^{3}He dating of shallow groundwater. *Earth Planet. Sci. Lett.*, **89**, 353–362.

Schonhofer, F. and Henrich, E. (1987). Recent progress and application of low-level liquid scintillation counting. *J. Radioanal. Nucl. Chem.*, (*Articles*), **115**, 317–333.

Schubel, J. R. (1974). Effects of tropical storm Agnes on the suspended solids of the northern Chesapeake Bay. In Gibbs, R.J. (Ed.) *Suspended Solids in Water*: 113–132. Plenum Press, New York.

Schubel, J. R. and Carter, M. M. (1977). Suspended sediment budget for Chesapeake Bay. In Wiley, M. (Ed.) *Estuarine Processes*, vol. II: 48–67. Academic Press, New York.

Schultz, V. and Klement, A. W. (1963). *Radioecology. Proceedings of the First National Symposium, Colorado State University, 1961*. Reinhold, New York.

SCOPE. (1989). Pittock, A. B., Ackerman, T. P., Crutzen, P. J., MacCracken, M. C., Shapiro, C. S. and Turco, R. P. (Eds) *Environmental Consequences of Nuclear War*, Vol. I, *Physical and Atmospheric Effects*, 2nd edn 1. John Wiley & Sons Ltd. Chichester.

Scott Russell, R. S. (1966). *Radioactivity and Human Diet*. Pergamon Press, Oxford.

Sedunov, Y. S., Borzilov, V. A., Klepikova, N. V., Troyanova, N. I. and Chernokozhin, E. V. (1990). A set of models for radionuclide transport and its application for estimating source parameters and describing contamination after the Chernobyl accident. Proc. 1st Int. Workshop on past severe accidents and their consequences, 30 Oct.–3 Nov. 1989, Sochi, USSR.

Sehmel, G. A. (1973a). Particle eddy diffusivities and deposition velocities for isothermal flow and smooth surfaces. *Aerosol Sci.*, **4**, 125–138.

Sehmel, G. A. (1973b). Particle resuspension from an asphalt road caused by car and truck traffic. *Atmos. Environ.*, **7**, 291–309.

Sehmel, G. A. (1980a). Particle resuspension: A review. *Environ. Int.*, **4**, 107–127.

Sehmel, G. A. (1980b). Particle and gas dry deposition: A review. *Atmos. Environ.*, **14**, 983–1011.

Sehmel, G. A. (1984). Deposition and resuspension. In Randerson, D. (Ed.) *Atmospheric Science and Power Production*: 533–583. US Department of Commerce, Springfield, Virginia.

Sharma, P., Gardner, L. R., Moore, W. S. and Bollinger, M. S. (1987). Sedimentation and bioturbation in a salt marsh as revealed by ^{210}Pb, ^{137}Cs, and ^{7}Be studies. *Limnol. Oceanogr.*, **32**, 313–326.

Shaw, G. and Bell, J. N. B. (1989). The kinetics of caesium absorption by roots of winter wheat and the possible consequences for the derivation of soil-to-plant transfer factors for radiocaesium. *J. Environ. Radioact.*, **10**, 213–233.

Shaw, G. and Bell, J. N. B. (1991). Competitive effects of potassium and ammonium on caesium uptake kinetics in wheat. *J. Environ. Radioact.*, **13**, 283–296.

Shen, G. T., Sholkovitz, E. R. and Mann, D. R. (1983). The coagulation of dissolved ^{239,240}Pu in estuaries as determined from a mixing experiment. *Earth Planet. Sci. Lett.*, **64**, 437–444.

Shleien, B., Cochran, J. A. and Magno, P. J. (1970). Strontium-90, strontium-89, plutonium-

239 and plutonium-238 concentrations in ground level air, 1964–1969. *Environ. Sci. Technol.*, **4**, 598–602.

Sholkovitz, E. R. and Mann, D. R. (1984). The pore water chemistry of ^{239,240}Pu and ^{137}Cs in sediments of Buzzards Bay, Massachusetts. *Geochim. Cosmochim. Acta*, **48**, 1107–1114.

Sholkovitz, E. R. and Mann, D. R. (1987). ^{239,240}Pu in estuarine and shelf waters of the north-eastern United States. *Est. Coast. Shelf Sci.*, **25**, 413–434.

Sholkovitz, E. R., Boyle, E. and Price, W. B. (1978). The removal of dissolved humic acids and iron during estuarine mixing. *Earth Planet. Sci. Lett.*, **40**, 130–136.

Sholkovitz, E. R., Carey, A. E. and Cochran, J. K. (1982). Aquatic chemistry of plutonium in seasonally anoxic lake waters. *Nature*, **300**, 159–161.

Sickle, J. V., Weimer, W. C. and Larsen, D. P. (1983). Mixing rates in Shagawa Lake, Minnesota, sediments as determined from ^{106}Ru. *Geochim. Cosmochim. Acta*, **47**, 2189–2197.

Singh, N. P. (1988). Radiochemical procedures for the determination of actinides in biological and environmental samples. In Garcia-Leon, M. and Madurga, G. (Eds) *Proc. 1st International Summer School on Low-Level Measurements and their Application to Environmental Radioactivity*: 471–490. World Scientific, La Rabida, Spain.

Slinn, S. A. and Slinn, W. G. N. (1980). Predictions for particle deposition on natural waters. *Atmos. Environ.*, **14**, 1013–1016.

Sokolov, V. E., Ryabov, I. N., Ryabtsev, I. A., Tikhomirov, F. A., Shevchenko, V. A. and Taskaev, A. I. (1990). Ecological and genetic consequences of the Chernobyl atomic power plant accident. Paper RL.30.90 presented at SCOPE–RADPATH First Case-study Meeting, 26–30 March. University of Lancaster.

Sombre, L., Vanhouche, M., Thiry, Y., Ronneau, C., Lambotte, J. M. and Myttenaere, C. (1990). Transfer of radiocaesium in forest ecosystems resulting from a nuclear accident. In Desmet, G., Nassimebeni, P and Belli, M. (Eds) *Transfer of Radionuclides in Natural and Semi-natural Environments*: 74–83. Elsevier, London.

Spirin, D. A., Romanov, G. N., Tikhomirov, F. A., Smirnov, E. G., Suvorova, L. T. and Shevehenko, V. A. (1990). The consequences of the Kyshtym accident for flora and fauna. Seminar on Comparative Assessment of the Environmental Impact of the Radionuclides Released During Three Major Nuclear Accidents: Kyshtym, Windscale, Chernobyl. CEC, Luxembourg. EUR 13574.

Sprugel, D. G. and Bartelt, G. E. (1978). Erosional removal of fallout plutonium from a large Mid-Western watershed. *J. Environ. Qual.*, **7**, 175–177.

Stadler, J. and Gowan, J. W. (1964). Observations on the effects of continuous irradiation over ten generations on reproductivities of different strains of mice. In Carlson, W. D. and Gassner, F. X. (Eds) *Effects of Ionizing Radiation on the Reproductive System*: 111–122. Macmillan, New York.

Stanners, D. A. and Aston, S. R. (1981). An improved method of determining sedimentation rates by the use of artificial radionuclides. *Est. Coast. Shelf Sci.*, **13**, 101–106.

Stegen de Schrieck, van der, J. and Myttenaere, C. (1992) The K and Ca biochemical cycles in forest ecosystems: a review. *Trends Soil Sci.* (Ed.) Council of Scient. Res. Int. (In press).

Stewart, N. G. and Crooks, R. N. (1958). Long range travel of the radioactive cloud from the accident at Windscale. *Nature*, **182**(4636), 627–628.

Sumerling, T. J. and Darby, S. C. (1981). Statistical aspects of the interpretation of counting experiments designed to detect low levels of radioactivity. National Radiological Protection Board Report, Chilton, UK, NRPB-R113.

Swenson, M. J. (1975). *Duke's Physiology of Domestic Animals*, 8th edn. Comstock Publishing Associates, Ithaca and London.

Tawil, J. J., Bold, F. C., Harrer, B. J. and Currie, J. W. (1985). Off-site consequences for radiological accidents: methods, costs and schedules for decontamination. NUREG/CR-3413, PNL-4790, Pacific Northwest Laboratory.

Ternovskij, I. A., Romanov, G. N., Fedorov, E. A., Teverovskij, E. N. and Kholina, Yu. B. (1989). Radioactive cloud trace formation dynamics after the radiation accident in the South Urals in 1957: migration processes, Paper presented to IAEA, Vienna, November.

Thiry, Y. and Myttenaere, C. (1991) Study on the bioavailability of radiocaesium following contamination of the forest floor. *Environmental Impact of Radionuclides Released During Three Major Nuclear Accidents: Kyshtym, Windscale, Chernobyl*. CEC, EUR 13574, Luxembourg.

Thiry, Y., Sombre, L., Ronneau, C., Myttenaere, C., Kutlahmedov, Y. and Davidchuck, V.S. (1990). Behaviour of Cs-137 in forested polygones of the Chernobyl contamination zone. All-Union Conference Geochemical Pathways of Artificial Radionuclides in Biosphere, 15–19 October, Gomel, USSR.

Thomas, G. W. and Swoboda, A. R. (1970). Anion exclusion effects on chloride movement of soils. *Soil Sci.*, **110**, 163–166.

Thomas, J. M., Hanson, W. C. and Eberhardt, L. L. (1982). Evaluating the compatibility of simulation model output and observed data. *Kybernetes*, **11**, 195–203.

Thommeret, Y. and Huynh Ngoc, L. (1990). Le ^{37}Cs et le ^{134}Cs dans les Sediments Marins Superficiels de la Zone cotière Cannes-Monaco et du Canal de Corse. *Rapport CIESM*, **32**(1), 320.

Thorne, M. C. and Coughtrey, P. J. (1983). Dynamic models for radionuclides in forage and domestic animals. In Coughtrey, P. J., Bell, J. N. B. and Roberts, T. M. (Eds) *Ecological Aspects of Radionuclide Release*: 127–140. Blackwell Scientific Publications, Oxford.

Tikhomirov, F. A. (1990). Urals and Chernobyl accidents: Radioecological impact on forest ecosystems. *Comparative Assessment of the Environmental Impact of Radionuclides Released During Three Major Nuclear Accidents: Kyshtym, Windscale, Chernobyl*. CEC, EUR 13574. Luxembourg.

Tomar, K. P. (1987). Chemistry of pedogenesis in Indo-Gangetic alluvial plains. *Soil Sci.*, **38**, 405–14.

Toole, J., Baxter, M. S. and Thompson, J. (1987). The behaviour of uranium isotopes with salinity change in three U.K. estuaries. *Est. Coast. Shelf Sci.*, **25**, 283–297.

Toole, J., Mackrell, A., Cook G. T. and Baxter, M. S. (1990). An assessment of inland transfer of radioactivity near Dounreay, Scotland. *J. Environ. Radioact.*, **9**, 299–329.

Torgersen, T., Top, Z., Clarke, W. B., Jenkins, W. J. and Broecker, W. S. (1977). A new method for physical limnology—tritium–helium-3–ages—results for Lakes Erie, Huron, and Ontario. *Limnol. Oceanogr.*, **22**, 181–193.

Trabalka, J. R. and Auerbach S. I. (1991). One Western perspective of 1957 soviet nuclear accident. *Comparative Assessment of the Environmental Impact of Radionuclides Released during Three Major Accidents: Kyshtym, Windscale and Chernobyl*: 41–69. Luxembourg, 1–5 October 1990, Vol. 1, Report EUR 13574.

Trabalka, J. R., Auerbach, S. I. and Eyman, L. D. (1980). Technical note: The 1957–1958 Soviet nuclear accident in the Urals. *Nucl. Safety*, **21**(1), 94–99.

Tracy, B. L., Prantl, F. A. and Quinn, J. M. (1984). Health impact of radioactive debris from the satellite Cosmos 954. *Health Phys.*, **47**, 225–233.

Tukey, H. B., Wittwer, S. H. and Tukey, H. B. (1958). Leaching of nutrients from plant foliage as determined by radioisotopes. In *Radioisotopes in Scientific Research*, Vol. 4: 304–322. Pergamon Press, Oxford.

Turner, D. R., Knox, S., Penedo, F., Titley, J. G., Hamilton-Taylor, J., Kelly, M. and Williams, G. (1991). Surface complexation modelling of plutonium absorption on sediments of the Esk Estuary, Cumbria. In Kershaw, P. J. and Woodhead, D. S. (Eds) *Radionuclides in the Study of Marine Processes*: 165–174. Elsevier, London.

Turner, F. B. (1975). Effects of continuous irradiation on animal populations. In Augenstein,

L.G., Mason, R., and Zelle, M. (Eds.) *Advances in Radiation Biology*, vol. 5: 83–144. Academic Press, New York.

Tveten, U. (Ed). (1990a). Environmental consequences of releases from nuclear accidents. A Nordic perspective. Final Report of the NKA Project AKTU-200, 261.

Tveten, U. (1990b). Environmental consequences of releases from nuclear accidents—a Nordic perspective. Nordic Liaison Committee for Atomic Energy. NORD 1990:46. Copenhagen, Denmark.

Tveten, U. (1991). *Radionuclide Behavior in the Environment—The Post-Chernobyl Experience in Norway*. NUREG/CR-5304 (SAND90-7116). Washington DC.

Ugedal, O. and Blakar, I. (1991). Radiocesium in fish and sediments from lakes in Trondelag, Norway. To be published. (In Norwegian.)

Uncles, R. J. (1979). A comparison of the axial, distributions of salt and ^{137}Cs in the Severn Estuary during August 1974. *Est. Coast. Mar. Sci.*, **9**, 585–594.

Underwood, B. Y. (1987). Dry deposition to an urban complex. *Radiat. Prot. Dos.*, **21**, 21–32.

UNSCEAR (1977a). *Natural Sources of Radiation*. United Nations Scientific Committee on Effects of Atomic Radiation. Report, Annex B.

UNSCEAR (1977b). Developmental effects of irradiation *in utero*. In *Sources and Effects of Ionizing Radiation*: 655–725. United Nations Scientific Committee on the Effects of Atomic Radiation 1977 Report to the General Assembly, with annexes. United Nations, New York.

UNSCEAR (1977c). Genetic effects of radiation. In *Sources and Effects of Ionizing Radiation*: 425–564. United Nations Scientific Committee on the Effects of Atomic Radiation 1977 Report to the General Assembly, with annexes. United Nations, New York.

UNSCEAR (1982a). Radiation-induced life shortening. In *Ionizing Radiation: Sources and Biological Effects*: 665–725. United Nations Scientific Committee on the Effects of Atomic Radiation 1982 Report to the General Assembly, with annexes. United Nations, New York.

UNSCEAR (1982b). Non-stochastic effects of irradiation. In *Ionizing Radiation: Sources and Biological Effects*: 571–654. United Nations Scientific Committee on the Effects of Atomic Radiation 1982 Report to the General Assembly, with annexes. United Nations, New York.

UNSCEAR (1982c). Genetic effects of radiation. In *Ionizing Radiation: Sources and Biological Effects*: 425–569. United Nations Scientific Committee on the Effects of Atomic Radiation 1982 Report to the General Assembly, with annexes. United Nations, New York.

UNSCEAR (1986). Biological effects of pre-natal irradiation. In *Genetic and Somatic Effects of Ionizing Radiation*: 263–366. United Nations Scientific Committee on the Effects of Atomic Radiation 1986 Report to the General Assembly, with annexes. United Nations, New York.

UNSCEAR (1988a). *Sources, Effects and Risks of Ionizing Radiation*. United Nations Scientific Committee on the Effects of Atomic Radiation. United Nations, New York.

UNSCEAR (1988b). Exposures from the Chernobyl accident. In *Sources, Effects and Risks of Ionizing Radiation*: 309–374. United Nations Scientific Committee on the Effects of Atomic Radiation 1988 Report to the General Assembly, with annexes. United Nations, New York.

Unsworth, E. F., Pearce, J., McMurray, C. H., Moss, B. W., Gordon, F. J. and Rice, D. (1989). Investigations on the use of clay minerals and Prussian Bluc in rcducing the transfer of dietary radiocaesium to milk. *Sci. Tot. Environ.*, **85**, 339–347.

US DOE (1980). In Hanson, W. C. (Ed.) *Transuranic Elements in the Environment*. DOE/TIC-22800, US Department of Energy, NTIS, Springfield, VA.

US DOE (1985). In Howard, W. A. and Fuller, R. G. (Eds) *The Dynamics of Transuranics and Other Radionuclides in Natural Environments*. NVO-272, DE87014456, U.S. Department of Energy, NTIS, Springfield, VA.

US Geological Survey (1987). *Subsurface Transport of Radionuclides in Shallow Deposits of the Hanford Nuclear Reservation, Washington—Review of Selected Previous Work and Suggestions for Further Study*. US Geological Survey Open-File Report 87–222, Tacoma, Washington.

USSR State Committee on the Utilisation of Atomic Energy (1986). *The Accident at the Chernobyl Nuclear Power Plant and its Consequences*. Safety Series No. 75-INSAG-1. IAEA, Vienna.

Van Middlesworth, L. (1958). Iodine-131 in sheep before and after a nuclear reactor accident. *Nature*, **181**(4640), 256.

Vernon, A. J., Emose, P. N. and Mudakiar, T. (1976). Coconut fertilizer yield trials in Fiji 1970–75. *Fiji Agric. J.*, **38**(2), 49–60.

Veska, E. and Tracy, B. L. (1986). The migration of reactor-produced tritium in Lake Huron. *J. Environ. Radioact.*, **4**, 31–38.

Voitsekhovitch, O. V., Borzilov, V. A. and Konoplyov, A. V. (1991). Hydrological and physico-chemical aspects of the contamination of water bodies after the Chernobyl accident. *Environmental Impact of Radionuclides Released During Three Major Accidents: Kyshtym, Windscale and Chernobyl*. CEC, EUR 13574, Luxembourg.

Voris, Van P. and Dahlman, R.C. (1976). Flood plain data: Ecosystem characteristics and Cs-137 concentrations in biota and soil. USAEC Report ORNL/TM-5526.

Voris, van P., Cowan, C. E., Cataldo, D. A., Wildung, R. E. and Shugart, H. H. (1990). Chernobyl case study: Modelling the dynamics of long-term cycling and storage or Cs-137 in forested Ecosystems. In Desmet, G., Nassimebeni, P. and Belli, M. (Eds) *Transfer of Radionuclides in Natural and Semi-natural Environments*: 61–73. Elsevier, London.

Waber, U., von Gunten, H. R. and Krahenbuhl, U. (1987). The impact of the Chernobyl accident on a river/groundwater aquifer. *Radiochim. Acta*, **41**, 191–198.

Wahlberg, J. S. and Fishman, M. J. (1962). Absorption of caesium on clay minerals. *US Geol. Surv. Bull.* **1140-A**, Washington, DC.

Waldman, J. M., Munger J. W., Jacob, D. J. and Hoffman M. R. (1983). Fogwater composition in Southern California. In Pruppacher, H. R., Semonin, R. G. and Slinn, W. G. N. (Eds) *Precipitation Scavenging*: pp. 101–123. Elsevier, New York.

Walker, M. I., McKay, W. A., Pattenden, N. J. and Liss, P. S. (1986). Actinide enrichment in marine aerosols. *Nature*, **323**(6084), 141–143.

Wallace, A., Romney E. M., Wood, R. A. and Alexander, G. V. (1983). Influence of potassium on uptake and distribution of caesium in bush beans. *J. Plant Nutr.*, **6**(5), 397–403.

Walling, D. E. and Bradley, S. B. (1988). Transport and redistribution of Chernobyl fallout radionuclides by fluvial processes: some preliminary evidence. *Environ. Geochem. Health*, **10**, 35–39.

Ward, G. M. and Johnson, J. E. (1986). Validity of the term transfer coeffient. *Health Phys.*, **50**, 411–414.

Ward, G. M. and Johnson, J. E. (1989). Assessment of milk transfer coefficients for use in prediction models of radioactivity transport. *Sci. Tot. Environ.*, **85**, 287–294.

Warming, L. (1982). Weathering and decontamination of radioactivity deposited on asphalt surfaces. Risø-M-2273. Risø National Laboratory.

Warner, F. E. (1987). Chernobyl—Ethical and environmental considerations. *IEE Proceedings*, **134** (Pt. A10), 834–840.

WASH-1400 (1975). *Reactor Safety Study: An Assessment of Accident Risks in U.S. Commercial Nuclear Power Plants*. NUREG 75/014, Washington, DC.

Watkins, B. M. (1990). Interception, retention absorption and transolocation of radionuclides by vegetation following application in irrigation waters. NSS R221. UK Nirex Ltd, Harwell.

Wheatcroft, R. A., Jumars, P. A., Smith, C. R. and Nowell, A. R. M. (1990). A mechanistic view of the particulate biodiffusion coeffcient: Step lengths, rest periods and transport directions. *J. Mar. Res.*, **48**, 177–207.

Whicker, F. W. and Fraley, L. (1974). Effects of ionizing radiation on terrestrial plant communities. In Augenstein, L.G., Mason, R., and Zelle, M. (Eds.) *Advances in Radiation Biology*, vol. 4; 317–366. Academic Press, New York.

Whicker, F. W. and Schultz, V. V. (1982). *Radioecology, Nuclear Energy and the Environment*, Vols. 1 and 2. CRC Press, Ohio.

White, J. C. and Angelovic, J. W. (1966). Tolerances of several marine species to Co-60 irradiation. *Chesapeake Sci.*, **7**, 36–39.

White, G. C., Hakonson, T. E. and Ahlquist, A. J. (1981) Factors affecting radionuclide availability to vegetables grown at Los Alamos. *J. Environ. Qual.*, **10**, 294–299.

Wilkins, B. T. (1987). The retention behaviour of radiocaesium on common building materials under natural outdoor conditions. *Radiat. Prot. Dos.*, **21**, 69–73.

Winkelmann, I., *et al.* (1987). *Radioactivity Measurements in the Federal Republic of Germany After the Chernobyl Accident.* Institut für Strahlenhygiene des Bundesgesundheitamtes, Neuherberg, Munich.

Winsor, T. F. and Whicker, F. W. (1980). Pocket gophers and redistribution of plutonium in soil. *Health Phys.*, **39**, 257–262.

Wirth, E., Koehler, H. and Burkhardt, J. (1985). The suitability of transfer coefficients used for stochastic calculation in radioecology. *Health Physics*, **49**, 1165–1172.

Witherspoon, J. P. and Taylor, F. G. Jr. (1969). Retention of a fallout simulant containing Cs-134 by pine and oak trees. *Health Phys.*, **17**, 825–829.

Witkamp, P. and Van der Drift, J. (1961). Behaviour of forest litter in relation to environmental factors. *Plant Soil*, **XV**(4), 295–311.

Wogman, N. A. (1981). Design and environmental applications of an ultra-low background, high efficiency intrinsic Ge gamma-ray spectrometer. In: Methods of Low-Level Counting and Spectrometry. IAEA, Vienna, STI/PUB/592, 15–29.

Wolff, W. J. (1980). Biotic aspects of the chemistries of estuaries. In Olausson, E. and Cato, I. (Eds) *Chemistry and Biogeochemistry of Estuaries*: 263–295. John Wiley & Sons, Chichester.

Woodard, K. (1979). Assessment of off-site radiation doses from the Three Mile Island Unit 2 accident. General Public Utilities Report TDR-TMI-116.

Woodhead, D. S. (1970). Assessment of the radiation dose to developing fish embryos due to the accumulation of radioactivity by the egg. *Radiat. Res.*, **43**, 582–597.

Woodhead, D. S. (1971). The biological effects of radioactive waste. *Proc. Roy. Soc. Lond. B*, **177**, 423–437.

Woodhead, D. S. (1973). The radiation dose received by plaice (*Pleuronectes platessa*) from the waste discharged into the northeast Irish Sea from the fuel reprocessing plant at Windscale. *Health Phys.*, **25**, 115–121.

Woodhead, D. S. (1974). The estimation of radiation dose rates to fish in contaminated environments, and the assessment of the possible consequences. In *Population Dose Evaluation and Standards for Man and his Environment*: 555–575. IAEA, Vienna.

Woodhead, D. S. (1977). The effects of chronic irradiation on the breeding performance of the guppy, *Poecilia reticulata* (Osteichthyes: Teleostei). *Int. J. Radiat. Biol.*, **32**, 1–22.

Woodhead, D. S. (1979). Methods of dosimetry for aquatic organisms. In *Methodology for Assessing Impacts of Radioactivity in Aquatic Ecosystems.* Technical Report Series No. 190: 43–96. IAEA, Vienna.

Woodhead, D. S. (1984). Contamination due to radioactive materials. In Kinne, O. (Ed.)

Marine Ecology, vol. 5: *Ocean Management*: Part 3: 11111287. John Wiley & Sons, Chichester.

Woodhead, D. S. (1986). The radiation exposure of the black-headed gull (*Larus ridibundus*) in the Ravenglass estuary, Cumbria, UK: a preliminary assessment. *Sci. Tot. Environ.*, **58**, 273–281.

Woodhead, D. S. (1988). Mixing processes in near-shore marine sediments as inferred from the distribution of radionuclides discharged into the northeast Irish Sea from BNFL, Sellafield. In Guary, J. C., Gueguenіat, P. and Pentreath, R. J. (Eds) *Radionuclides: A Tool for Oceanography*: 331–340. Elsevier, London.

Woodwell, G. M. and Rebuck, A. L. (1967). Effects of chronic gamma radiation on the structure and diversity of an oak-pine forest. *Ecol. Monogr.*, **37**, 53–69.

Yaron, B., Dagan, G. and Goldshmid, J. (1984). *Pollutants in Porous Media. The Unsaturated Zone Between Soil Surface and Groundwater*. Springer-Verlag, Berlin.

Young, P. C. (1990). Quantitative systems methods in evaluation of environmental pollution problems. In Harrison, R.M. (Ed.) *Pollution: Causes, Effects and Control*: 367–388. Royal Society of Chemistry, Cambridge.

Yu, Yu-Fu, Bjornstad, H. E. and Salbu, B. (1991). Determination of low-level alpha and beta emitters using liquid-liquid extraction and a liquid scintillation spectrometer. In Garcia-Leon, M. and Madurga, G. (Eds) *Low-level Measurements of Man-made Radionuclides in the Environment*: 176–186, World Scientific, Singapore.

Zifferero, M. (1988). A post-Chernobyl view. In Harley, J. H., Schmidt, G. D. and Silini, G. (Eds) *Radionuclides in the Food Chain*: 1–9. Springer Verlag, New York.

Zucker, C. L., Olsen C. R., Larsen, I. L. and Cutshall, N. H. (1984). Inventories and sorption-desorption trends of radiocaesium and radiocobalt in James River Estuary sediments. *Environ. Geol. Water Sci.*, **6**, 171–182.

Index

Index compiled by G. Jones